Computational Theoretical Organic Chemistry

# NATO ADVANCED STUDY INSTITUTES SERIES

*Proceedings of the Advanced Study Institute Programme, which aims at the dissemination of advanced knowledge and the formation of contacts among scientists from different countries*

The series is published by an international board of publishers in conjunction with NATO Scientific Affairs Division

| | | |
|---|---|---|
| A | Life Sciences | Plenum Publishing Corporation |
| B | Physics | London and New York |
| C | Mathematical and Physical Sciences | D. Reidel Publishing Company Dordrecht, Boston and London |
| D | Behavioural and Social Sciences | Sijthoff & Noordhoff International Publishers |
| E | Applied Sciences | Alphen aan den Rijn and Germantown U.S.A. |

**Series C – Mathematical and Physical Sciences**

*Volume 67 – Computational Theoretical Organic Chemistry*

# Computational Theoretical Organic Chemistry

*Proceedings of the NATO Advanced Study Institute held at Menton, France, June 29 - July 13, 1980*

edited by

I. G. CSIZMADIA
*Department of Chemistry, University of Toronto, Toronto, Ontario, Canada*

and

R. DAUDEL
*Centre de Mécanique Ondulatoire Appliquée,*
*Centre National de la Recherche Scientifique, Paris, France*

D. Reidel Publishing Company

Dordrecht : Holland / Boston : U.S.A. / London : England

Published in cooperation with NATO Scientific Affairs Division

Library of Congress Cataloging in Publication Data

Main entry under title:

Computational theoretical organic chemistry.

(NATO advanced study institutes series: Series C, mathematical and physical sciences; v. 67)

Includes index.

1. Chemistry, organic-data processing–Congresses. I. Csizmadia, I. G. II. Daudel, Raymond. III. NATO advanced study institute. IV. Series.

QD255.5.E4C65 547'.12'02854 81-4619

AACR2

ISBN-13: 978-94-009-8474-5 e-ISBN-13: 978-94-009-8472-1

DOI: 10.1007/978-94-009-8472-1

Published by D. Reidel Publishing Company
P.O. Box 17, 3300 AA Dordrecht, Holland

Sold and distributed in the U.S.A. and Canada
by Kluwer Boston Inc.,
190 Old Derby Street, Hingham, MA 02043, U.S.A.

In all other countries, sold and distributed
by Kluwer Academic Publishers Group,
P.O. Box 322, 3300 AH Dordrecht, Holland

D. Reidel Publishing Company is a member of the Kluwer Group

TABLE OF CONTENTS

## PREFACE

As a general rule any interdisciplinary subject and that includes Computational Theoretical Organic Chemistry (CTOC) incorporates people from the two overlaping areas. In this case the overlaping areas are Computational Theoretical Chemistry and Organic Chemistry. Since CTOC is a relatively young science, people continue to shift from their major discipline to this area. At this particular time in history we have to accept in CTOC people who were trained in Computational Theoretical Chemistry and do not know very much about Organic Chemistry, but more often the opposite case is operative Experimental Organic Chemistry who have not been exposed to Computational Theoretical Chemistry. This situation made NATO Advanced Study Institute in the field of CTOC necessary.

The inhomogenity outlined above was present in the NATO Advanced Study Institute, held at Menton in July 1980, and to some degree it is noticable from the content of this volume. This book contains 20 contributions. The first contribution is an Introduction chapter in which the initiated experimental chemists are briefed about the subject matter. The last chapter describes very briefly the "Computational Laboratory" that was designed to help people with an experimental back ground in order to obtain some first hand experience. Between the first and the last chapters there are 18 contributions. These contributions were arranged in a spectrum from the exclusively method oriented papers to the applications of existing computational methods to problems of interest in Organic Chemistry.

ACKNOWLEDGEMENTS

The Editors are grateful to the NATO (Scientific affairs Division) for providing a grant for the organization of a NATO Advanced Study Institute on which this volume is based. The financial assistance of CNRS for the organization of a Computational Laboratory that accompaning the NATO ASI is gratefully acknowledged. The Editors have very much appreciated, as all participants, the kind hospitality of the Mayor of Menton : General Aubert.

The names of the people who helped the ASI to be successful are too numerous to mention, however, it is impossible not to express our gratitude to the secretarial staff of CMOA Madam Escuillié, Madam Le Denmat, Madam Piton and Madam Schuhé whos effort beyond the call of duty both before and during A.S.I. as well as in the preparation of the manuscripts helped to made this volume possible.

# SOME FUNDAMENTALS OF COMPUTATIONAL THEORETICAL CHEMISTRY

I. G. Csizmadia
Department of Chemistry
University of Toronto
Toronto, Ontario, Canada

## INTRODUCTION

Although the title of this NATO Advanced Study Institute (ASI) specified the subject matter as "Computational Theoretical Organic Chemistry", certain fundamentals used explicitely or implicitely are valid regardless whether the molecule is classified as organic, inorganic or biological. For this reason the term "Organic" is omitted from the title of this chapter.

The lecturers of this ASI are grouped into three categories, with a number of sections in each of the categories. In brief one may break down the overall content in the following simplified fashion.

a) Data Generation. Calculation of Molecular Wavefunction and Density. Calculation of Molecular Properties.

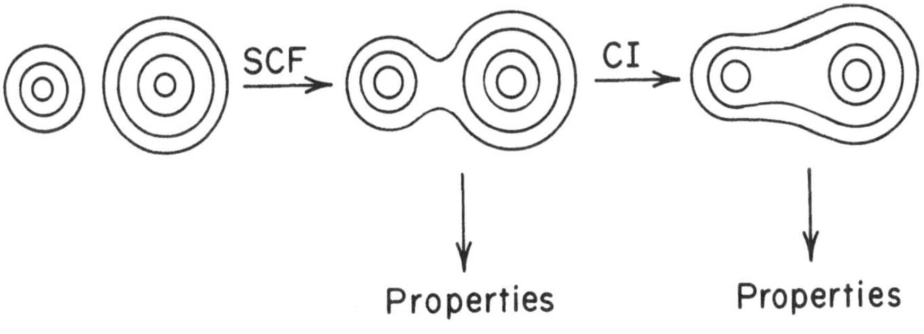


Figure 1. A schematic representation of Data Generation.



*I. G. Csizmadia and R. Daudel (eds.), Computational Theoretical Organic Chemistry, 1–14.*

b) Data Analysis. Analysis of Wavefunction or Density. Analysis of Properties.

c) Data Correlation. Theoretical Concepts.
Experimental Concepts.

## DEMYTHOLOGIZATION OF QUANTUM CHEMISTRY

Throughout the years some magic is attached to anything that is theoretical in chemistry and by now certain amount of mythology is built around Quantum Chemistry as it may be applied to study organic problems. For this reason some demythologization is in order. This is hoped to be achieved in five brief statements.

Statement no. 1. The term "orbital" is a synonym for the term "one-electron function (OEF)".

Statement no. 2. A single-centred OEF is synonymous for "Atomic Orbital (AO)". A multi-centered OEF is synonymous for "Molecular Orbital (MO)".

Figure 2. A schematic representation of Single (AO) and Multi (MO) centered one-electron functions.

An orbital (AO or MO) has as much to do with reality as the functions

$e^x$

$\{x^n\}$

$\{\sin k\alpha, \cos k\alpha\}$

do. Nevertheless these functions (AO and MO) enable us to construct molecular wavefunctions that may be used to compute molecular electron density and molecular properties. It is axiomatic that these latter quantities must have a one-to-one correspondence to the corresponding physical properties of the molecule as determined experimentally.

Statement no. 3. There are three ways to express a mathematical function:
(i) Explicitely in analytic form

$$f(x) = e^x$$

the hydrogen-like AO are usually expressed in this form.
(ii) As a table of numbers

| x | f(x) |
|---|---|
| 0.0 | 1.000 |
| 0.1 | 0.905 |
| . | . |
| . | . |
| . | . |

The Hartree-Fock type AO are usually expressed in this numerical form.
(iii) In the form of an expansion

$$f(x) \equiv e^{+x} = \frac{f(0)}{0!}x^0 + \frac{f'(0)}{1!}x^1 + \frac{f''(0)}{2!}x^2 + \frac{f'''(0)}{3!}x^3 + \cdots\cdots$$

which is analogous to the expression of MO in terms of a set of AO:

$$\phi = C_0\eta_0 + C_1\eta_1 + C_2\eta_2 + C_3\eta_3 + \cdots\cdots$$

Statement no. 4. The generation of MO ($\phi$) from AO ($\chi$) is equivalent to the rotation of an N-dimensional vector space to another N-dimensional vector space where $\{\eta\}$ is the original set of non-orthogonal functions. After orthogonalization the orthogonal set $\{\chi\}$ is rotated to the other orthogonal set $\{\phi\}$

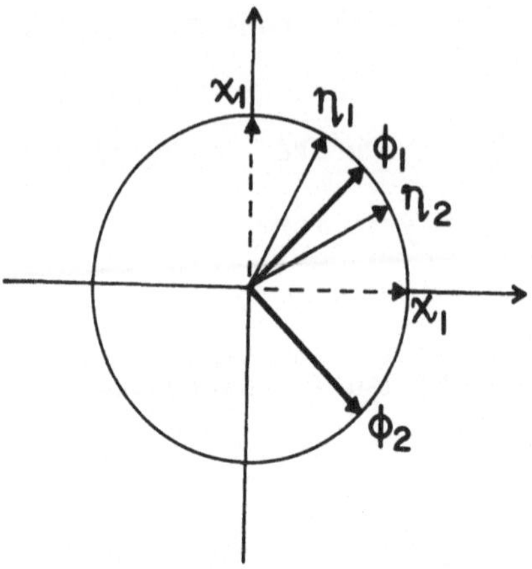


Figure 3. Two dimensional Vector Model of AO→MO transformation.

$$\underset{\text{AO}}{\underline{\eta}} \xrightarrow{\text{orthogonalization}} \underset{\text{AO}}{\underline{\chi}} \xrightarrow{\text{SCF}} \underset{\text{MO}}{\underline{\phi}}$$

Statement no. 5. There are certain differences between the shape of numerical Hartree-Fock atomic orbitals (HF-AO), the analytic Slater type orbitals (STO) and the analytic Gaussian type functions (GTF) but these differences are irrelevant to the final results as the MO can be expanded in terms of any of these sets to any desired degree of accuracy.

## ATOMIC ORBITAL BASIS SETS

The generation of MO from AO requires the generation and transformation of the Fock-Matrix into diagonal form. The elements of the Fock-Matrix are assembled from integrals in the following fashion:

$$f_{ij} = \langle i|\hat{h}|j\rangle + \sum_{k}^{N}\sum_{l}^{N}\{2[ij|kl] - [ik|jl]\}$$

where the first term is a one-electron integral and the latter terms are two-electron integrals and they have the following form

One-Electron Integrals

$$\langle i|\hat{h}|j\rangle \equiv \langle\eta_i|\hat{h}|\eta_j\rangle \equiv \int_i \eta_i(1)\hat{h}_1\eta_j(1)d\tau_1$$

Two-Electron Integrals

$$[ij|kl] \equiv \langle \eta_i \eta_k | \frac{1}{r_{12}} | \eta_l \eta_j \rangle \equiv \iint_{12} \eta_i(1)\eta_k(2)\frac{1}{r_{12}}\eta_l(2)\eta_j(1)d\tau_2 d\tau_1$$

As the running indices range from 1 to N the number of one- and two-electron integrals are calculable by the following formulas

number of 1-electron integrals $p = \frac{N(N+1)}{2}$

number of 2-electron integrals $q = \frac{p(p+1)}{2}$

The table below illustrates how rapidly the number of one- and two-electron integrals grows with the basis set size N.

Table 1. The increase of the number of one-electron (p) and two-electron (q) integrals with increasing basis set size (N).

| N | P | q |
|---|---|---|
| 1 | 1 | 1 |
| 10 | 55 | 1,540 |
| 50 | 1,275 | 814,725 |
| 100 | 5,050 | 12,751,250 |
| 150 | 11,325 | 64,133,475 |

For the atomic orbitals $\{\eta\}$ two types of analytic functions are used in molecular computations.
1) Slater-type orbitals (STO) or Exponential type functions (ETF).
2) Gaussian-type orbitals (GTO) or Gaussian-type functions (GTF).
Some characteristics of these two sets of functions are illustrated in Figure 4. As far as STO are concerned the integral evaluation is very slow but a relatively small N gives fairly accurate results.

The GTF are more popular as it is possible to compute the integrals over Gaussians very quickly, but only a relatively large N gives accurate results.

These AO basis sets need to be optimized for molecular calculation. This may be achieved by minimizing the electronic energy with respect to all orbital exponents. Figure 5 illustrates the

STO or ETF

GTO or GTF

no cusp as
r→0

decays too
rapidly
as r→∞

$$\eta_{STO} = \text{polynomial} \cdot e^{-\xi r}$$

$$\eta_{GTF} = \text{polynomial} \cdot e^{-\alpha r^2}$$

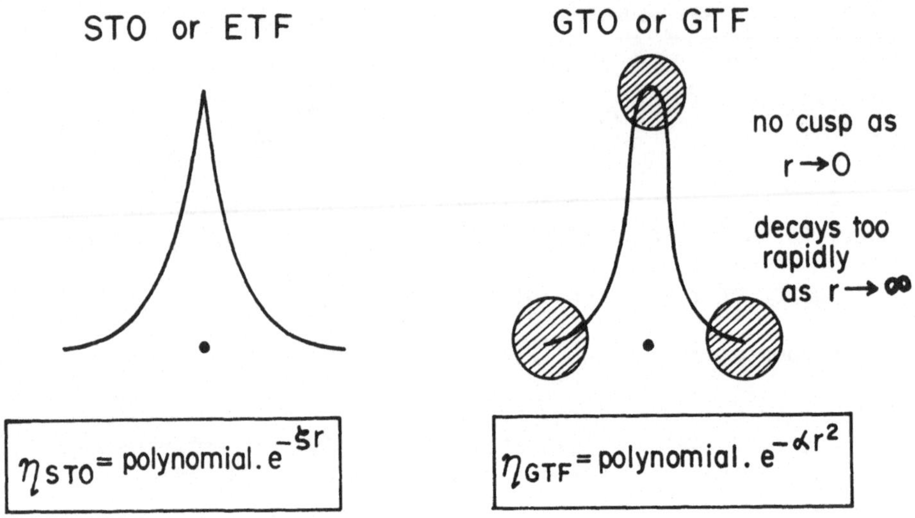

Figure 4.   Some characteristics of STO and GTF.

situation for a two-orbital case:

$$\eta_1 = \text{polynomial}_1 \cdot e^{-\alpha_1 r^2}$$

$$\eta_2 = \text{polynomial}_2 \cdot e^{-\alpha_2 r^2}$$

$$\frac{\partial E}{\partial \alpha_1}$$

$$\frac{\partial E}{\partial \alpha_2}$$

$E(\alpha_1 \alpha_2)$

$\alpha_2$

$\alpha_1$

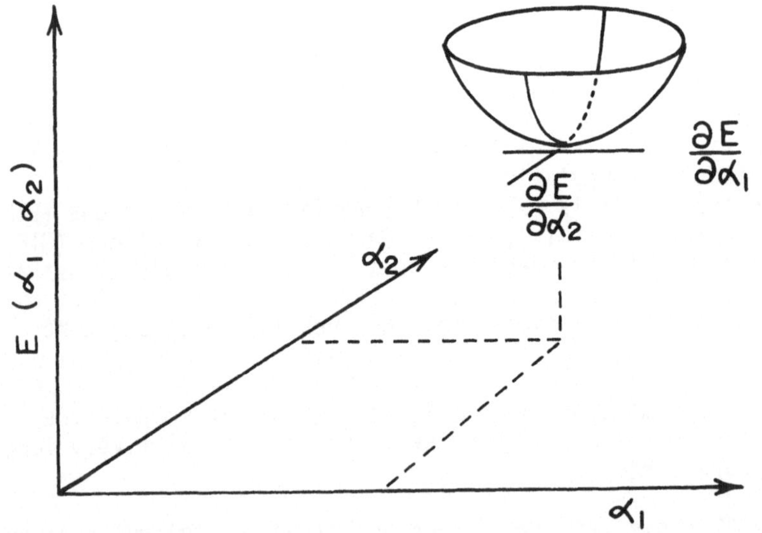

Figure 5.   Electronic Energy Surface as a function of orbital exponents $\{\alpha\}$.

Very often STO are expanded in terms of GTF, or, in other words, the GTF basis set is contracted. The contraction of three Gaussian to a Slater is very popular (STO-3G).

$$\eta_i^{STO} = a_{i1}\eta_1^{GTO} + a_{i2}\eta_2^{GTO} + a_{i3}\eta_3^{GTO}$$

This means that the total number of integrals evaluated can be reduced by contraction as illustrated below:

$$\langle\eta_i^{STO}|\hat{h}|\eta_j^{STO}\rangle =$$

$$= \langle a_{i1}\eta_1^{GTO} + a_{i2}\eta_2^{GTO} + a_{i3}\eta_3^{GTO}|\hat{h}|a_{j1}\eta_1^{GTO} + a_{j2}\eta_2^{GTO} + a_{j3}\eta_3^{GTO}\rangle =$$

$$= a_{i1}a_{j1}\langle\eta_1^{GTO}|\hat{h}|\eta_1^{GTO}\rangle$$
$$+ a_{i2}a_{j1}\langle\eta_2^{GTO}|\hat{h}|\eta_1^{GTO}\rangle$$
$$+ a_{i3}a_{j1}\langle\eta_3^{GTO}|\hat{h}|\eta_1^{GTO}\rangle$$
$$\vdots$$
$$+ a_{i3}a_{j3}\langle\eta_3^{GTO}|\hat{h}|\eta_3^{GTO}\rangle$$

LOCALIZED MOLECULAR ORBITALS

<u>CMO ($\phi$).</u> Molecular orbitals that produce a Fock-Matrix in canonical (diagonal) form are known as canonical molecular orbitals

$$\underline{\underline{F}}^{\phi} = \begin{pmatrix} \langle\phi_1|\hat{F}|\phi_1\rangle & 0 & 0 & \cdots \\ 0 & \langle\phi_2|\hat{F}|\phi_2\rangle & 0 & \cdots \\ 0 & 0 & \langle\phi_3|\hat{F}|\phi_3\rangle & \cdots \\ \vdots & \vdots & \vdots & \end{pmatrix}$$

These CMO are delocalized all over the molecule.

<u>LMO ($\psi$).</u> Localized molecular orbitals produce a non-diagonal real symmetric Fock-Matrix

$$\underline{\underline{F}}^{\psi} = \begin{pmatrix} \langle\psi_1|\hat{F}|\psi_1\rangle & \langle\psi_1|\hat{F}|\psi_2\rangle & \langle\psi_1|\hat{F}|\psi_3\rangle & \cdots \\ & \langle\psi_2|\hat{F}|\psi_2\rangle & \langle\psi_2|\hat{F}|\psi_3\rangle & \cdots \\ & & \langle\psi_3|\hat{F}|\psi_3\rangle & \cdots \\ & & & \ddots \end{pmatrix}$$

The unitary matrix $\underline{\underline{U}}$ that transforms $\underline{\phi}$ to $\underline{\psi}$:

$$\underline{\psi} = \underline{\phi}\, \underline{\underline{U}}$$

will change $\underline{\underline{F}}^{\phi}$ to $\underline{\underline{F}}^{\psi}$ via a similarity transformation:

$$\underline{\underline{F}}^{\psi} = \underline{\underline{U}}^{+}\, \underline{\underline{F}}^{\phi}\, \underline{\underline{U}}$$

The generation of $\underline{\phi}$ has the unique condition that $\underline{\underline{F}}$ be a diagonal matrix, but the generation of $\underline{\psi}$ has no such condition.

Consequently $\underline{\underline{U}}$ could be any unitary matrix and therefore there are literally an infinity of possible sets of LMO.

The concept of localization means the separation of electron pairs and therefore the definition of separation means the definition of $\underline{\underline{U}}$.

Separation has been defined by Boys in stereochemical terms and by Edmiston-Ruedenberg in energetic terms.

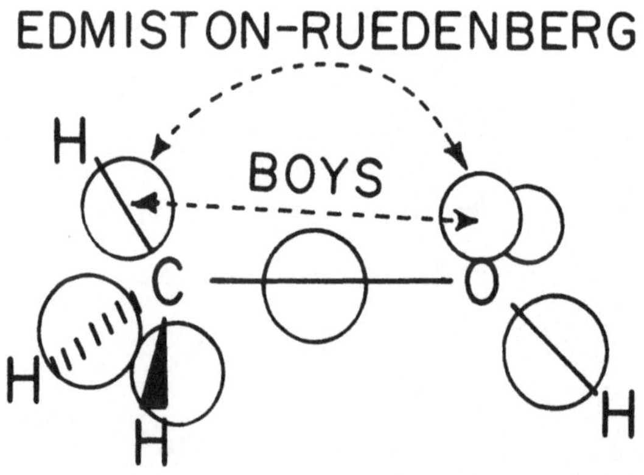


Figure 6. A schematic illustration of the Boys and Edmiston-Ruedenberg localization.

In either case the degree of degeneracy of the diagonal elements of $\underline{\underline{F}}^{\psi}$ will follow the equivalency of the chemical bonds. This is illustrated for $CH_4$ in Figure 7.

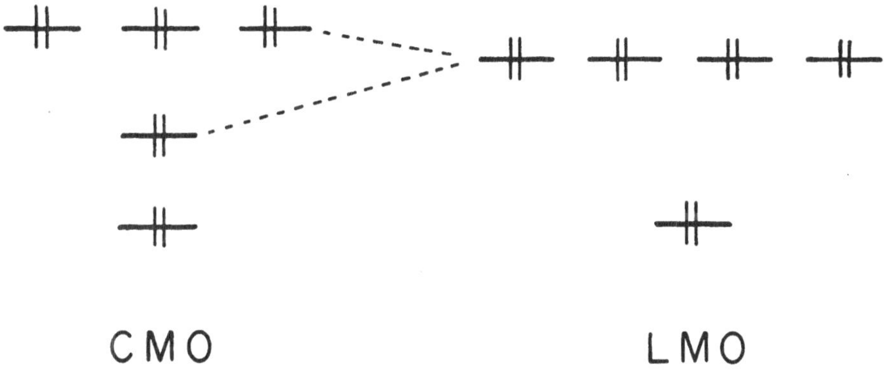


Figure 7. Schematic illustration of CMO and LMO energies for $CH_4$.

PRIMARY MOLECULAR PROPERTIES AS OBSERVABLE

The following statements are axiomatic in Quantum Mechanics. Many electron state wavefunction:

$$\Phi_{\nu} = \Phi_{\nu}(1, 2, 3, \ldots)$$

Quantum Chemical Operator:

$$\hat{\Omega} = \hat{\Omega}_0 + \sum_i \hat{\Omega}_i + \sum_{ij}{}' \hat{\Omega}_{ij}$$

Observable:

$$\Omega_{\nu} = \langle \Phi_{\nu}(1, 2, 3, \ldots) | \hat{\Omega} | \Phi_{\nu}(1, 2, 3, \ldots) \rangle$$

The energy and dipole moment is used to illustrate the computation of observables.

<u>Energy</u> (a two-electron property)

$$E = \langle \Phi | \hat{H}_0 + \sum_i \hat{H}_i + \sum_{ij}{}' \hat{H}_{ij} | \Phi \rangle =$$

$$= \langle \Phi | \hat{H}_0 | \Phi \rangle + \langle \Phi | \sum_i \hat{H}_i | \Phi \rangle + \langle \Phi | \sum_{ij}{}' \hat{H}_{ij} | \Phi \rangle$$

$$= E_0 \qquad + \qquad \sum_i E_i \qquad + \qquad \sum_{ij}{}' E_{ij}$$

| no electron energy | one electron energy | two electron energy |
|---|---|---|

Dipole Moment (a one-electron energy)

$$\mu = \langle\Phi|\hat{\mu}_0 + \sum_i \hat{\mu}_i|\Phi\rangle = \langle\Phi|\hat{\mu}_0|\Phi\rangle + \langle\Phi|\sum_i \hat{\mu}_i|\Phi\rangle =$$

$$= \mu_0 + \sum_i \mu_i$$

The matrix below gives the dipole moments over electronic states. The 1,1 element is the dipole moment for the electronic ground state and the other diagonal elements are dipole moments of the various electronic excited states. The off-diagonal elements are frequently referred to as transition moments.

$$\underline{\underline{\mu}} = \begin{pmatrix} \langle\Phi_0|\hat{\mu}|\Phi_0\rangle & \langle\Phi_0|\hat{\mu}|\Phi_1\rangle & \langle\Phi_0|\hat{\mu}|\Phi_2\rangle & \cdots \\ & \langle\Phi_1|\hat{\mu}|\Phi\rangle & \langle\Phi_1|\hat{\mu}|\Phi_2\rangle & \cdots \\ & & \langle\Phi_2|\hat{\mu}|\Phi_2\rangle & \cdots \\ & & & \ddots \end{pmatrix}$$

ELECTRON PROPERTIES FOR GROUND STATE MO WAVEFUNCTIONS

The dipole moment discussed above falls in the category of one-electron properties

$$P = \langle\Phi_0|\hat{P}|\Phi_0\rangle$$

simply because the operator is of the following form

$$\hat{P} = \hat{P}_0 + \sum_i \hat{P}_i$$

the MO wavefunction is an antisymmetrized orbital product.

$$\Phi_0 = \mathcal{A}[\phi_1(1)\alpha(1)\phi_1(2)\beta(2)\ldots\phi_M(2M-1)\alpha(2M-1)\phi_M(2M)\beta(2M)]$$

After substituting $\Phi_0$ and $\hat{P}$ into P and integrating over the spin variable one obtains the following relationship

$$P = P_0 + 2 \sum_{i=1}^{M} \langle\phi_i|\hat{P}_i|\phi_i\rangle$$

The expression under the $\Sigma$ sign is in fact the property over the MO basis and it may be CMO or LMO.

The following Multipole Moments are routinely computed nowadays.

First Moment $(\hat{r})$ ... Dipole Moment

Second Moment $(\hat{r}^2)$ ... Quadrupole Moment

Third Moment $(\hat{r}^3)$ ... Octupole Moment

## THE CHEMICAL SIGNIFICANCE OF CORRELATION ENERGY

In MO Calculations the improvement in basis set will lower the total molecular energy to a limiting value called the Hartree-Fock limit (HFL). Using a single Slater determinant as the molecular wavefunction, one cannot obtain lower energy values. Yet, within the framework of non-relativistic quantum mechanics the ultimate limit is the non-relativistic limit (NRL) that is approachable only by using the most sophisticated wavefunction. The difference between these two limits is normally referred to as correlation energy $(\Delta E_{cor})$

$$\Delta E_{cor} = E_{NRL} - E_{HFL}$$

The origin of this discrepancy $(\Delta E_{cor})$ may be looked upon as a systematic error as Hartree-Fock wavefunctions do not correlate properly. Strictly speaking electrons of like spin are properly correlated as they must be assigned to different region of space, ie, to different MO (Pauli exclusion principle) but electrons of opposite spin <u>may</u> occupy the same region of space due to the double occupancy inherently assumed in a Hartree-Fock MO wavefunction.

One traditionally popular way to compensate for this systematic error is to carry out Configuration Interaction (CI) calculations in which configurations where the double occupancy is relieved, as illustrated for the 2-orbital 2-electron case in Figure 9, are taken in a suitable linear combination

$$\Psi_0(1,2) = C_0\Phi_0(1,2) + C_1\Phi_1(1,2) + C_2\Phi_2(1,2)$$

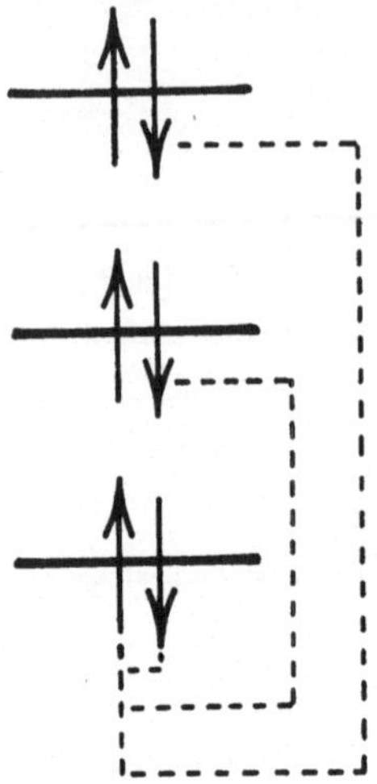

Figure 8. Schematic representation of the correlation of electrons with opposite spin.

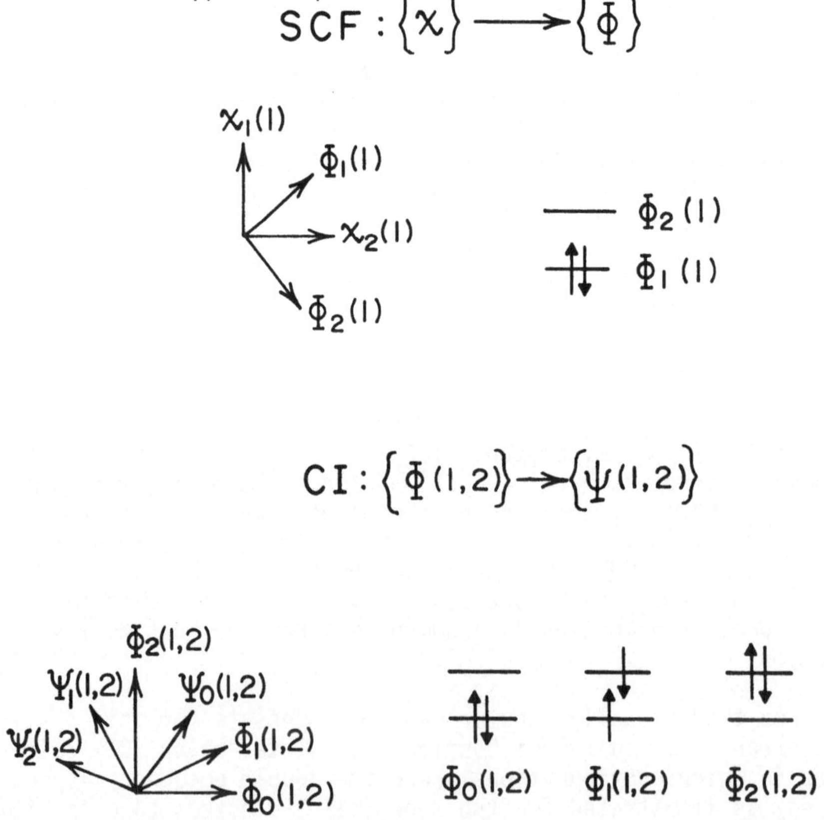


Figure 9. A schematic comparison of SCF and CI calculations.

In practice a limited basis set SCF followed by a limited CI (say including no more than double excitations within the valence shell) is expected to recover less than 50% of the correlation energy. Certain advantages may be achieved by using an LMO basis instead of a CMO basis, that may also be augmented with polarization functions centred at the centroid of charge of each LMO.

The energetics of all of these concepts ($E_{exp}$, HFL, NRL, $\Delta E_{cor}$, $E_{SCF}$, etc...) are illustrated in Figure 10.

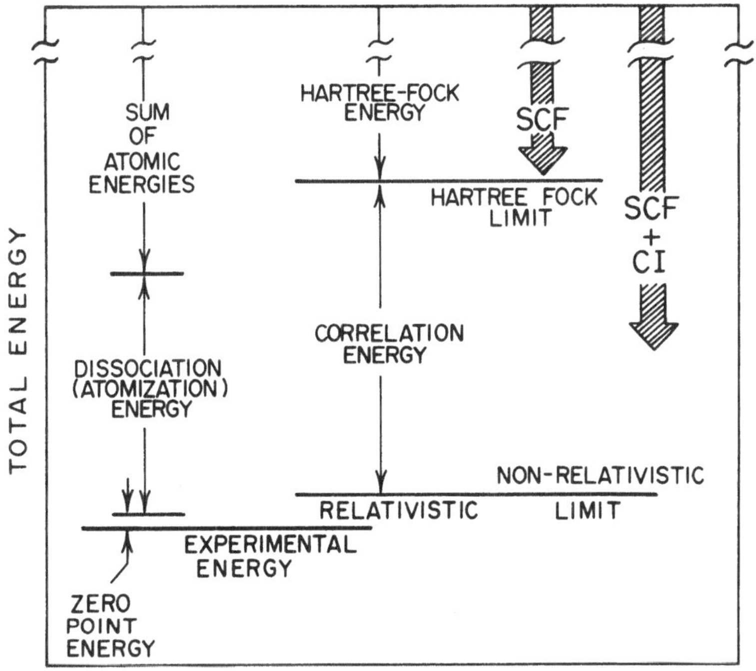


Figure 10. A schematic illustration of experimental and theoretical energy values for a molecular system.

Because correlation energy is associated with electron pairing, it is clear that any process where electrons are unpaired (e.g. ionization, electronic excitation) the correlation energy assumes chemical significance. This must be so by virtue of the fact that one has a different number of electron pairs in the final state than in the initial state, which implies a difference in correlation energy that in turn leads to a systematic error in the computed thermodynamic stability.

In certain chemical processes where the electron distribution is dramatically altered, even though the pairing scheme is formally unchanged, the difference in correlation energy may be significant (e.g. 5 kcal/mol). If the barrier is small than such a change of correlation energy will cause a quantitative difference that sometime is sufficient to lead to a qualitative difference as illustrated by Figure 11.

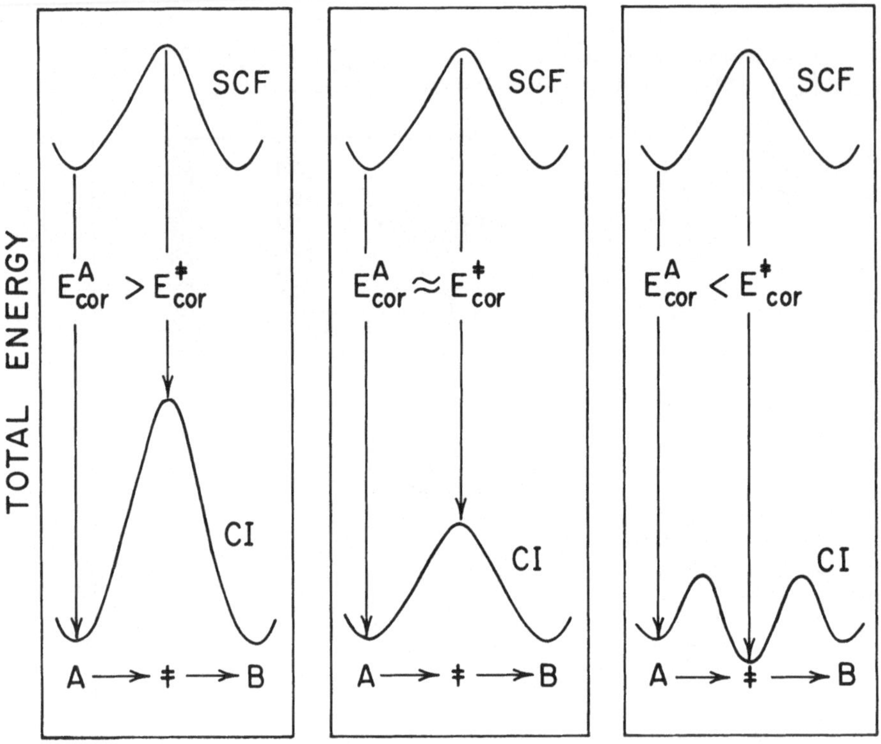


Figure 11. The chemical significance of correlation energy as illustrated in three different hypothetical chemical reactions.

The situation at the right hand side of Figure 11 indicates that under certain circumstances the difference in correlation energy may alter the energy profile obtained at the SCF level sufficiently such that a transition state in fact becomes a reaction intermediate. Such a case has in fact been observed in the 1,2 hydride shift of $C_2H_3^+$.

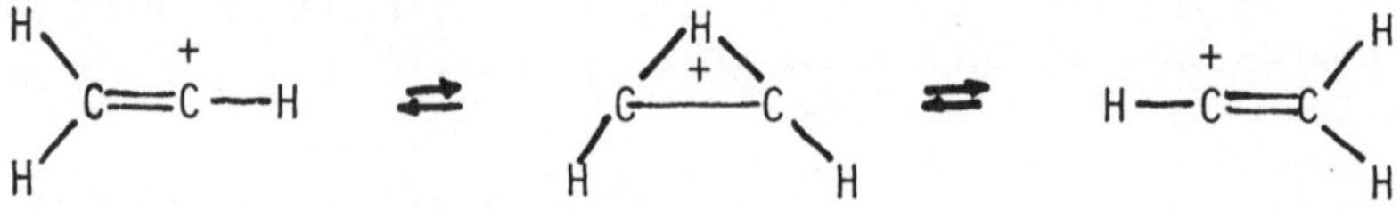

# GAUSSIAN BASIS SETS

Raymond A. Poirier

Centre de Mécanique Ondulatoire Appliquée
Conseil National de Recherche Scientifique
23, rue du Maroc
75940, PARIS Cédex 19, France

All calculations performed at the school are of the LCAO-MO-SCF or LCAO-MO-SCF-CI type. The calculations were performed using a program (MONSTERGAUSS (1)) which uses as atomic orbitals (AO) or basis functions, the Gaussian function.

The general Gaussian, first proposed by Boys in 1950 (2) as a possible basis set orbital for molecular calculations, may be written as,

$$\eta_i^A = x_A^l y_A^m z_A^n \exp\{-\alpha_i[x_A^2+y_A^2+z_A^2]\}$$

$$= x_A^l y_A^m z_A^n \exp\{-\alpha_i r_A^2\} \qquad (1)$$

where,

$$x_A = x-A_x$$

$$y_A = y-A_y$$

$$z_A = z-A_z \qquad (2)$$

and the A's are the coordinates at which the Gaussian is centered. The $\alpha_i$ is the orbital exponent, a parameter to be determined. For example for an s-type Gaussian function (l=m=n=o), and dropping the A's,

$$\eta(\alpha_i^s) = \exp\{-\alpha_i^s r^2\} \qquad (3)$$

and for a $p_x$-type Gaussian function (l=1, m=n=o),



*I. G. Csizmadia and R. Daudel (eds.), Computational Theoretical Organic Chemistry, 15–20.*

Table I. Gaussian Basis Sets Available in MONSTERGAUSS.

| BASIS SET | ATOMS | REFERENCES |
|---|---|---|
| STO-NG | | |
| N=1,6 | H to Kr | 3-6 |
| STO-NG*[a] | | 14 |
| N=1,6 | Na to Ar | 14 |
| NM-LG | | |
| 3-21G | H to Ar | 16, 17 |
| 4-21G | H, B to F | 15 |
| 5-21G | Li, Be, B | 12, 13 |
| 6-21G | H to Ar | 16, 17 |
| 4-31G | H, B to F, P, S, Cl | 7, 9, 10 |
| 5-31G | C to F | 8 |
| 6-31G | H, Li to F | 8, 12, 13 |
| NM-LG* (and) | NM-LG**[a] | |
| 4-31G* | Li to F | 11 |
| 6-31G* | Li to F | 11 |

[a] The superscript * implies the addition of d-polarization functions, ** would also include p-polarization on H.

$$\eta(\alpha_j^x) = x\exp\{-\alpha_j^x r^2\} \tag{4}$$

Therefore, a Gaussian basis set is a set of individual Gaussian functions referred to as the primitive Gaussians or simply primitives. A Gaussian basis set with k primitives ($\alpha_i$, i=1, k) may have s-type, x-type, y-type...etc. Gaussian functions. However, in general the primitives of a given type are grouped together or contracted. A basis function then becomes a linear combination of more than one primitive Gaussian function,

$$\chi_s = \sum_{i=1}^{n} d_{s,i}\ \eta(\alpha_i^s) \tag{5}$$

In this case two sets of parameters must be determined, the exponents $\alpha$'s and contraction coefficients d's.

Two types of Gaussian basis sets exist in MONSTERGAUSS which are referred to as minimal STO-NG (2-6) and extended NM-LG (7-17). A list of the basis sets available in MONSTERGAUSS (1) is given in Table I.

In order to illustrate some concepts about basis sets in general, two basis sets will be described, the minimal STO-3G and the extended 3-21G basis set. These two basis sets are presented for a general first-row atom in Table II. The STO-3G and 3-21G basis sets are similar in two ways. First, both are s=p basis sets (as are all other basis sets in MONSTERGAUSS), that is, the exponents of the Gaussian function ($\alpha$'s), representing the 2s and 2p AO for example, are constrained to have the same value ($\alpha_{2s}=\alpha_{2p}$). Secondly, both basis sets are 3G since there is a total of 3 Gaussian functions (primitives) for each AO (1s, 2s and 2p's for first-row atoms).

These basis sets are also different in a number of ways. First, the two basis sets are contracted in different ways, that is, the STO-3G has minimal contraction (also known as single zeta) or only one basis function, represented by 3 Gaussians (the '3' in STO-3G), for each AO. The 3-21G has a similar minimal contraction in the core ('3' in 3-21G) and split valence part (double zeta) where two basis functions describe the valence part (2s and 2p), one with two Gaussians ('2' in 21G) and the other a single Gaussian ('1' in 21G). The extended or split valence type basis sets have more flexibility in the valence region and therefore can adjust more freely to the molecular environment.

These two basis sets are also different in the way the parameters of Table II, $\xi$'s, $\alpha$'s and d's are determined. In the case of the STO-3G basis set the Gaussian exponents ($\alpha$'s) and contraction coefficients (d's) are determined by fitting a Slater or exponential type function ($\eta'=e^{-\gamma r}$) with an exponent of $\gamma$= 1.0 and

Table II. Representation of a General First-row Atom STO-3G and 3-21G Basis Set.

| | ATOMIC ORBITAL | SCALE | EXPONENT | CONTRACTION S | COEFFICIENTS P |
|---|---|---|---|---|---|
| STO-3G | 1s | $\xi_1$ | $\alpha_1$ | $d_1$ | |
| | | | $\alpha_2$ | $d_2$ | |
| | | | $\alpha_3$ | $d_3$ | |
| | 2sp | $\xi_2$ | $\alpha_4$ | $d_4$ | $d_7$ |
| | | | $\alpha_5$ | $d_5$ | $d_8$ |
| | | | $\alpha_6$ | $d_6$ | $d_9$ |
| 3-21G | 1s | $\xi_1$ | $\alpha_1$ | $d_1$ | |
| | | | $\alpha_2$ | $d_2$ | |
| | | | $\alpha_3$ | $d_3$ | |
| | 2sp | $\xi_2$ | $\alpha_4$ | $d_4$ | $d_6$ |
| | | | $\alpha_5$ | $d_5$ | $d_7$ |
| | 2sp' | $\xi_3$ | $\alpha_6$ | 1.0 | 1.0 |

minimizing the following sum of squares,

$$\varepsilon_s = \int(\eta'_{1s}-\chi_s)^2 d\tau$$

$$\varepsilon_{sp} = \int(\eta'_{2s}-\chi_s)^2 d\tau - \int(\eta'_{2p}-\chi_p)^2 d\tau \tag{6}$$

The Gaussian expansions thus obtained are then uniformly scaled such that for each $\alpha$ of equation 5,

$$\alpha'_k = \xi^2 \alpha_k, \; k=1,N \tag{7}$$

and where in this case optimizing the parameter $\xi$ corresponds to optimizing the exponents of the Slater fitted by the Gaussian expansion ($\xi \equiv \gamma$).

The scale factors (or Slater exponents) for the valence of the STO-NG basis sets have been chosen by minimizing the total energy for a standard set of molecules (3,4) with the core fixed at the best atom values (5).

The 3-21G basis set on the other hand is optimized (d's and $\alpha$'s) in two steps. First a 6-21G basis set is optimized by minimizing the total energy with respect to $\alpha$'s and d's. The 6G core representation prevents (16) the valence region (21G part) from "falling inward", that is, preventing the valence basis functions from adjusting to represent the core where the largest gain in energy can be obtained. Once the atomic energy-optimized 6-21G basis set is obtained, the 6G core is replaced by a 3G representation which is than energy-optimized for the atom keeping the 21G valence part fixed. The resulting 3-21G basis set is then left unscaled ($\xi$=1.0) except for H($\xi$=1.10).

## REFERENCES


1. MONSTERGAUSS: M. R. Peterson and R. A. Poirier, Department of Chemistry, University of Toronto, Toronto, Canada.

2. S. F. Boys: 1950, Proc. Roy. Soc. (London), A200, pp. 542.

3. W. J. Hehre, R. F. Stewart and J. A. Pople: 1969, J. Chem. Phys., 51, pp. 2657.

4. W. J. Hehre, R. Ditchfield, R. F. Stewart and J. A. Pople: 1970, J. Chem. Phys., 52, pp. 2769.

5. E. Clementi and D. R. Raimondi: 1963, J. Chem. Phys., 38, pp. 2686.

6. R. F. Stewart: 1969, J. Chem. Phys., 52, pp. 431.

7. R. Ditchfield, W. J. Hehre and J. A. Pople: 1971, J. Chem. Phys., 54, pp. 724.

8. W. J. Hehre, R. Ditchfield and J. A. Pople: 1972, J. Chem. Phys., 56, pp. 2257.

9. W. J. Hehre and J. A. Pople: 1972, J. Chem. Phys., 56, pp. 4233.

10. W. J. Hehre and W. A. Lathan: 1972, J. Chem. Phys., 56, pp. 5255.

11. P. C. Hariharan and J. A. Pople: 1973, Theoret. Chim. Acta (Berl.), 28, pp. 213.

12. J. D. Dill and J. A. Pople: 1975, J. Chem. Phys., 62, pp. 2921.

13. J. S. Binkley and J. A. Pople: 1977, J. Chem. Phys., 66, pp. 879.

14. J. B. Collins, P. v. R. Schleyer, J. S. Binkley and J. A. Pople: 1976, J. Chem. Phys., 64, pp. 5142.

15. P. Pulay, G. Fogarasi, F. Pang and J. E. Boggs: 1979, J. Am. Chem. Soc., 101, pp. 2550.

16. J. S. Binkley, J. A. Pople and W. J. Hehre: 1980, J. Am. Chem. Soc., 102, pp. 939.

17. M. S. Gordon, J. S. Binkley, J. A. Pople, W. J. Pietro and W. J. Hehre: 1980, J. Am. Chem. Soc., 102, pp. xx.

# SINGLE- AND MULTI-CONFIGURATION SELF CONSISTENT FIELD METHODS

Michael A. Robb and Richard H. A. Eade

Department of Chemistry
Queen Elizabeth College
London, W8 7AH.

## INTRODUCTION

The basic concept in the modern theory of molecular structure is the molecular orbital(MO). The primary tool for the computation of MO is the self consistent field theory(SCF). Today one can routinely perform SCF calculations on stable closed shell systems and many open shell systems where the wavefunction is well represented by a single configuration of electrons distributed among a set of MO.

In situations where electrons undergo rearrangements such as making and breaking bonds, one must go to a multi-configuration description to obtain even a qualitatively correct description of molecular structure. However, in this case the orbital concept and SCF survive in the form of the multi-configuration SCF method (MC-SCF). The principle conceptional difference is that some of the orbitals now have non-integral occupation numbers. From the computational point of view one now has the problem of the simultaneous optimization of both the linear variation coefficients in the multi-configuration expansion of the many electron wavefunction and orbitals which occur in all of those configurations.

Although the theory of MC-SCF was first derived in its most general form in 1955 by McWeeny [1] it is only very recently that MC-SCF methods have begun to be applied in a routine manner. (Space limitations preclude a full historical development. For two comprehensive bibliographies of the literature up to 1979 the reader is referred to references [2] and [3]). The development of SCF methods that are applicable in the case of the general multi-configuration wavefunction has been hampered by the lack of



*I. G. Csizmadia and R. Daudel (eds.), Computational Theoretical Organic Chemistry, 21–54.*

an effective formalism for the unified treatment of both the CI problem and the orbital optimization problem. The introduction of the Unitary Group formalism into quantum chemistry by Paldus [4] in 1965 and subsequent applications [5,6,7,8,9] provides the framework in which both of these problems can be discussed succinctly and also provides the most efficient methodology for implementation at the computational level.

## BASIC MATHEMATICAL FORMALISM OF SCF THEORY

It is our intention in this introductory section to develop SCF theory in a very general way so that various approaches can be easily discussed later.

We are concerned first with the Configuration Interaction (CI) expansion for the wavefunction $\psi$ for some state (for a comprehensive review of the CI method the reader is referred to reference [10]).

$$\psi = \sum_K C_K |K\rangle \tag{1}$$

where the states $|K\rangle$ are referred to as configuration state functions (CSF) and are, in general, linear combinations of Slater determinants built from a suitable set of orthogonal orbitals $\{\phi_i\}$. The expansion coefficients $C_K$ are linear variation coefficients obtained from the solution of the usual CI matrix eigenvalue problem

$$HC = EC \tag{2}$$

with

$$H_{KL} = \langle K|\hat{H}|L\rangle \tag{3}$$

and

$$C = \begin{pmatrix} C_1 \\ \vdots \\ C_K \\ \vdots \\ C_N \end{pmatrix} \tag{4}$$

where $\hat{H}$ is the full many-electron hamiltonian operator.

The matrix elements which occur in equation 3 will in general be combinations of one and two electron integrals over the set of orthogonal orbitals $\{\phi_i\}$. Thus we can write equation 3 as

$$\langle K|\hat{H}|L\rangle = \sum_{ij} \langle i|\hat{h}|j\rangle A_{ij}^{KL} + \frac{1}{2}\sum_{ijkl} [ij|kl]\, B_{ijkl}^{KL} \tag{5}$$

with

$$\langle i|\hat{h}|j\rangle = \int \phi_i(1)\, h\, \phi_j(1)\, dr_1 \tag{6}$$

$$[ij|kl] = \iint \phi_i(1)\,\phi_j(1)\,\frac{1}{r_{12}}\,\phi_k(2)\,\phi_l(2)\,dr_1\,dr_2 \qquad (7)$$

The coefficients $A_{ij}^{KL}$ and $B_{ijkl}^{KL}$ are often called the symbolic one and two electronic density matrices. The first $\gamma_{ij}$ and second order $\Gamma_{ijkl}$ density matrices are obtained by contraction with the eigenvector $C$ .

$$\gamma_{ij} = \sum_{KL} C_K C_L A_{ij}^{KL} \qquad (8)$$

$$\Gamma_{ijkl} = \sum_{KL} C_K C_L B_{ijkl}^{KL} \qquad (9)$$

It is clear that the central problem of the CI method is the determination of these coefficients $A_{ij}^{KL}$ and $B_{ijkl}^{KL}$ for arbitrary CSF $|K\rangle$ and $|L\rangle$ . Further, as we shall presently demonstrate, the SCF conditions involve these coefficients as well via the density matrices $\gamma$ and $\Gamma$ . Thus it is essential to use an effective formalism for discussing these coefficients. The Unitary Group method provides this formalism and we now digress to give some very elementary discussion.

It is possible to regard the coefficients $A_{ij}^{KL}$ as the matrix elements of a one electron operator so that we have

$$A_{ij}^{KL} = \langle K|\hat{E}_{ij}|L\rangle \qquad (10)$$

where $\hat{E}_{ij}$ is defined as

$$\hat{E}_{ij} = |\phi_i\rangle\langle\phi_j| + |\bar{\phi}_i\rangle\langle\bar{\phi}_j| \qquad (11)$$

We use the bar to denote an orbital with $\beta$ spin. The operator $\hat{E}_{ij}$ when it acts on a determinant will remove orbital $\phi_j$ wherever it occurs and replaces it with orbital $\phi_i$ since

$$\hat{E}_{ij}|\phi_j\rangle = |\phi_i\rangle\langle\phi_j|\phi_j\rangle + |\bar{\phi}_i\rangle\langle\bar{\phi}_j|\phi_j\rangle \qquad (12)$$

where $\langle\phi_j|\phi_j\rangle = 1$ and $\langle\bar{\phi}_j|\phi_j\rangle = 0$ . Also, if orbital $\phi_j$ is not contained in the determinant, the result of $\hat{E}_{ij}$ is zero. It is also possible to prove the following relationships

$$[\hat{E}_{ij}, \hat{E}_{kl}] = \delta_{jk}\hat{E}_{il} - \delta_{il}\hat{E}_{kj} \qquad (13)$$

$$\langle K|\hat{E}_{ij}|L\rangle = \langle L|\hat{E}_{ji}|K\rangle \qquad (14)$$

where we use the [] to mean the commutator ([AB] = AB −BA). Finally, the usefulness of these operators $\hat{E}_{ij}$ arises because the $B_{ijkl}^{KL}$ can also be expressed in terms of them as

$$B_{ijkl}^{KL} = \langle K|\hat{E}_{ij}\hat{E}_{kl} - \delta_{jk}\hat{E}_{il}|L\rangle \qquad (15)$$

By virtue of equations 10 and 15 we can write the hamiltonian operator as

$$\hat{H} = \sum_{ij}\langle i|\hat{h}|j\rangle\hat{E}_{ij} + \frac{1}{2}\sum_{ijkl}[ij|kl](\hat{E}_{ij}\hat{E}_{kl} - \delta_{jk}\hat{E}_{il}) \qquad (16)$$

It is easy to verify that the use of $\hat{H}$ as defined by equation 16 together with equation 11 and 13 is completely equivalent to the use of the usual rules for the matrix elements of one and two electron operators. The advantage of the present formalism is that we can perform manipulations at the operator level using equation 13 and then form the resulting matrix elements of these operators with the CSF $|k\rangle$ and $|L\rangle$ . Further, the required matrix elements of these operators are very easy to compute for arbitrary CSF (see reference 4 for a complete discussion).

Let us now turn to the problem that will be our central concern when we come to MC-SCF theory. We shall consider the energy obtained from the CI problem (eq.2) when the orbitals are subjected to a unitary transformation.

$$\phi' = U\,\phi \tag{17}$$

$$U^{+} = U^{-1} \tag{18}$$

We shall assume that the $C_K$ are fixed and the form of the CSF remains the same except that the orbitals $\phi$ are replaced by $\phi'$ where ever they occur. Further, we shall limit our discussion to the case where the orbitals are real (so we take $+$ to mean transpose rather than transpose complex conjugate).

It is possible to write a unitary (orthogonal) transformation in terms of a parameter matrix A as

$$U = \exp A \tag{19}$$

where A is skew symmetric so that

$$A^{+} = -A \tag{20}$$

The matrix U is found formally by expanding the exponential

$$U = 1 + A + \tfrac{1}{2}A^2 + \cdots\cdots \tag{21}$$

Clearly we also have

$$U^{+} = \exp(-A) \tag{22}$$

This "trick" allows us to view the energy variation as a function of $\frac{n(n-1)}{2}$ independent parameters which can be chosen without constraint.

It is now useful to rewrite equation 19 in terms of the operators $\hat{E}_{ij}$ that were introduced previously. If we define an operator $\hat{A}$ by

$$\hat{A} = \sum_{ij} A_{ij}\hat{E}_{ij} \tag{23}$$

then we have

$$\hat{U} = \exp \hat{A} \tag{24}$$

and the matrix elements of U are given as

$$U_{ij} = \langle \phi_i | \hat{U} | \phi_j \rangle \tag{25}$$

It is clear that a transformation of the orbitals defined by equations 17-25 must induce a transformation of the wavefunction itself so that

$$\psi' = \exp \hat{A} \, |\psi\rangle \tag{26}$$

Alternatively, a transformation of the orbitals transforms H and using the Hausdorf[11] expansion we can write the transformed hamiltonian as

$$\hat{H}' = e^{-A} \hat{H} e^{A} = \hat{H} + [\hat{H},\hat{A}] + \frac{1}{2}[[\hat{H},\hat{A}],\hat{A}] + \ldots \tag{27}$$

Note that equation 27 is valid both for the operators and corresponding matrices; however, if we use the operators as defined in equation 3, then one may perform the manipulations at the operator level using equation (13). The energy associated with $\hat{H}'$ can now be written as

$$\begin{aligned} E &= \langle \psi | \hat{H}' | \psi \rangle \\ &= \langle \psi | \hat{H} | \psi \rangle + \langle \psi | [\hat{H},\hat{A}] | \psi \rangle + \frac{1}{2} \langle \psi | [[\hat{H},\hat{A}],\hat{A}] | \psi \rangle + \ldots \end{aligned} \tag{28}$$

Thus the total energy is a function of the parameters $A_{ij}$. One can now rewrite equation 28 as a quadratic form in the $A_{ij}$ as

$$E = E_0 + FA + \frac{1}{2} A^{+} G A \tag{29}$$

with

$$F_{ij} = \langle \psi | [\hat{H}, \hat{M}_{ij}] | \psi \rangle \quad i > j \tag{30}$$

$$G_{ijkl} = \langle \psi | [\hat{M}_{ij} \; \hat{H} \; \hat{M}_{kl}] | \psi \rangle \quad ij > kl \tag{31}$$

and

$$\hat{M}_{ij} = \hat{E}_{ij} - \hat{E}_{ji} \tag{32}$$

where we have used the symmetric form of the double commutator

$$[\hat{M}_{ij} \hat{H} \hat{M}_{kl}] = \frac{1}{2}[[\hat{M}_{ij},\hat{H}],\hat{M}_{kl}] + \frac{1}{2}[\hat{M}_{ij},[\hat{H},\hat{M}_{kl}]] \tag{33}$$

In equation 29 the lower half of the matrix A is written as a column vector

$$A = \begin{pmatrix} A_{11} \\ A_{21} \\ \vdots \\ A_{ij} \\ \vdots \end{pmatrix} \quad i > j \tag{34}$$

We can thus identify $F_{ij}$ and $G_{ijkl}$ as being the first and second derivatives of the energy with respect to the $A_{ij}$.

At this stage we can state the central ideas of SCF theory in a very general way. One wishes to find that transformation $\mathcal{U}$ which makes E defined by equation 29 stationary. Clearly we have

$$F + GA = 0 \tag{35}$$

and the condition for optimum orbitals is that

$$F_{ij} = 0 \tag{36}$$

Thus to find the optimum orbitals for a given wavefunction $\psi$ defined by equation 1, one need only compute the $F_{ij}$ and $G_{ijkl}$ in equation 35, and solve for the $A_{ij}$. Since one has used a truncated (not exact) expression for E (Eq.28) one must iterate until all the $F_{ij}$ are less than some threshold.

It is now useful to examine equation 30 in more detail, in order to be able to interpret the significance of equation 36. If we expand the commutator, substituting for $\hat{M}_{ij}$ we have

$$\begin{aligned} &\langle\psi|[\hat{H},\hat{E}_{ij}-\hat{E}_{ji}]|\psi\rangle = \\ &\quad -\langle\hat{E}_{ji}\psi|\hat{H}|\psi\rangle + \langle\psi|\hat{H}|\hat{E}_{ij}\psi\rangle - \langle\psi|\hat{H}|\hat{E}_{ji}\psi\rangle + \langle\hat{E}_{ij}\psi|\hat{H}|\psi\rangle \end{aligned} \tag{37}$$

where we have used equation 14 to write

$$\langle\psi|\hat{E}_{ij}\hat{H}|\psi\rangle = \langle\hat{E}_{ji}\psi|\hat{H}|\psi\rangle \tag{38}$$

Clearly the first and third terms on the right hand side of equation 37 are equal, as are the second and fourth, so that we can rewrite the conditions (36) as

$$2\{\langle\psi|\hat{H}|\hat{E}_{ij}\psi\rangle - \langle\psi|\hat{H}|\hat{E}_{ji}\psi\rangle\} = 0 \tag{39}$$

$$\langle\psi|\hat{H}|(\hat{E}_{ij}-\hat{E}_{ji})\psi\rangle = 0 \tag{40}$$

The conditions (40) are just the Brillouin conditions for a general CI wavefunction $\psi$. These are usually referred to as the generalized Brillouin conditions [12,13]. Their meaning becomes clear with an example. Let us suppose that orbital $\phi_j$ is occupied in some of the CSF $|k\rangle$ but orbital $\phi_i$ is not. The operator $\hat{E}_{ij}-\hat{E}_{ji}$ produces some state $|{}^i_j\rangle$ not contained in $\psi$

$$|{}^i_j\rangle = (\hat{E}_{ij} - \hat{E}_{ji})|\psi\rangle \tag{41}$$

The state $|{}^i_j\rangle$ differs from state $|\Psi\rangle$ in that orbital j is replaced wherever it occurs by orbital i (since the effect of $\hat{E}_{ji}$ gives zero by definition). The state $|{}^i_j\rangle$ is a type of singly excited state and condition (40) implies that for SCF orbitals, these matrix elements will vanish at SCF convergence.

As a final step in this introductory section it is necessary to evaluate the commutator

$$[\hat{H}, \hat{E}_{ij}] = \hat{H}\hat{E}_{ij} - \hat{E}_{ij}\hat{H} \tag{42}$$

which has appeared in many of the expressions so far. We use the expression for $\hat{H}$ defined in equation 16 and then apply the commutation relations (13). The result is then easily obtained as

$$[\hat{H}, \hat{E}_{ij}] = 2\Big\{\sum_a (\langle i|h|a\rangle \hat{E}_{aj} - \langle j|h|a\rangle \hat{E}_{ai}) + \sum_{bc} [ab|ci](\hat{E}_{ab}\hat{E}_{cj} - \delta_{bc}\hat{E}_{aj}) - \sum_{bc} [ab|cj](\hat{E}_{ab}\hat{E}_{ci} - \delta_{bc}\hat{E}_{ai})\Big\} \tag{43}$$

and with equation 36, we have the most general formulation of the SCF conditions possible. When the energy is optimized with respect to unitary transformations of the orbitals in $\Psi$ then we have

$$\langle\Psi|[\hat{H}, \hat{E}_{ij} - \hat{E}_{ji}]|\Psi\rangle = 0$$

for all i and j. In the following subsection we shall explore these conditions for the simple case of closed shell single configuration SCF. In later subsections the problems will be discussed for general MC-SCF.

## AN ELEMENTARY TREATMENT OF SINGLE CONFIGURATION SCF

In this section we will present the familiar closed shell single configuration SCF within the framework just discussed. Our objectives are to familiarize the reader with the methodology presented in the previous section and also to demonstrate the simplicity of the formalism.

We are concerned with a single configuration of doubly occupied MO

$$\Psi = |\phi_\alpha(1)\,\bar{\phi}_\alpha(2)\,\ldots\ldots\ldots\ldots\,\phi_\gamma(N-1)\,\bar{\phi}_\gamma(N)| \tag{44}$$

where we have a set of N/2 doubly occupied MO $\{\phi_\alpha\}$ and a set of (m − N/2) virtual or unoccupied MO $\{\phi_r\}$. (We use subscripts $\alpha\ \beta\ \gamma$ for doubly occupied orbitals and r s t u for virtual orbitals, reserving subscripts i j k l for orbitals of either set.).We must now consider the matrix elements $F_{ij}$ for the wavefunction $\Psi$. Fortunately the only non-zero matrix elements of $\hat{E}_{ij}$ are those for which i = j = an occupied orbital. Thus we have

$$\langle \Psi | \hat{E}_{\gamma\gamma} | \Psi \rangle = 2 \tag{45}$$

for all $\gamma$ occupied. However, in equation 43 we also have bi-linear terms in the $\hat{E}_{ij}$ of the form

$$\langle \Psi | \hat{E}_{ij} \hat{E}_{kl} - \delta_{jk} \hat{E}_{il} | \Psi \rangle$$

However because of equation 45 the only non-zero elements are

$$\langle \Psi | \hat{E}_{\gamma\gamma} \hat{E}_{\delta\delta} | \Psi \rangle = 4 \tag{46}$$

and

$$\langle \Psi | \hat{E}_{\gamma\delta} \hat{E}_{\delta\gamma} - \hat{E}_{\gamma\gamma} | \Psi \rangle = - \langle \Psi | \hat{E}_{\gamma\gamma} | \Psi \rangle = -2 \tag{47}$$

Thus if we evaluate $\langle \Psi | [\hat{H}, \hat{E}_{ij} - \hat{E}_{ji}] | \Psi \rangle$ in equation 43, it is clear that the summations for a, b and c can be limited to the occupied orbitals only and that either i or j must be a doubly occupied MO. Thus there are two possible cases; either i and j are both occupied (case a) or i is occupied and j is virtual (case b). For case (a) we have from equation 43,

$$\begin{aligned}\langle \Psi | [\hat{H}, \hat{E}_{\alpha\beta} - \hat{E}_{\beta\alpha}] | \Psi \rangle &= 2 \langle \Psi | \Big\{ \sum_{\mu} \big( \langle \alpha | h | \mu \rangle \hat{E}_{\mu\beta} - \langle \beta | h | \mu \rangle \hat{E}_{\mu\alpha} \big) \\ &+ \sum_{\eta\chi} [\mu\eta | \chi\alpha] \big( \hat{E}_{\mu\eta} \hat{E}_{\chi\beta} - \delta_{\eta\chi} \hat{E}_{\mu\beta} \big) - \sum_{\eta\chi} [\mu\eta | \chi\beta] \big( \hat{E}_{\mu\eta} \hat{E}_{\chi\alpha} - \delta_{\eta\chi} \hat{E}_{\mu\alpha} \big) \Big\} | \Psi \rangle\end{aligned} \tag{48}$$

However from equations 45, 46 and 47 the only non-zero terms are given by

$$\begin{aligned}\langle \Psi | [\hat{H}, \hat{E}_{\alpha\beta} - \hat{E}_{\beta\alpha}] | \Psi \rangle = 2 \langle \Psi | \Big\{ &\langle \alpha | h | \beta \rangle \hat{E}_{\beta\beta} - \langle \beta | h | \alpha \rangle \hat{E}_{\alpha\alpha} \\ &+ \sum_{\eta} \big( [\eta\eta | \beta\alpha] \hat{E}_{\eta\eta} \hat{E}_{\beta\beta} - [\eta\beta | \eta\alpha] \hat{E}_{\beta\beta} \big) \\ &- \sum_{\eta} \big( [\eta\eta | \alpha\beta] \hat{E}_{\eta\eta} \hat{E}_{\alpha\alpha} - [\eta\alpha | \eta\beta] \hat{E}_{\alpha\alpha} \big) \Big\} | \Psi \rangle\end{aligned} \tag{49}$$

Now when we take expectation values of the $\hat{E}_{ij}$ in equation 49 it is clear that the right hand side of equation 49 is identically zero. Thus the SCF conditions hold trivially when one considers mixing between doubly occupied MO.

Turning now to case b we have

$$\begin{aligned}\langle \Psi | [\hat{H}, \hat{E}_{r\alpha} - \hat{E}_{\alpha r}] | \Psi \rangle &= 2 \langle \Psi | \Big\{ \sum_{\mu} \big( \langle r | h | \mu \rangle \hat{E}_{\mu\alpha} - \langle \alpha | h | \mu \rangle \hat{E}_{\mu r} \\ &+ [\mu\mu | \alpha r] \hat{E}_{\mu\mu} \hat{E}_{\alpha\alpha} - [\mu\alpha | \mu r] \hat{E}_{\alpha\alpha} - [\mu\mu | r\alpha] \hat{E}_{\mu\mu} \hat{E}_{rr} + [\mu r | \mu\alpha] \hat{E}_{rr} \big) \Big\} | \Psi \rangle\end{aligned} \tag{50}$$

However, since $\langle\Psi|\hat{E}_{ij}|\Psi\rangle=0$ unless i and j are equal and occupied we have

$$\langle\Psi|[H,\hat{E}_{r\alpha}-\hat{E}_{\alpha r}]|\Psi\rangle = 4\Big(\langle r|h|\alpha\rangle + \sum_{\mu}\big(2[\mu\mu|\alpha r]-[\mu r|\mu r]\big)\Big) \quad (51)$$

$$= 4\langle\alpha|h^F|r\rangle$$

where

$$h^F(1) = h(1) + \sum_{\mu} 2J_{\mu}(1) - K_{\mu}(1) \quad (52)$$

Thus equation 51 and 52 define the usual Hartree-Fock SCF operator. It is clear that for the optimum orbitals the matrix element of $h^F$ between occupied and virtual orbitals will vanish. This condition will hold if the occupied orbitals are eigenfunctions of $h^F$ so that

$$h^F|\alpha\rangle = \varepsilon_{\alpha}|\alpha\rangle \quad (53)$$

and it is in this form that one usually solves the closed shell SCF problem. Nevertheless it is clear that one may transform either the occupied or virtual orbitals amongst themselves without affecting the energy and thus the orbitals need not be eigenfunctions of $h^F$.

At this stage one may derive the usual LCAO form of the HF equation by substituting for the occupied MO in equations 52 and 53 to obtain

$$FY = ESY \quad (54)$$

with

$$|\alpha\rangle = \sum_{A} C_{A\alpha}|A\rangle \quad (55)$$

$$F_{AB} = \langle A|h^F|B\rangle \quad (56)$$

$$S_{AB} = \langle A|B\rangle \quad (57)$$

and

$$\langle A|h^F|B\rangle = \langle A|h|B\rangle + \sum_{CD}\sum_{\mu}\big\{2[AB|CD]-[AC|BD]\big\}c_{C\mu}c_{D\mu} \quad (58)$$

Further, in equation 54 we have collected together $c_{A\mu}$ for all A in each column of Y. One may then use any of the standard methods to solve equation 54 at each iteration. Usually one transforms F to an orthongonal basis defined by

$$VSV^{+} = 1 \quad (59)$$

and solves

$$\bar{F}U = EU \quad (60)$$

with

$$\bar{F} = VFV^{+} \quad (61)$$

so that

$$Y = V^{+}U \tag{62}$$

In practice one usually uses the matrix Y from the previous iteration so that

$$V^{i} = Y^{(i-1)} \tag{63}$$

Thus at each iteration, the Fock operator is transformed to the basis of the MO of the previous iteration so that

$$\bar{F}_{ij}^{(k)} = \langle \phi_i^{(k-1)} | h^F | \phi_j^{(k-1)} \rangle \tag{64}$$

and

$$Y^{(k)} = Y^{(k-1)} U^{(k)} \tag{65}$$

with

$$F^{(k)} U^{(k)} = E^{(k)} U^{(k)} \tag{66}$$

Thus each step of the SCF process generates a $U^{(k)}$ which transforms the previous set of MO $\phi^{(k-1)}$. At convergence $U^{(k)} = 1$ and $\bar{F}_{\alpha r}^{(k)} = 0$ for all $\alpha$ and $r$.

It is now instructive to compare this procedure with the second order process defined by equation 35. In order to do this we must understand the relationship between the diagonalization of the Fock matrix and the solution of the equation system 35. It is clear that we can write a unitary transformation as a product of 2 x 2 Jacobi rotations

$$U = \prod_{ij} U^{(ij)} = \exp\left(\sum_{ij} A_{ij}\right) \tag{67}$$

with

$$U^{(ij)} = \begin{array}{c|ccccc|} & & i & & j & \\ & 0 & \vdots & 0 & \vdots & 0 \\ i & \cdots & \cos A_{ij} & \cdots & \sin A_{ij} & \cdots \\ & 0 & \vdots & 0 & \vdots & 0 \\ j & \cdots & \sin A_{ij} & \cdots & \cos A_{ij} & \cdots \\ & 0 & \vdots & 0 & \vdots & 0 \end{array} \tag{68}$$

and

$$A_{ij} = \begin{array}{c|ccccc|} & & \vdots & & \vdots & \\ i & \cdots & 0 & \cdots & A_{ij} & \cdots \\ & & \vdots & & \vdots & \\ j & \cdots & A_{ij} & \cdots & 0 & \cdots \\ & & \vdots & & \vdots & \end{array} \tag{69}$$

Thus the "angles" $A_{ij}$ in each Jacobi rotation are just the "parameters" of the matrix A. In order to diagonalise F we must choose $A_{ij}$ so that

$$\tan 2A_{ij} = \frac{2F_{ij}}{F_{ii} - F_{jj}} \qquad (70)$$

and if $A_{ij}$ is sufficiently small we have

$$A_{ij} = \frac{F_{ij}}{F_{ii} - F_{jj}} \qquad (71)$$

Now, if we solve equation 35 formally we have

$$A_{ij} = -F_{ij} / G_{ijij} \qquad (72)$$

Thus in ordinary SCF theory we approximate the second derivative matrix by $F_{ii} - F_{jj}$. As we shall see, in multi-configuration methods this approximation may not be adequate.

Even in ordinary closed shell SCF, convergence may be slow. However, if we recognize that we are using an approximate form of equation 35 it is possible to understand the reasons. The matrix G must be positive definite in order to obtain convergence. However, this can be ensured by modifying G to $G + \lambda$ where $\lambda$ is a level shift parameter[14]. Thus if one of the occupied orbitals is of higher energy than one of the virtual orbitals in closed shell SCF, the process will diverge. However, this can be circumvented by shifting the occupied orbital energies to lower values using a level shift. In general, since we are using an approximate $G$ anyway, we can replace equation 71 by

$$A_{ij} = \beta_{ij} F_{ij} / (F_{ii} - F_{jj} + \lambda_{ij}) \qquad (73)$$

where $\beta$ and $\lambda$ are chosen empirically to obtain the best convergence. The parameter $\beta$ is chosen as the damp factor. If $\lambda$ are chosen large enough and $\beta$ small enough then the closed shell SCF is guaranteed to converge.

Before moving to a general discussion of multi-configuration SCF theory it is useful to summarize the main features of closed shell SCF theory. The formalism is very simple because the matrix elements of $\hat{E}_{ij}$ are very simple (cf. equations 45, 46 and 47). When formulated in the basis of the trial MO the SCF conditions reduce to

$$F_{\alpha r} = 0$$

(i.e. all Fock matrix elements between occupied and virtual orbitals are zero). If the SCF procedure is carried out by successive diagonalisations of the Fock operator, it can be seen that one is using an approximate second order procedure where the inverse second derivative matrix is approximated as the difference in the diagonal Fock matrix elements.

OBJECTIVES OF MC-SCF CALCULATIONS.

The state of our understanding about electronic structures that can occur in the course of chemical reactions (when molecular systems are in transition from one equilibrium configuration to another) is still modest. This is because concepts that are familiar from static equilibrium configurations do not carry over since the standard closed shell SCF approximation yields results which are not even qualitatively correct. Rather one requires multi-configurational wavefunctions that include those configurations essential for correctly generating the essential features of a specific reaction as distinct from those configurations that aim to recover correlation effects. In addition one must optimize the orbitals. Thus the objective of the MC-SCF approach is the simultaneous optimization of both CI coefficients and orbital expansion coefficients within the variation method. It is clear then, that we have two problems - one must develop an effective computational strategy and then one must apply ones chemical knowledge to the choice of configurations to be used in the calculations. In this section we aim to give some discussion of the latter and also to attempt to give some physical insight into the process of orbital optimization.

Ultimately, ones choice of configurations for a limited CI reduces to a choice of possible orbital occupancies. Thus we should begin our discussion with a very general form of CI expansion. Let us partition the orbital space of our problem into "core" orbitals ($\alpha\beta\gamma\delta$) "valence" orbitals ($\mu\nu\chi\eta$) and virtual or unoccupied orbitals ($rstu$). The "complete" MC-SCF-CI space (to which we shall refer as the "reference space") consists of all possible arrangements of valence electrons in valence orbitals (i.e. a full valence shell CI). In this reference space all the "core" orbitals are doubly occupied. Thus the reference space is defined by the partition of the occupied MO set into orbitals that are doubly occupied and those which, at any point on the reaction surface may have occupation numbers less than one. In addition to the complete MC-SCF-CI space we have the complementary "secondary" space which contains all possible configurations with core and valence orbitals replaced by virtual orbitals, and core orbitals replaced by valence and virtual orbitals. We refer to these secondary states as single excitations if they differ from a reference configuration by one orbital, and double excitations if they differ by two.

We can again write the expansion for the wavefunction in terms of the $\hat{M}_{ij}$ as

$$\Psi' = e^{T}|\Psi\rangle \qquad (74)$$

with $$T = T_1 + T_2 + \ldots\ldots \tag{75}$$

$$T_1 = \sum_{ij} A_{ij} \hat{M}_{ij} \tag{76}$$

$$T_2 = \sum_{ijkl} D_{ijkl} \hat{M}_{ij} \hat{M}_{kl} \tag{77}$$

where $T_1$ generates the single excitations, $T_2$ generates the doubles and $A, D$ are coefficients to be determined variationally. For the optimum orbitals the $A_{ij}$ go to zero and thus the single excitations are removed. Note also that because of the products which occur on the expansion of the exponential, one has removed the major components of the triple excitations as well. Thus the MC-SCF orbitals are the optimum starting point for a computation of the "dynamic" (double excitation part) of the correlation energy. However, we are concerned with the non-dynamic part of the correlation energy in the reference space. The single excitations are thus implicitly included when the orbitals are optimum.

Provided we use a full CI in a set of valence orbitals - complete MC-SCF-CI - our chemical intuition need only be applied to determine which orbitals shall be allowed to have partial occupancy - the valence orbitals. In most cases this choice is rather obvious. If we are breaking or making bonds we require the bonding and antibonding MO's for the particular bonds to have partial occupancy. In other words we require that those orbitals that are "open shell" in the fragments to be partially occupied. In other examples, an examination of a qualitative MO correlation diagram for orbital energy crossings tells one which orbitals may be partly occupied at some point on the energy curve.

In practice, one may not be able to use the complete MC-SCF-CI reference space. Rather one may be forced to select configurations from this space (due to the large dimension). The use of the complete MC-SCF-CI space has the advantage that the energy is invariant to a unitary transformation of the valence orbitals. Thus in this case the valence orbitals may be chosen either delocalized (symmetry adapted) or localized on the fragment (for ease of intepretation). If one selects only certain valence configurations, then one must be certain that the neglected configurations do not contribute at any geometry. Further, there is no guarantee that a transformation among the valence orbitals which destroys the symmetry will not lower the energy. In other words, for less than a full valence shell CI, symmetry is a constraint on the wavefunction.

## COMPUTATIONAL STRATEGY IN MC-SCF THEORY.

### (i) Orbital Optimization.

In this section we shall discuss the orbital optimization

problem in more detail using the methodology developed in the introductory section to extend the method for the single configuration case. This formalism is much more powerful than the original Lagrangian multiplier method proposed by Roothaan for open shell systems.

There are essentially three distinct methods that are now in common use for the solution of the orbital optimization problem - the Newton-Raphson method (and various approximations to it) which involve the solution of equation 35, the so-called Super CI method (see references [15-19] for a detailed exposition) and the older Iterative Natural Orbital method (INO) first proposed by Davidson [20]. The Newton-Raphson method is the simplest computationally and we shall devote most of our discussion to this.

In principle, the MC-SCF equations in the Newton-Raphson method are as easy to set up and solve as in the case of closed shell SCF. One must evaluate the matrix elements of $\mathcal{F}$ and $\mathcal{G}$ as defined by equations 30 (or equation 43) and 31 and solve 35 by some stable method. The parameter matrix A is then used to generate the unitary transformation for that iteration. Ideally, one would like to approximate $\mathcal{G}$ in some manner as in closed shell SCF; however, it is not clear how one should do this in the general case. The major practical difficulty arises because the MC-SCF orbitals cannot be written as eigenfunctions of a single Fock operator. Thus in order to discuss these problems in detail we must give the form of the matrix elements of the Fock operator $F_{ij}$ in the general case.

As in the single configuration SCF case the $F_{ij}$ are obtained by taking the expectation value of equation 43. Using the equations 45-47 ,we can show that

$$\langle \Psi | \hat{E}_{\mu\nu} \hat{E}_{\gamma\gamma} | \Psi \rangle = 2\, V_{\mu\nu} \tag{78}$$

and

$$\langle \Psi | \hat{E}_{\gamma\mu} \hat{E}_{\nu\delta} | \Psi \rangle = - \langle \Psi | \hat{E}_{\nu\delta} \hat{E}_{\gamma\mu} | \Psi \rangle - \langle \Psi | \hat{E}_{\nu\mu} | \Psi \rangle$$
$$= - V_{\nu\mu} \tag{79}$$

where $\gamma\delta$ are core orbitals and $\mu\nu$ are valence orbitals. Thus we have the remarkable simplification that we can express the Fock matrix elements $F_{ij}$ in terms of the density matrices of the valence MO alone. We need only two independent Fock operators defined as

$$(F_1)_{ij} = 2\Big[\langle i | h^c | j \rangle + \sum_{\mu\nu} \big\{ [\mu\nu | ij] - \tfrac{1}{2} [\mu i | \nu j] \big\} V_{\mu\nu} \Big] \tag{80}$$

$$(F_2)_{i\eta} = \sum_{x} \langle i | h^c | x \rangle V_{x\eta} + \sum_{\mu\nu} [x\mu | \nu i] \, \Gamma_{x\mu\nu\eta} \tag{81}$$

for arbitrary orbitals i and j ($x\mu\nu$ are valence orbitals). We then have 4 distinct types of $F_{ij}$ matrix element depending upon the orbitals :

$$F_{r\alpha}^{cu} = \langle r \,|\, F_1 \,|\, \alpha \rangle \qquad (82)$$

$$F_{r\eta}^{vu} = \langle r \,|\, F_2 \,|\, \eta \rangle \qquad (83)$$

$$F_{\alpha\eta}^{cv} = \langle \eta \,|\, F_1 \,|\, \alpha \rangle - \langle \alpha \,|\, F_2 \,|\, \eta \rangle \qquad (84)$$

$$F_{\eta\chi}^{vv'} = \langle \eta \,|\, F_2 \,|\, \chi \rangle - \langle \chi \,|\, F_2 \,|\, \eta \rangle \qquad (85)$$

with

$$h^c = h + \sum_{\alpha} 2 J_\alpha - K_\alpha \qquad (86)$$

The superscripts c u, v u, c v, and v v′ refer to matrix elements between closed and unoccupied orbitals, valence and unoccupied orbitals, core and valence orbitals, and finally different valence orbitals. The final matrix F has the form

$F =$ (87)

| | c | v | u |
|---|---|---|---|
| c | | $F^{cv}$ | $F^{cu}$ |
| v | | $F^{vv'}$ / $F^{vv'}$ | $F^{vu}$ |
| u | | | |

It must be emphasized that the matrices $F^{cc}$, $F^{uu}$ and the diagonal elements of $F^{vv'}$ are not defined at this stage (as in the closed shell case). If one were to proceed as in the closed shell case and use an approximate Newton-Raphson method (i.e. diagonalize F) one must assume some form for the diagonal elements of F. The most obvious choice is to use the operator $h^c$ to compute the diagonal elements.

If we adapt an approximate Newton-Raphson method using diagonalization of the F matrix, then the SCF procedure is formally identical to the closed shell case - each iteration generates a matrix $U^{(i)}$ which transforms the orbitals from iteration i-1. The only practical problem is the determination of the starting set of orthogonal MO. The simplest solution is to use orbitals from closed shell SCF calculations or to diagonalize $h^c$ in the original atomic orbital basis.

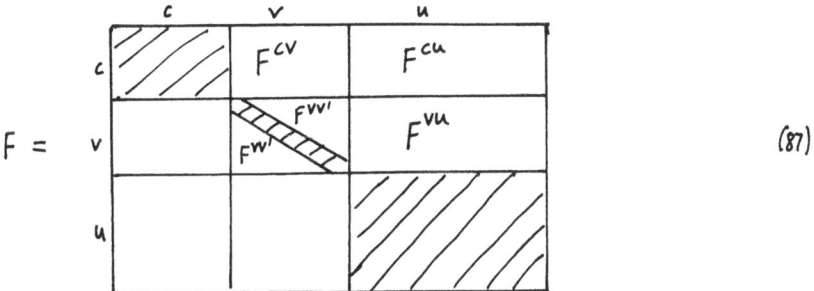

It is clear that in many cases the approximate Newton-Raphson method just described will not converge sufficiently rapidly. One must therefore consider using the full Newton-Raphson method as defined by equation 35. However, the quadratic convergence must be paid for in terms of the much more involved calculations of the matrix elements $G$ at each iteration (for the explicit formulae of the $G$ matrix elements the reader is referred to reference 21). The alternative, is to regard the $G_{ijkl}$ as empirical parameters and to make judicious use of the level shift and damp factors as defined in equation 73. In doing this one may hope to gain enough experience so that the full $G$ need not be computed.

Let us now turn our attention to the so-called "Super CI" method which has proved to be a popular alternative to the full Newton-Raphson method. In the super CI method one still satisfies the Brillouin conditions (equation 40); however, the second derivative matrix $G$ corresponds to a more truncated expansion of the energy than in the full Newton-Raphson method. As in our previous discussion, the wavefunction that results from a unitary transformation of the orbitals is written in terms of the operator $\hat{A}$ defined by equations 23 and 26. However, we now truncate the expansion of $\psi'$ at two terms so that we have

$$\psi' = (1 + \hat{A})\,|\psi\rangle \tag{88}$$

The energy associated with this wave function can now be written as

$$E' = \langle\psi'|\hat{H}|\psi'\rangle \,/\, \langle\psi'|\psi'\rangle \tag{89}$$

Note that in our derivation of the Newton-Raphson method $\langle\psi'|\psi'\rangle$ is equal to the unit matrix for every order in A, but this is not true if the wavefunction expansion is truncated as in equation 88 above. Thus we have on substition for $\psi'$

$$\langle\psi'|\hat{H}|\psi'\rangle = \langle\psi|\hat{H}|\psi\rangle + 2\langle\psi|\hat{H}|\hat{A}\psi\rangle + \langle\hat{A}\psi|\hat{H}|\hat{A}\psi\rangle \tag{90}$$

and

$$\langle\psi'|\psi'\rangle = 1 + \langle\hat{A}\psi|\hat{A}\psi\rangle \tag{91}$$

Thus using the following expansion truncated at the second term

$$\frac{1}{1 + \langle\hat{A}\psi|\hat{A}\psi\rangle} = 1 - \langle\hat{A}\psi|\hat{A}\psi\rangle + \langle\hat{A}\psi|\hat{A}\psi\rangle^2 + \cdots \tag{92}$$

we can write equation 28 as

$$E' = \langle\psi|\hat{H}|\psi\rangle + 2\langle\psi|\hat{H}|\hat{A}\psi\rangle + \langle\hat{A}\psi|\hat{H}|\hat{A}\psi\rangle - \langle\psi|\hat{H}|\psi\rangle\langle\hat{A}\psi|\hat{A}\psi\rangle \tag{93}$$

If we again subtitute for $\hat{A}\psi$ we have an equation similar to equation 29 except that the $F_{ij}$ and $G_{ijkl}$ now have the form

$$F_{ij} = \langle\psi|\hat{H}\hat{M}_{ij}|\psi\rangle$$

$$G_{ijkl} = \langle\psi\hat{M}_{ij}|\hat{H}|\hat{M}_{kl}\psi\rangle - \langle\psi|\hat{H}|\psi\rangle\langle\psi\hat{M}_{ij}|\hat{M}_{kl}\psi\rangle \tag{94}$$

Now we have already shown that the $F_{ij}$ as defined by equation 30 differ from the $F_{ij}$ defined above only by a factor of 1/2(cf. equations 37-40). Thus it is the form of $G_{ijkl}$ that differ fundamentally in the two methods.

From a practical point of view, in the Super CI method, one does not attempt to solve the linear equation system of the form given by equation 35. Instead one solves the complete equivalent "partitioned" eigenvalue problem.

$$\begin{pmatrix} \langle\Psi|\hat{H}|\Psi\rangle & \langle\Psi|\hat{H}|\hat{m}_{ij}\Psi\rangle \\ \langle\Psi\hat{m}_{ij}|\hat{H}|\Psi\rangle & \langle\Psi\hat{m}_{ij}|\hat{H}|\hat{m}_{kl}\Psi\rangle \end{pmatrix}\begin{pmatrix} 1 \\ A_{ij} \end{pmatrix} = \varepsilon\begin{pmatrix} 1 & 0 \\ 0 & \langle\Psi\hat{m}_{ij}|\hat{m}_{kl}\Psi\rangle \end{pmatrix}\begin{pmatrix} 1 \\ A_{ij} \end{pmatrix} \qquad (96)$$

where $\varepsilon$ is just

$$\varepsilon = \langle\Psi'|\hat{H}|\Psi'\rangle / \langle\Psi'|\Psi'\rangle$$

However, this is just a CI calculation using an expansion of the form given by equation 88, and hence the name Super CI.

At this point we should comment briefly on the relative merits of the full Newton-Raphson method and the Super CI method since they are the most commonly used second order methods. Firstly, in the full Newton-Raphson method the energy is correct to second order in the $A_{ij}$ whereas in the Super CI, the wavefunction is truncated at terms that are linear in the $A_{ij}$ . As a consequence, the Super CI must be regarded as an approximate second order method. Secondly, the Newton-Raphson method has overwhelming advantages from a practical point of view in terms of the ease of evaluation of the $G$ . Obviously the metric matrix $\langle\Psi'|\Psi'\rangle$ does not need to be constructed in the Newton-Raphson method. However, there is considerable cancellation that takes place in the evaluation of $G$ in Newton-Raphson that does not occur in the corresponding terms in Super CI. Indeed in Newton-Raphson the expectation values of the double commutators reduce to expressions which involve only second order density matrices. In contrast, in Super CI one has the difficult practical task of constructing third order density matrices (for a discussion of third order density matrices see reference 22). In spite of the fact that Super CI may be computationally inefficient, because of the relationship of this optimization procedure to a well defined CI procedure, it does offer conceptional advantages in discussing divergence problems, we shall return to this in the next subsection.

The last optimization method to be discussed is the INO method. Although this is an older method, it is perhaps best thought of as an extension of the super CI method to the case where the CSF coefficient variations are not assumed to be independent of orbital variation. Let us begin by considering the full variational space of the super CI wavefunction defined by equation 88. Let us use the symbols $|I\rangle\,|k\rangle$ to denote the CSF appearing in the MC-SCF function $\psi$ . Now if we allow $\hat{A}$ to operate on $\psi$ we generate an orthogonal space spanned by CSF consisting of states where an occupied orbital in $\psi$ is replaced by an unoccupied one. We denote the states in this orthogonal space by $|R\rangle\,|S\rangle$ etc. Thus for example if we consider

$$\hat{M}_{ij}\,|\psi\rangle = \sum_{k} c_k\,\hat{M}_{ij}\,|k\rangle \tag{97}$$

one sees that $\hat{M}_{ij}$ produces a linear combination of secondary space states

$$\hat{M}_{ij}\,|k\rangle = \sum_{R} \langle k\,|\,\hat{M}_{ij}\,|\,R\rangle\,|R\rangle \tag{98}$$

The complete CI expansion in the super CI space then has the form

$$\psi' = \sum_{K} c_K\,|k\rangle + \sum_{R} c_R\,|R\rangle \tag{99}$$

In super CI we assume that the $c_K$ are fixed and thus the $c_R$ are completely determined by the $A_{ij}$ so that

$$\begin{aligned} c_R &= \langle R\,|\,\psi'\rangle \\ &= \sum_{K}\sum_{ij} \langle K\,|\,\hat{M}_{ij}\,|R\rangle\, c_K\, A_{ij} \end{aligned} \tag{100}$$

where we have used equation 88 to expand $\psi'$. On the other hand, if we now allow both the $c_K$ and $c_R$ to be determined variationally the $A_{ij}$ are given as

$$\begin{aligned} A_{ij} &= \langle \psi\,|\,\hat{M}_{ij}\,|\,\psi'\rangle \\ &= \sum_{KR} \langle R|\,\hat{M}_{ij}\,|k\rangle\, c_K\, c_R \end{aligned} \tag{101}$$

where we have used equation 99 for the expansion of $\psi'$ this time. Since at most one of the two contributions from $\hat{M}_{ij}$ (cf. eq.32) will be non-zero (or else they will both be equal), we have

$$A_{ij} = \gamma_{ij} \tag{102}$$

where $\gamma$ is the first order density matrix of the complete expansion given by equation 8. Thus the parameter matrix $A_{ij}$ can be obtained from the first order density matrix for the complete Super CI expansion.

The "natural orbitals" are defined by the unitary transformation which diagonalizes $\gamma$ . Thus one could generate a unitary transformation from the $A_{ij}$ by diagonalization (cf. eq. 67-71). This is the origin of the name iterative natural orbital method. Alternatively, the $A_{ij}$ could be used to generate U directly.

It should be clear that INO will be computationally intractable in general because of the large CI expansion generated in equation 99. One would expect it to have much the same convergence properties as super CI itself unless the orbital variations were strongly coupled to the variation of the $C_K$.

It is possible, however, to take into account the coupling of the $C_K$ with the orbital variation in the Newton-Raphson method (see reference 3 for a full discussion). If this refinement proves necessary the Newton-Raphson method is to be preferred for the same reason as discussed previously.

Finally, we should make a few remarks on the computer implementation of MC-SCF. The main computational bottleneck involves the transformation of the integrals which appear in the formulae for the $F_{ij}$ and $G_{ijkl}$. In closed shell SCF one need only compute matrix elements of the closed shell Fock-operator. However in MC-SCF one needs certain transformed molecular integrals as well. If we do not use the full Newton-Raphson method, then the Fock operators required in equations 80 to 86 can be constructed from the coulomb and exchange operators constructed from the valence orbitals.

$$(J_{\mu\nu})_{AB} = \sum_{CD} [AB|CD]\, C_{C\mu} C_{D\nu} \tag{103}$$

$$(K_{\mu\nu})_{AB} = \sum_{CD} [AC|BD]\, C_{C\mu} C_{D\nu} \tag{104}$$

Thus the computation time will be approximately $N_V^2$ times the computation time of a closed shell calculation where $N_V$ is the number of valence orbitals. In contrast, for the evaluation of the second derivative matrix one needs all molecular integrals except those which contain more than 2 virtual orbital indices. A detailed discussion of this problem has been given by Ruedenberg et al [19]. However, even though fewer molecular integrals are required than in the full 4-index transformation, the computation time for this step will be approximately proportional to $N^4$ as opposed to $N^5$ for the full transformation. Thus it is apparent that the full Newton-Raphson method becomes prohibitive for large basis sets, while the approximate method depends primarily on the number of valence (open shell MO). However, the approximate Newton-Raphson method may converge very slowly.

Let us now briefly summarize the orbital optimization problem for general MC-SCF. It is clear that if one works in the parameter space of the $A_{ij}$ one has no need for the complicated coupling operator method of the earlier open shell SCF theories. Indeed, if one does not require the rigorous determination of the second derivative matrix $G_{ijkl}$ the SCF theory for MC-SCF is very similar to closed shell methods. The only new features are the appearance of the first and second order density matrices of the valence CI expansion and as we shall discuss in the next section, these are easly determined. It is clear that if the full Newton-Raphson method can be approximated so that the $G_{ijkl}$ need not be computed explicitly there is a considerable computational advantage. However, if one must use a second order method the Newton-Raphson method has overwhelming computational advantage over the super CI method.

## ii-The CI Problem

A general MC-SCF procedure must embody an iterative alternation between the optimization of the MO (just discussed) and the determination of the CI coefficients. The latter procedure basically requires the solution of the standard eigenvalue problem (cf. equation 1-4). However, since one must use the results of the CI calculation in the orbital optimization step, this aspect of the calculation needs to be organised in such a manner as to represent this information in the most convenient fashion.

The general computational strategy is obvious from equations 1-16. If we use the Hamiltonian in the form given by equation 5, (using the definitions of equations 10 - 15). Then one must prepare a file of the matrices A and B defined by equations 10 - 15. There is one distinct matrix for each integral $\langle i|h|j\rangle$ or $[ij|kl]$ and the rows and columns are labelled by the configuration index K and L. This file is then processed with the valence MO integrals to form H in the CI step or alternatively it is processed with an eigenvector of the CI step to produce the one and two particle density matrices (defined by equations 8 and 9) which are required by the orbital optimization step.

In early MC-SCF and open shell calculations one required complicated tabulations of so called vector coupling coefficients (the $A_{ij}^{KL}$ and $B_{ijkl}^{KL}$ of equations 8 and 9). With the development of Unitary Group methods (see references 4-10), these coefficients can now be obtained in a completely automatic ("black box") way and the CI step becomes almost trivial.

There is however, a complication which arises in the CI step if one requires an excited state of the same symmetry as the ground state. In this case one has a constraint on both the orbital optimization step and the CI step - the excited state CI eigenvector must be orthogonal to the lower states of the same symmetry. However, if we optimize the orbitals for an excited state eigenvector, the lower eigenvectors will change and rigorous orthogonality with the optimum lower state eigenvector will be lost. Clearly, one could simply orthogonalize at each iteration. However, in practice, provided one includes the dominant configurations of both lower and excited eigenvectors in ones CI space orthogonalisation will be maintained approximately.

## SOME ILLUSTRATIVE NUMERICAL EXAMPLES.

In this subsection, we wish to discuss some numerical examples. The molecules were chosen not because of their chemical interest but rather because they illustrate the types of theoretical and practical problems encountered in performing calculations using MC-SCF theory.

### (i) $H_2$

Oddly, $H_2$ turns out to be a pathelogical system for MC-SCF, in the sense that for this simple chemical system one has severe convergence problems with MC-SCF. However, through the resolution of these difficulties one gains experience that is applicable to more complex systems.

Ones objective in an MC-SCF computation for $H_2$ is to use a CI expansion so that one obtains proper dissociation into two ground state H atoms. Thus at infinite separation the wavefunction must have the form

$$\Psi = \frac{1}{\sqrt{2}}\left(1\sigma_g^{\,2} - 1\sigma_u^{\,2}\right)$$

while at the equilibrium nuclear configuration the coefficient of the $(1\sigma_g)^2$ configuration approaches 1.0 and the coefficient of the $(1\sigma_u)^2$ configuration approaches zero. Thus we are dealing with a 2 valence orbital 2 valence electron problem

As one approaches the separated atoms, the MC-SCF procedure converges quickly. However, at the equilibrium separation one has convergence problems. As was first pointed out by Wood and Veillard [23], the simple approximate Newton-Raphson method (which involves diagonalization of the F matrix) converges to a saddle point on the energy surface which is above the starting point. Thus this is an excellent example to examine the various MC-SCF procedures described in earlier sections.

In table I we present the convergence of various optimization procedures for this two configuration MC-SCF calculation (equilibrium internuclear separation).

TABLE I. Convergence of MC-SCF calculations of $H_2$.

| METHOD / ITERATION | a | b | c | d |
|---|---|---|---|---|
| 0 | -1.1341 | -1.1340 | -1.1394 | -1.1394 |
| 1 | -1.1509 | -1.1454 | -1.1464 | -1.1439 |
| 2 | -1.1513 | -1.1509 | -1.1504 | -1.1443 |
| 3 | -1.1513 | -1.1511 | -1.1511 | -1.1445 |
| 4 | ⋮ | -1.1511 | -1.1512 | -1.1447 |
| | | ⋮ | ⋮ | ⋮ |
| 46 | | | | -1.1499 |

(a) Full Newton-Raphson method (ref. 21)
(b) Iterative Natural Orbital method.
(c) Approximate Newton-Raphson with level shifting and damp factors.
(d) Approximate Newton-Raphson with no level shifting or damp factors.

In column a we present the results using the full Newton-Raphson method obtained by Kendrick and Hillier [21]. In column b we have the results of the INO method. In both these computations, iteration 0 corresponds to the energy obtained with the SCF closed shell MO. In columns c and d we present the result of the approximate Newton-Raphson method. In these two cases the starting orbitals had to be chosen to be different from the closed shell SCF MO because of the divergence problem just discussed. In each case we did a preliminary computation optimizing the orbital for the $(1\sigma_u)^2$ configuration by itself (which converges rapidly). The purpose is merely to move the optimization procedure away from the local minimum encountered using the closed shell SCF-MO.

It can be seen that both the second order procedures (methods a and b) converge starting from the closed shell SCF MO. However, convergence of INO is very slow at the end. In contrast, the approximate Newton-Raphson method (method d) converges too slowly to be of any practical value. On the other hand, method c, where one has used empirical damp factors and a level shift, converges at an acceptable rate. We thus have encountered two separate problems which we need to examine in more detail - the initial divergence with the closed shell SCF MO and the apparently slow convergence of the simple approximate Newton-Raphson method.

If we examine the wavefunction in the complete variational space (reference space + single excitations) the origin of the initial MC-SCF divergence becomes clear. The dominant terms are given as

$$\psi^{(0)} = 0.997(1\sigma_g)^2 + 0.059(1\sigma_u)^1(2\sigma_u)^1 + 0.037(1\sigma_u)^2$$

whereas at convergence we have

$$\psi^{(F)} = 0.993(1\sigma_g)^2 - 0.117(1\sigma_u)^2 + 0.0000(1\sigma_u)^1(2\sigma_u)^1$$

The closed shell $\sigma_u$ orbital is rather poor and the $G$ matrix will be indefinite due to the presence of the $(1\sigma_u)^1(2\sigma_u)^1$ configuration resulting in convergence to a saddle point for the approximate method. It is clear that if we had chosen a reference space with 3 valence orbitals ($1\sigma_g\, 2\sigma_u\, 2\sigma_u$) the problem would be avoided by including the configuration $(1\sigma_u)^1(2\sigma_u)^1$ in the reference space. However, simple physical considerations enable one to avoid the initial divergence as we shall now discuss.

If one examines the closed shell SCF $\sigma_u$ MO one observes that it is very diffuse. This is as expected since the virtual orbitals are shielded by the doubly occupied $\sigma_g$ MO. However, ones intuition says that the $\sigma_u$ MO ought to be similar to that for the $^2\Sigma_u$ state of $H_2^+$. Thus one ought to start the calculation with an $\sigma_u$ orbital appropriate to $H_2^+$. However, since we have a two electron system it could be argued that one should optimize the virtual orbitals for the closed shell problem for the $(\sigma_u)^2$ state. This was the choice made here. It can be seen that this choice for $\sigma_u$ is only slightly better than the SCF MO on energetic grounds (see the energies for iteration 0 in Table I). Nevertheless, the approximate Newton-Raphson method is convergent with this choice.

Let us turn our attention to the slow convergence of the approximate Newton-Raphson method (method D in Table I). Since the wavefunction is dominated by the $(1\sigma_g)^2$ configuration the energy is very insensitive to the optimization of the $\sigma_u$ orbital. Further, the $\sigma_g$ orbital will be almost optimal if it was obtained from a closed shell SCF computation. This implies that the first and second derivatives of the energy will be small in magnitude. In the absence of quantitative information about the second derivatives, one may use damp factors which scale the $F_{ij}$ as in equation 73. In the present example, after a little experimentation, the optimum damp factor for mixing the $\sigma_u$ orbitals is around 50.

Thus for this example it can be seen that the second order methods (Newton-Raphson method or INO method b) give rapid convergence. In contrast, if one avoids the explicit construction of $G$ , one must choose the starting orbitals carefully and make judicious use of damp factors. However, the time per iteration is significantly slower in the second order methods. In our calculations the iteration time for INO was ten times slower than for method c.

(ii) $F_2$ Dissociation

One of the earliest MC-SCF calculations was performed on the $F_2$ molecule by Das and Wahl in 1971 [24]. Since this is a particularly "delicate" problem we have re-run these calculations in order to illustrate some of the ideas discussed in the first sections of this article. This problem illustrates dramatically the improper formal description of molecular formation and dissociation at the SCF level and the increase in the correlation energy with molecular formation provided by the creation of a new electron pair.

The electronic configuration of $F_2$ is

$$\text{core } (1\pi_u)^4 (1\pi_g)^4 (3\sigma_g)^2 \; 3\sigma_u \; 4\sigma_g \; 4\sigma_u \; 2\pi_g \; 2\pi_u$$

where the core is

$$(1\sigma_g)^2 (1\sigma_u)^2 (2\sigma_g)^2 (2\sigma_u)^2$$

In dissociation or molecule formation one must allow the $3\sigma_g$ and $3\sigma_u$ orbitals to have variable occupancy just as in the $H_2$ problem. Thus one has a 2 valence orbital 2 valence electron problem and the MC-SCF calculation is formally very similar to $H_2$ .

In order to decide which additional configurations to include in the MC-SCF space one must think in terms of localized atomic orbitals of the F atom. It is clear that our 2 configuration reference space contains the following configurations written in terms of the atomic orbitals.

$$\text{I } (2p_A\sigma)^1 (2p_B\sigma)^1 \qquad \text{II } (2p_A\sigma)^2 \qquad \text{III } (2p_B\sigma)^2$$

At infinite separation

$$\frac{1}{\sqrt{2}} (3\sigma_g)^2 - \frac{1}{\sqrt{2}} (3\sigma_u)^2 = (2p_A\sigma)^1 (2p_B\sigma)^1$$

however at finite separations the "ionic" terms II and III contribute as well. However since the atomic state

$$(2p_A\pi)^4 (2p_A\sigma)^1 \quad {}^2P$$

has 2 other components arising from

$$(2p_A\pi)^3 (2p_A\sigma)^2$$

these should be included in the reference space as well. Similarly one must include the corresponding "ionic" states as well. Thus we must enlarge the valence space to include the $1\pi_{gu}$ orbitals by adding all the configurations of the form

$$(1\pi_{gu})^2 \longrightarrow (3\sigma_u)^2$$

Thus we now have a 6 valence orbital 10 valence electron problem. This would give us 21 configurations in all. However, because of the high symmetry most of these do not contribute to the energy.

Finally, one would like to add those configurations that correspond to the change in dynamic correlation during bond formation. Again, this analysis is most easily done in terms of the configurations of the separated atoms. Since it is the ionic configurations II and III, that increase in importance during bond formation one must include correlation effects in these two configurations by including the effects of terms like

$$(2p_A)^6 \longrightarrow (2p_A)^5(3p_A)^1$$

One may think about this type of excitation as recovering the extra atomic correlation introduced in atom A due to the extra electron transferred to it. In terms of the MO , one must include terms of the form

$$1\pi_{ug}\, 3\sigma_g \longrightarrow 3\sigma_u\, 2\pi_{ug}$$

At this level, one interprets this configuration as "semi-internal" correlation. Hence one electron is excited within the valence orbital space and one electron is excited from the valence orbital space into the virtual orbital space. This type of excitation is expected to be the dominant effect in multi-reference correlation effects. The case where one excites two electrons into the virtual manifold should be much less important.

Let us now turn to the results which are summarized in table II. In the top half of the table we show the convergence of the MC-SCF for the two configuration calculation using the approximate Newton-Raphson method both with (method c) and without (method d) level shifting/damp factors. In the second half of the table we give the results for the 14 configuration 10 valence orbitals computation MC-wavefunction. Here we have given the INO results (method b) as an indication of the performance of a second order method.

The results are in general agreement (considering the rather modest Gaussian basis used here consisting of double zeta sp orbitals only) with the earlier calculations of Das and Wahl. Considering the results of method c we see one recovers only a small portion of the binding energy with the two valence orbital MC-SCF; however, when the valence orbital space is increased to 10 orbitals (14 configurations) one recovers about 2/3 of the binding energy. The semi-internal component ($1\pi_{ug}3\sigma_g \rightarrow 3\sigma_u\ 2\pi_{ug}$ ) is estimated to be around .003 A.U. by Das and Wahl. Thus it is apparent that the choice of the valence orbital space is critical to the success of MC-SCF computations.

Turning now to the problem of convergence of the approximate Newton-Raphson procedure, we encountered similar problems to $H_2$ .

Table II Convergence of MC-SCF procedures for $F_2$ .

| | $r = 2.68$, $E_{SCF} = -198.7128$ | | | $r = 6.00$, $E_{SCF} = -198.4408$ |
|---|---|---|---|---|
| Iteration Number | b | c | d | d |
| | | | 2 Configurations | |
| 0 | | -198.7624 | -198.7624 | -198.7554 |
| 1 | | -198.7847 | -198.7696 | -198.7751 |
| 2 | | -198.7898 | -198.7747 | -198.7818 |
| 3 | | -198.7900 | -198.7788 | -198.7841 |
| 4 | | -198.7901 | -198.7819 | -198.7849 |
| Final(Iteration) | | -198.7902(6) | -198.7898(15) | -198.7854(8) |
| | | | 14 Configurations | |
| 0 | -198.7830 | -198.8168 | -198.8154 | -198.7867 |
| 1 | -198.8201 | -198.8178 | -198.8165 | -198.7867 |
| 2 | -198.8252 | -198.8184 | -198.8169 | |
| 3 | -198.8269 | -198.8189 | -198.8171 | |
| 4 | -198.8278 | -198.8192 | -198.8173 | |
| Final(Iteration) | | -198.8276(20) | -198.8132(11) | -198.7867(2) |

At the equilibrium internuclear separation, convergence is slow with method a because the wavefunction is dominated by the $(3\sigma_g)^2$configuration and the $(3\sigma_u)$ orbital is very poor. Clearly one must use a damp factor for the mixing of $3\sigma_u$ with the remainder of the virtual orbitals. We have made no attempt to optimize this factor; however, as can be seen from table II one obtains a reasonable rate of convergence with a damp factor of around 20. The INO method probably gives an approximate indication of the convergence of a second order approach. Here we see that convergence is rapid. Nevertheless it is very much slower per iteration. Careful optimization of the damp factor would clearly improve the convergence rate of the approximate Newton-Raphson method.

It is clear that the technique used here should be adequate for all first row homo-nuclear diatomics. For example, in $N_2$ the valence space would still be $1\pi_u 3\sigma_{gu}$. However since one has fewer electrons (the $\pi_g$ shell is formally empty), there will be many more configurations (175 in all, of which 32 contribute due to symmetry arguments).

Table III. Timing Data.

| | CPU + I/O TIME/ITERATION. |
|---|---|
| SCF | 1.0 |
| 2 Configuration MC-SCF | 4.0 |
| 14 Configuration MC-SCF | 50.0 |
| INO | 288.3 |
| Full orbital transformation (for full Newton-Raphson) | 70.0 |

Finally, we should comment very briefly on the computer time required for these computations. In Table III we have summarized the computer time relative to the time required to perform a single SCF iteration. Also we have included the time required to perform a full orbital integral tranformation. (This quantity should give an indication of the time required for each iteration of a full second order Newton-Raphson). The INO time is dominated by the time required to solve the CI problem with 768 configurations

While we would not pretend that our programs are terribly efficient, the timing data should give an order of magnitude estimate for the various approaches. As expected for the MC-SCF calculation the time is approximately proportional to the square of the number of valence orbitals. For the 14 configuration calculation we have a total of 10 valence orbitals so the computation time rises dramatically from the 2 configuration case.

Since the total number of orbitals is small the full transformation is fast but this will not be the case in general. It is clear that the INO method is useful only to check the occasional problem encountered in MC-SCF.

(iii) Dimerization of two Methylenes

Basch's [25] calculation in 1972 of the least motion co-planar approach of two $CH_2$ molecules represented the first application of the MC-SCF method to a polyatomic molecule of chemical interest. This problem also turns out to be a "delicate" one in the sense that if one chooses the MC-SCF configuration space incorrectly one obtains a result that is qualitatively correct. In the present work we have repeated these computations to show how the problem of configuration selection can be aleviated using localized fragment orbitals.

Naive orbital correlation diagrams suggest that at the equilibrium geometry one has the configuration

$$\text{I} \quad (\sigma)^2(\pi)^2$$

while at infinite separation one will have the configuration

$$\text{II} \quad (\sigma)^2(\sigma^*)^2$$

corresponding to two methylenes with spins singlet paired in the $\sigma$ orbitals. Thus there will be an allowed orbital crossing of the $\sigma^*$ and $\pi$ orbitals leading to an avoided crossing of configurations I and II and an associated barrier. Further one expects the correlation energy of the $\pi$ electrons to be dominated by the configuration

$$\text{III} \quad (\sigma)^2(\pi^*)^2$$

Indeed, if one performs this 3 configuration MC-SCF one finds the expected avoided crossing and activation barrier.

However, the above reasoning is manifestly incorrect since the separated methylenes will have a total of 4 unpaired electrons (for the tripet state). Basch was able to obtain a qualitatively correct result using the additional configurations.

$$\text{IV} \quad (\pi^*)^2(\pi)^2$$

$$\text{V} \quad (\sigma^*)^2(\pi)^2$$

$$\text{VI} \quad (\pi^*)^2(\sigma^*)^2$$

By analogy with our considerations on $F_2$ such a configuration set is reasonable. One is breaking two bonds - a $\sigma$ and a $\pi$ bond. Thus one clearly needs configurations I, III and IV to describe this process as well as configuration V for the $\sigma$ correlation. However Basch found that the wavefunction at infinite separation had the form

$$\psi = |\mathrm{I}\rangle - |\mathrm{III}\rangle - |\mathrm{IV}\rangle + |\mathrm{VI}\rangle$$

Thus configuration VI, a quadruple excitation, was of vital importance. The importance of the quadruple excitation VI is typical of computations for more than 2 valence electrons. In general one requires all possible products of double excitations to correctly describe dissociation.

From the above discussion, we see that the choice of configurations is critical for this type of computation. However, the choice of configurations I-VI above is not at all obvious from the outset. It is clear that one must include in ones calculation all the low lying states of the separated fragments. Thus there may be advantages in formulating the expansion directly in terms of fragment orbitals. We shall now give some discussion of this possibility.

Our general philosophy will be as follows. We choose a set of valence orbitals for each molecular fragment. We may then define the "fragments in molecules configurations" (FIMC) to be those configurations that correctly describe the ground and lowest excited states of the fragments. The valence orbitals of the combined system will be the union of the sets of valence orbitals for each fragment. At every internuclear separation, we first perform an MC-SCF computation using the FIMC configurations only. This provides the reference point for discussing binding. Finally, we perform a complete MC-SCF which includes a full valence shell CI.

It is clear that the energy of an MC-SCF calculation is invariant to a unitary transformation among the valence orbitals if one performs a full valence shell CI. However, if we use a subset of the full set of valence configurations, this is not true and in general if we break the symmetry the energy will converge to a lower value. In particular, for the subset of configurations corresponding to the FIMC, the orbitals will naturally localize on the fragments. These localized orbitals can then be used to interpret the full calculation.

For $(CH_2)_2$ the FIMC set must contain a minimum of 2 configurations : the configuration denoted T-T corresponding to two triplet methylenes coupled to a singlet and S-S corresponding to two singlet methylenes. We have two valence orbitals from each methylene giving rise to a 4 valence orbital, 4 valence electron calculation, (20 configurations), for the full calculation.

The results of the calculations using these fragment orbitals are given in Table IV. The symbols T-T and S-S have just been

Table IV. Calculations on $(CH_2)_2$ using Fragment MO.

| r | $E^{FIMC}$ | $E_0^{MC-SCF}$ | | $E_\infty^{MC-SCF}$ | | | $E^{pt}$ |
|---|---|---|---|---|---|---|---|
| | | T-T | T-S | T-T | S-S | T-S | T-T |
| 1.30 | -77.1698 | -77.9239 | -77.7769 | -77.9532 | -77.3233 | -77.7937 | -77.9837 |
| 2.13 | -77.6372 | -77.7923 | -77.7671 | -77.8050 | -77.5836 | -77.7781 | -77.8138 |
| 2.66 | -77.7024 | -77.7387 | -77.7307 | -77.7432 | -77.6043 | -77.7342 | -77.7545 |
| 3.20 | -77.7192 | -77.7278 | -77.7251 | -77.7280 | -77.6224 | -77.7257 | -77.7351 |
| 9.54 | -77.7242 | -77.7242 | -77.7242 | -77.7242 | -77.6354 | -77.7242 | -77.7301 |

discussed. We use the symbol T-S to denote the triplet state formed by combination of triplet and singlet methylene. In the first column we give the energy of the 2 configuration MC-SCF formed by the FIMC. In columns 2 and 3 we give the energy of the 20 configuration CI calculation performed in the basis of the fragment orbitals (i.e. the 0th iteration of the full MC-SCF). In columns 4, 5, and 6 we give the energy obtained for the full MC-SCF (where the orbitals have been optimized for the T-T eigenvector). Finally in column 7 we give the results of a perturbation calculation, (for details of the method see reference [26]), in the basis of the orbitals optimized for the FIMC. In addition in table V we have given the coefficient of the T-T configuration in the full calculation at each internuclear separation. (The calculations were carried out in the standard STO 4-31G basis with a fixed H-C-H angle).

Table V. Eigenvector Components for $(CH_2)_2$ singlet calculations using Fragment Orbitals.

| r | Coefficient of Triplet-Triplet Methylene Configuration |
|---|---|
| 1.30 | .54 |
| 2.13 | .79 |
| 2.66 | .94 |
| 3.20 | .99 |
| 9.54 | 1.00 |

Since we have carried out the calculation in terms of the optimum orbitals for FIMC, at infinite separation the values of $E^{FIMC}$, $E_0^{MC-SCF}$ (T-T, T-S) and $E_\infty^{MC-SCF}$ (T-T, T-S) should be equal; and $E_\infty^{MC-SCF}$ (S-S) - $E_\infty^{MC-SCF}$ (T-T) should correspond to twice the singlet-triplet splitting of methylene. An examination of the result for 9.54 A shows that this is true. Further, from table V the coefficient of T-T goes to 1.00 at infinite separation.

If we compare the results tabulated in columns 2 and 4, where we give the results of the calculation with and without re-optimization of the FMO, it can be seen that, even for the equilibrium internuclear separation, the FMO are excellent approximations to the final orbitals. The binding energy computed with the FMO ($E_0^{MC-SCF}$) is only in error by 15%. As a consequence the binding can be easily interpreted in terms of the addition of charge transfer configurations to $E^{FIMC}$. As expected, $E^{FIMC}$ is strongly repulsive. Clearly the state S-S that correlates with two singlet methylenes is strongly repulsive ($E_\infty^{MC-SCF}$ column 5 table IV). As expected the FMO are not as good for describing the T-S triplet state because of the very different spin coupling.

Finally, almost as an aside we should make a brief comment on the perturbation theory calculation shown in table IV because it demonstrates the utility of the FMO. In this calculation, rather than re-optimize the FMO for the full valence shell MC-SCF, we have included all of the single excitations (that would be included in an INO calculation) including the semi-internal configurations, plus all valence orbital double excitations. It is reasonable to include these effects by perturbation theory because, as we have just pointed out, the effects of allowing the FMO to relax is very small. The binding energy computed in this fashion is 20% larger than $E_{\infty}^{MC-SCF}$. This is reasonable due to the contribution of semi-internal effects.

(iv) $Be_2$

As a final example we would like to briefly mention calculations on an exceedingly delicate problem - the $Be_2$ molecule. In this case one has empty low lying valence orbitals (approximate 2s 2p degeneracy).

The valence space for the MC-SCF computation consists of 8 valence orbitals for 4 valence electrons which gives rise to 336 configurations for the complete MC-SCF calculation. This example is interesting because if one freezes the 1s orbitals one can perform the full CI using 13,790 configurations.

Table VI. MC-SCF and CI calculations on $Be_2$.

| r | $E_0^{MC-SCF}$ | $E_{\infty}^{MC-SCF}$ | $E^{CI}$ |
|---|---|---|---|
| 4.0 | -28.8471 | -28.8911 | -28.9078 |
| 5.0 | -28.8511 | -28.9107 | -28.9176 |
| 6.0 | -28.8507 | -28.9154 | -28.9205 |
| 7.0 | -28.8512 | -28.9132 | -28.9225 |
| 8.0 | -28.8515 | -28.9191 | -28.9233 |
| 9.0 | -28.8516 | -28.9193 | -28.9234 |
| 10.0 | -28.8516 | -28.9192 | -28.9232 |
| . | . | . | . |
| . | . | . | . |
| . | . | . | . |
| 15.0 | -28.8506 | -28.9190 | -28.9231 |

The results of the MC-SCF and full CI calculations are summarized in table VI. In column 1 we have given the energy obtained with the closed shell SCF orbitals, in column 2 the MC-SCF (336 configurations) results and finally the CI results (13,790 configurations). The CI results should be very close to the basis limit. The agreement between MC-SCF and CI results is remarkable. However, the effects of full orbital optimization is critical as can be seen by the erratic behaviour of the results obtained with the closed shell SCF orbitals.

## ACKNOWLEDGEMENTS

The paper was prepared using the FORMAT text program developed by S.Cardy and P.N.Jackson. The authors are grateful for the help of P.N.Jackson in preparing the final version of the text of the manuscript.

One of us (RHAE) is grateful to the Science Research Council for a studentship.

## REFERENCES

1). R. McWeeny, Proc.R.Soc. A232, 114 (1955).
2) R. Carbo and J. M. Riera, A General SCF Theory, (Springer-Verlag, Berlin 1978), Vol. 5.
3) E. Dalgaard and P. Jorgensen, J.Chem.Phys. 69, 3833 (1978).
4) a) J. Paldus, J.Chem.Phys. 61, 5321 (1975). b) J. Paldus, Theoretical Chemistry : Advances and Perspectives, Vol. 2, ed. H. Eyring and D. G. Henderson (Academic Press), 1976.
5) M. Downward and M. Robb, Theor.Chim.Acta 46, 129 (1977).
6) I. Shavitt, Int.J. Quant.Chem. S11, 131 (1977), S12, 5, (1978).
7) B. R. Brooks, and H. F. Schaeffer, J.Chem.Phys. 70, 5092 (1979).
8) D. Hegarty and M. A. Robb, Mol.Phys. 38, 1795 (1979).
9) P. E. M. Sieghbahn, J.Chem.Phys. 70, 5391, (1979).
10) I. Shavitt, Methods of Electronic Structure Theory, ed. H. F. Schaeffer (Plenum) 1977.
11) R. M. Wilcox, J.Math.Phys. 8, 962, (1967).
12) L. Brillouin, Act.Sci.Inst. 71, 159 (1933).
13) B. Levy and G. Berthier, Int.J.Quantum Chem. 2, 307 (1968).
14) V. R. Saunders, and M. F. Guest, Quantum Chemistry : the State of the Art, Procedings of the SRC ATLAS Symposium No. 4 p.119.
15) A. Banerjee and F. Grein, Int. J. Quantum Chem. 10, 123 (1976).
16) F. Grein and T. C. Chang, Chem.Phys.Lett. 12, 44 (1971).

17) F. Grein and A. Banerjee, Int. J. Quantum Chem. S9, 147 (1975); J.Chem.Phys. 66, 1054 (1977).
18) W. H. E. Schwarz and T. C. Chang, Int. J. Quantum Chem. S10, 91 (1976); Theoret.Chim.Acta. 44, 45 (1977).
19) K. Ruedenberg, L. M. Cheung and S. T. Elbert, Int. J. Quantum Chem. 16, 1069 (1979).
20) C. F. Bender and E. R. Davidson, J.Phys.Chem. 70, 2675 (1966).
21) J. Kendrick and I. H. Hillier, Chem.Phys.Lett. 41, 283 (1976).
22) B. Roos, P. R. Taylor and Per. E. M. Siegbahn, (Preprint).
23) M. H. Wood and A. Viellard, Mol.Phys. 26, 595 (1972).
24) G. Das and A. C. Wahl, J.Chem.Phys. 56, 3532 (1972).
25) H. Basch, J.Chem.Phys. 55,1700 (1971).
26) D. Hegarty and M. A. Robb, Mol.Phys. 37, 1455 (1979).

# DEVELOPMENT OF A COMPUTATIONAL STRATEGY IN ELECTRONIC STRUCTURE CALCULATIONS: ERROR ANALYSIS IN CONFIGURATION INTERACTION TREATMENTS

Robert J. Buenker
Lehrstuhl für Theoretische Chemie, Universität Wuppertal
5600 Wuppertal 1, West Germany

Sigrid D. Peyerimhoff and Pablo J. Bruna
Lehrstuhl für Theoretische Chemie, Universität Bonn
5300 Bonn 1, West Germany

## I. INTRODUCTION

From a practical point of view one of the most serious problems in the field of atomic and molecular calculations is the need for estimating their reliability in a given application. In all but the simplest of cases an exact treatment is not possible, or at least is not economically feasible, and as a result it is imperative that certain approximations be introduced. The difficulties attendant upon such procedures thus lie not only in providing that the necessary computations be carried out in a technically optimum manner, but also in recognizing to what extent the resulting calculated findings deviate from experimental reality. For people who do not work in the computational field directly and whose main interest lies in the interpretation of experimental data, the second of these problems is by far the more critcal, particularly in view of the great variety of such theoretical treatments which have been brought forward in recent times.

The fact remains, however, that far more information can be found in the literature concerning the actual design of computational methods than regarding their reliability in specific applications. Nor is the reason for this state of affairs difficult to find, namely that while the methodology of a certain type of calculation is generally subject to precise mathematical analysis, the magnitude of the errors incumbant upon its use in practical investigations is often only determined after a great deal of numerical experience therewith becomes available. Arriving at such judgments in this manner is never as



*I. G. Csizmadia and R. Daudel (eds.), Computational Theoretical Organic Chemistry, 55–76.*

satisfactory as when a straightforward mathematical proof can be achieved, but as the complexity of the physical systems to be treated increases it becomes necessary to rely more and more on a careful observation of the results of explicit calculations in order to form a realistic appraisal of their overall accuracy. After sufficient numerical data are obtained it often becomes clear that a particular computational method is not sufficiently effective, and more importantly that by implementing a certain technical change the major portion of this deficiency can be removed.

In evaluating such numerical results it is quite helpful to develop an organizing principle which allows the various computational methods to be compared with one another on a quantitative basis. One means of accomplishing this objective is to consider the definitions of a few important theoretical techniques and to establish a clear relationship between them and an exact treatment, which in the present instance amounts to solution of the Schrödinger equation itself. For purpose of simplicity this discussion will contain itself within the clamped-nuclei or Born-Oppenheimer Approximation and will also disregard relativistic effects, that is, will deal with a purely electrostatic Hamiltonian operator, but the type of error analysis developed thereby will be seen to be generalizeable to situations in which even these theoretical restrictions are no longer acceptable.

## II. DEFINITIONS OF VARIOUS APPROXIMATE TREATMENTS

In order to understand the relationship between various approximate computational methods and the exact solution of the Schrödinger equation, it is helpful to consider the following arguments based on a matrix representation of the corresponding Hamiltonian operator. If an orthonormalized set of all the eigenfunctions of this operator is chosen as basis for the purpose of constructing the Hamiltonian matrix, the result would be a completely diagonal form with the corresponding energy eigenvalues appearing as the only non-zero elements. Although these solutions are not generally known, the fact that they form a complete set leads to a simple and at the same time very useful conclusion, namely that a Hamiltonian matrix formed in <u>any other complete orthonormal basis</u> differs by at most a unitary transformation from the above diagonal species. A more detailed discussion of the consequences of this observation may be found elsewhere (1,2), but for the present it suffices to say that diagonalization of such a transformed Hamiltonian matrix constitutes a formal solution of the corresponding Schrödinger equation, <u>equally valid for both the bound and the continuum states</u> of the physical system under discussion.

From a practical point of view a straightforward approximation to the above exact matrix method of solution consists in simply employing an incomplete set of many-electron functions to construct a representation of the Hamiltonian matrix, which upon diagonalization leads to hopefully accurate substitutes for the exact eigenstates of the problem. The resulting computational scheme is generally referred to as configuration interaction, or CI for short, and it is well known that the approximate energy values of this type of treatment, for excited species as well as the ground state, satisfy the variation principle. Clearly the only source of error in these calculations is the use of an incomplete basis, and the obvious means of improving on its results in a given application is through systematic expansion of this function set toward the goal of achieving a perfect representation of the asssociated Hamiltonian operator.

Over the past ten to fifteen years there has been a concerted effort to develop efficient computational techniques which will allow the CI method to become a powerful tool for theoretical chemistry. To obtain an overview of these developements it is important to recognize clearly various levels of approximations commonly employed in CI calculations. The many-electron basis functions used to form the required Hamiltonian matrix representation are the well-known Slater determinants or some variation thereof such as Young Tableaus or Gelfand States. To construct these species a series of one-electron functions are required, which are referred to as spin orbitals. In principle the many-electron basis does not have to be orthonormalized but the nature of the required computations is generally greatly complicated when this is not done. For this reason it is also quite convenient to work with an orthonormal spin orbital basis, which commonly consists of self-consistent field (SCF) solutions, some form of natural orbitals (NO's), or less commonly multi-configurational SCF orbitals.

The simplest form of configuration interaction to define and at the same time the most difficult to carry out in practice is the full CI method. The configuration space in this case consists of all possible distinct combinations (Slater products) of spin orbitals which can be formed from a given one-electron (AO) basis. If there are m spatial orbitals in the basis set and the system has n electrons, the total number of such spin orbital combinations is $\binom{2m}{n}$, which for even moderately large values of m is very large. A very useful property of a full CI is that its results are uniquely determined by the linear space spanned by the AO basis, i.e. they are *invariant to any non-singular basis transformation*, including one in which these functions are no longer mutally orthogonal. The eigenfunctions of a full CI transform according to the irreducible representations of the molecular point group of the system being treated because the Hamiltonian matrix is necessarily fully symmetric under all the operations contained therein, just as is the Hamiltonian operator itself; use of a non-symmetric one-electron basis does not alter this situation since again all results of the full CI are independent of such details.

In larger systems a full CI quickly becomes inpractical, whereupon a common variation is often introduced, namely a full valence CI. In this case a fixed one-electron basis for inner-shell or core orbitals is assumed and the CI basis consists of all Slater determinants in which these species are doubly occupied; commonly a null set of one-electron functions is also defined, which species are never occupied in the corresponding CI. The results of a full valence CI are therefore very much dependent on the constitutions of the core orbitals, although they remain independent of any non-singular linear transformation of the non-core (valence) species.

If the computations associated with a full CI in a given AO basis are too extensive to be practicable, experience has shown that it is useful to truncate the configuration space to include a series of reference species which appear with large expansion coefficients in one or more roots of interest, plus all singly and doubly excited (substituted) species thereto. The special significance of single and double excitations derives from the fact that since the electronic Hamiltonian contains at most two-particle terms, only matrix elements between Slater determinants constructed from an orthonormal one-electron basis which differ by no more than a double spin-orbital substitution can be non-zero in magnitude (3); use of a non-orthogonal one-electron basis destroys this simple relationship, however. The simplest case of this type of CI involves use of a single Hartree-Fock reference term plus all singly and doubly excited species formed from the (occupied plus virtual) SCF-MO's of the former configuration, such as has been employed in the work of Roos (4); traditionally this type of calculation has been referred to as an SD-CI, i.e. single- and double-excitation CI.

Since the great majority of electronic states are not adequately represented in this manner, it is very helpful to generalize upon this idea by taking all single and double excitations with respect to a series of all important leading terms in the root or roots of interest. This type of calculation has become very popular in the last few years (5-8) and is referred to as a multi-reference (single- and) double-excitation CI, or MRD-CI for short (D standing for both double and single substitutions together). The key feature of this approach is that by systematically expanding the size of the reference set the results of a full CI can be approached as closely as desired with considerable computational savings.

For large AO basis sets even the MRD-CI spaces quickly become too large to be treated directly by diagonalization of the corresponding Hamiltonian matrices and thus further truncation becomes advisable. The most common way of accomplishing this objective is to estimate the energy contributions of generated configurations by means of simple perturbation theory, and then to select only those species for inclusion in the actual CI calculations which are estimated to cause an energy lowering which exceeds

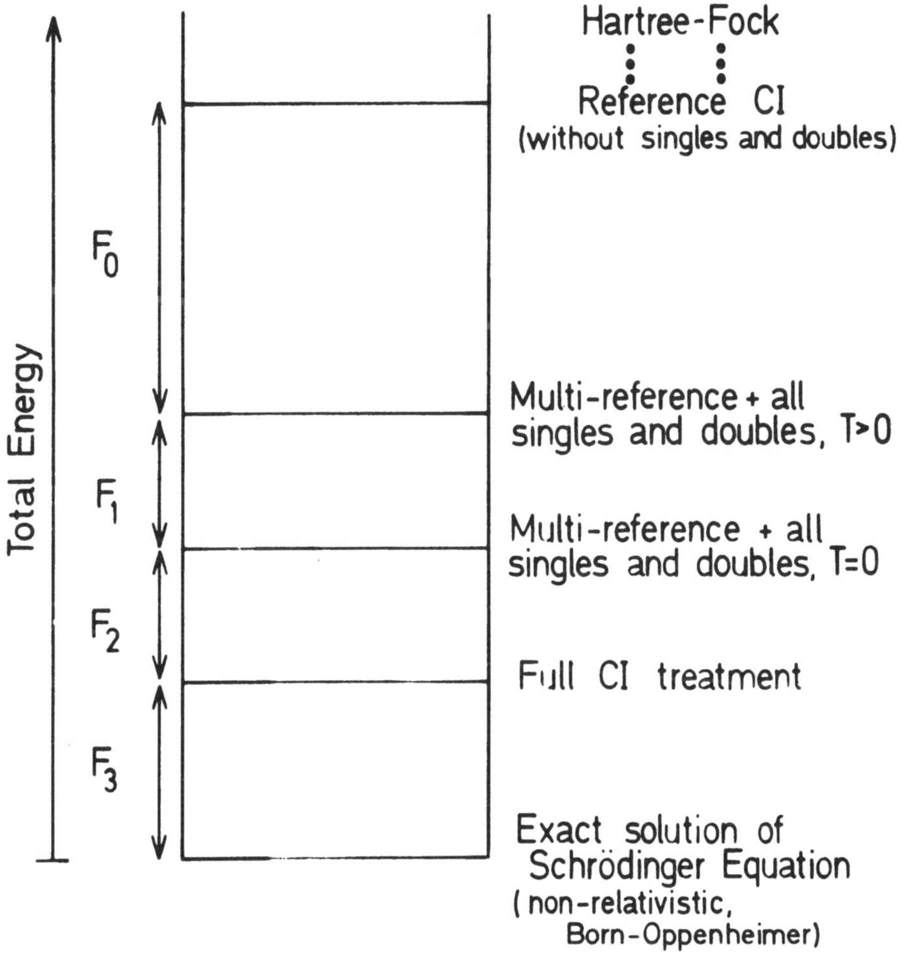


Fig. 1: Schematic diagram showing the variation of the total energy of a system with level of CI treatment.

some threshold value T. Another means of achieving an effective truncation is the Hartree-Fock interacting space technique of Bunge (9), which amounts to retaining only certain Slater determinants (spin assignments) within a given configuration based on their relationship to the reference species themselves. The threshold technique has the advantage of being quite flexible, with its results gradually approaching those of the entire MRD-CI space as the value of T is decreased to zero.

The different levels of CI discussed above can be summarized conveniently in terms of an energy diagram, as contained in Fig. 1. After the Hartree-Fock or SCF (single-configuration) treatment the simplest type of calculation is a CI for the series of reference configurations discussed above, without additional excited species. Then come respectively the truncated and untruncated MRD-CI treatments, followed by the corresponding full CI and finally the exact Schrödinger equation solution, as marked in the figure. The scale employed therein is not meaningful in itself, but in what follows the energy differences between the successive levles of treatment will be referred to as $F_o$-$F_3$ respectively. The magnitudes of these quantities clearly depend on the composition of the AO basis employed, as well as the nature of the reference configurations in the MRD-CI calculations and the threshold value T used to define the level of truncation thereby. In addition except for the full CI all other approximate treatments in the figure are dependent on the choice of the orbital transformation used to form the one-electron basis for the calculations. As with the full CI, variations of all the other treatments discussed can be defined by requiring certain core orbitals to always remain fully occupied and by excluding other one-electron species entirely from the excitation process.

## III. ERROR ANALYSIS FOR THE VARIOUS CI TECHNIQUES

The energy differences $F_i$ in Fig. 1 represent the errors made in CI calculations relative to exact (non-relativistic) results when a certain level of treatment is assumed. By varying the input parameters in a given CI procedure it is clearly possible to control the magnitudes of one or more of the $F_i$ quantities, and the goal in such calculations is to develop a strategy which balances computational expense against the degree of error minimization. To get an idea of the problems involved it is well to begin such an analysis with the full CI procedure itself. Since this type of treatment explicitly includes all configurations in the diagonalization procedure, the only deficiency in such calculations lies in the choice of the AO basis; in terms of Fig. 1 the error in the computed energy $\Delta E$ is simply equal to $F_3$. As pointed out in the previous section, however, a full CI leads to enormously large Hamiltonian matrices in all but the simplest of applications, and hence in practice for this procedure to be applicable a very restricted AO basis needs to be chosen, which is to say $F_3$ must remain a relatively large quantity.

Calculations of this type employing minimal AO basis sets have been carried out for a number of systems (10,11) and very useful results have been obtained with respect to the symmetry correlation of molecular electronic states in dissociation. From a more quantitative point of view, however, such treatments have been much less successful in obtaining agreement with measured data because the errors introduced through the use of such inflexible AO basis sets are too large. Even when one takes account of the fact that energy comparisons are involved in making such judgements, so that cancellation effects can be quite important, the experience is clear that for acceptable computational expense the AO basis errors are not sufficiently similar either for different electronic states or for different nuclear geometries in the same state to afford a high degree of accuracy for such full CI treatments.

With this background it is not difficult to understand the rationale behind the next type of treatment, the MRD-CI procedure. Because a more limited CI space is chosen in this case it is obvious that the computational times for a given AO basis will decrease substantiallly relative to that for a full CI. On the other hand this gain in speed has to be paid for at the price of introducing a second type of error into the calculations ($F_2$ in Fig. 1), in addition to the AO basis error $F_3$ already mentioned. The key point, however, is that it is no longer necessary to deal with such restricted AO basis sets in an MRD-CI calculation because the Hamiltonian matrices thereof possess much smaller orders than in a full CI. In other words, if at the same time that a more restrictive definition of the CI space is incorporated in the treatment, a significantly more flexible AO basis is introduced, the total error in the calculations can be substantially reduced without increasing the overall computational requirements, i.e. the sum of the two error types $F_2$ and $F_3$ can easily be less than $F_3$ for a full CI in a comparably expensive application. In absolute terms this statement simply means that considerably lower total energies are obtained in a given MRD-CI treatment than for a full CI requiring the same computation time, but more significantly for practical applications the result is that the relative errors between treatments of different states or nuclear geometries are smaller in the MRD-CI procedure.

Ultimately the reason that the MRD-CI approach is far more efficient can be traced to the fact that in this procedure attention is centered upon the description of a small number of the possible eigenstates of the system; in reality its results are superior to those of a comparably expensive full CI for only a few roots. Since in actual applications the goals in such calculations rarely involve a very large number of electronic states, however, this limitation is of little practical consequence.

Since limiting the type of configurations to be included in the diagonalizations has proven to be advantageous in the last example, the question arises whether an even more restrictive definition of the CI space might be called for. Even in the MRD-CI procedure the practical limits of such calculations are quickly approached as either the AO basis or the number of reference configurations is increased, for example, so an analogous situation presents itself as in the discussion of the full CI method. The next possibility as mentioned in Section II is to introduce some selection criterion into the MRD-CI framework so that only the most important of the singly and doubly excited configurations are included in the actual diagonalization. In terms of Fig. 1 a new type of error ($F_1$) is introduced thereby and the question thus becomes whether it can be overcome by achieving a reduction in the corresponding $F_2$ and $F_3$ quantities by expanding the AO basis and/or reference set relative to the original MRD-CI treatment without increasing the overall computation time thereby.

So long as the amount of selection is relatively moderate, as effected through the use of either a small T value in the perturbation technique or the employment of the Hartree-Fock interacting space approach, the experience with this computational strategy has generally been quite satisfactory. Because the transformation of electron repulsion integrals over the orthonormalized one-electron basis functions becomes an ever more critical factor as the AO basis increases, and since the size of the MRD-CI space increases in direct proportion to the number of reference configurations, however, it is often necessary to use a relatively high level of selection before any significant reduction in the $F_2$ and $F_3$ errors becomes beneficial from an economical standpoint, at which point the magnitude of $F_1$ therefore begins to be a dominant factor. In addition large-scale selection has a tendency to have a quite different effect on one state than on another (12,13), so that the benefits of error cancellation in relative energy comparisons are often rather small thereby.

## IV. ENERGY EXTRAPOLATION TECHNIQUES: ERROR MINIMIZATION

To this point in the discussion only simple diagonalization of the Hamiltonian secular matrices has been considered as a means of obtaining approximate solutions to the Schrödinger equation. In particular the errors in such calculations have been directly equated to the energy differences between the results of such diagonalization procedures and the corresponding exact findings. In a given application, however, the actual error in calculating a particular eigenvalue of the Schrödinger equation is clearly only the uncertainty in these energy differences, and this observation thus suggests that there is value in formulating a quantitative means of estimating the various $F_i$ quanties in Fig. 1. To illustrate this idea it is well to begin with a technique designed to predict the difference between truncated and T=0 MRD-CI energy results which has proven quite effective (14).

In a threshold type of selection each generated configuration is associated with a perturbation energy lowering $\Delta E_i$ and so a simple means of estimating the latter energy difference ($F_1$) is as the sum of these quantities for all unselected species in the T=0 treatment. This procedure amounts to an overestimation in most cases but it can be improved upon rather simply by introducing a proportionality factor: $F_1 = \lambda_1 \Sigma^1 \Delta E_i = \lambda_1 K_1$. The value for $\lambda_1$ can then be estimated by carrying out explicit truncated MRD-CI treatments at different thresholds and comparing the actual energy change in going from $T_1$ to $T_2$ with the corresponding variation in $K_1$ (see Fig. 2). Numerical experiments (14) have indicated that the uncertainty $\Delta F_1$ can be estimated to within about 5 % of the value of the correction term $K_1$ at the minimum threshold value in this manner.

With this extrapolation technique the errors connected with a given truncated MRD-CI treatment can therefore be substantially reduced without incurring any significant increase in computational expense. The $\Delta E_i$ values must be calculated anyway for the selection process itself and hence the only additonal effort involves summing these values up and in carrying out diagonalizations for smaller secular equations than that obtained at the minimum threshold; for the latter purpose no new Hamiltonian matrix elements need to be generated and eigenvector results obtained at one T value can be used to speed convergence at another.

These findings greatly increase the effectiveness of the truncated MRD-CI technique because the loss in accuracy relative to T=0 results is only the uncertainty in $F_1$ as a result of the extrapolation, thereby allowing selection to be applied more freely than without this procedure. It thus becomes possible to employ larger AO basis and reference sets in the calculations so that the total error of $\Delta E = \Delta F_1 + F_2 + F_3$ in the truncated CI is notably smaller than for a conventional (T=0) MRD-CI treatment of

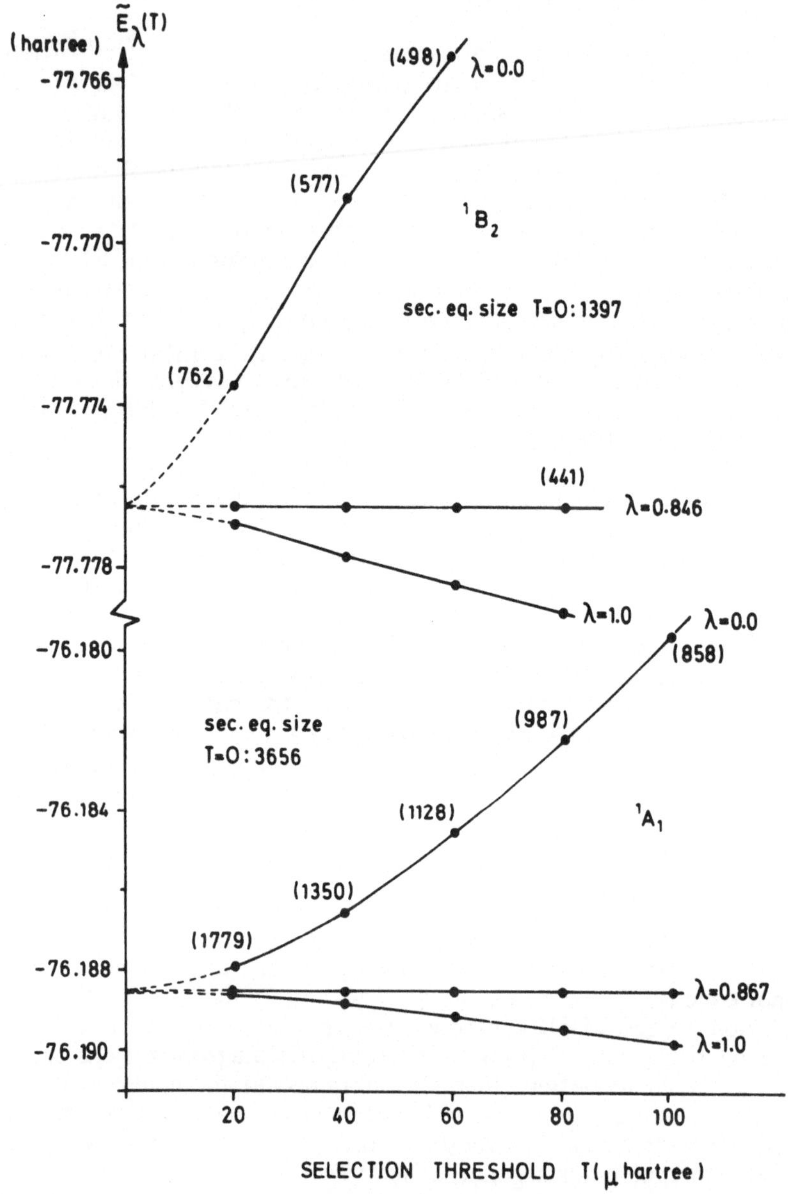


Fig. 2: Illustration of an MRD-CI extrapolation procedure for several electronic states of water; $E_\lambda(T) = E_{MRD-CI}(T) + \lambda \sum_i^1 \Delta E_i$.

comparable expense. As a result an analogous situation exists as before with the comparison of full CI and T=0 MRD-CI calculations in Sect. III, whereby a more restrictive definition of the CI space actually leads to increased accuracy for the same amount of computer time.

More recently progress has been made in developing other extrapolation corrections to further reduce the errors in CI calculations. On the basis of perturbation theory arguments (15) the energy lowering caused by adding all triple and quadruple excitations to an SD-CI can be estimated to be equal to $(1-c_o^2)\ (E_{CI}-E_{SCF})$, where $c_o$ is the CI coefficient of the Hartree-Fock reference function and $E_{SCF}$ is its diagonal energy. The resulting energy contribution is known as the Davidson correction and it can be looked upon as simply a linear extrapolation of the total correlation energy based on results obtained by taking single and double excitations with respect to a single leading term. From this point of view the result can be generalized (16) for a multi-reference function so that the quantity $F_2$ in Fig. 1 is estimated as $(1-\sum_i^{ref} c_i^2)\cdot(E_{MRD-CI}-E_{ref})$, where the sum is over all reference species and $E_{ref}$ is the energy of the desired root in the reference CI itself.

Just as in the first extrapolation procedure it is obvious that the above quantity by itself will not give a generally accurate estimate for the energy difference of interest; in the case of two-electron systems, for which any SD- or MRD-CI is a full CI, $F_2$ is clearly zero, for example, while the above estimate for it will not be as long as $\sum_i^{ref} c_i^2 \neq 1.0$ and $E_{ref} \neq E_{MRD-CI}$. Nevertheless just as before with $\Sigma^{1}\Delta E_i$ it is helpful to introduce a proportionality factor such that $F_2 = \lambda_2\ (1-\sum_i^{ref} c_i^2)\cdot(E_{MRD-CI}-E_{ref}) = \lambda_2 K_2$. If a series of MRD-CI calculations is carried out for a number of different reference sets, the value of $\lambda_2$ can be estimated by comparing the actual energy lowering obtained in going from one calculation to the next with the change in the $K_2$ perturbation quantity for the same variation. Alternatively one can simply plot the MRD-CI energy as a function of $\sum_i^{ref} c_i^2$ with and without the $K_2$ correction and note that both quantities must converge to the true full CI energy when $\sum_i^{ref} c_i^2 = 1$. An example of the latter plotting procedure is given in Fig. 3 for the $^3\Pi$ and $^1\Sigma^+$ states of $CN^+$ (17), from which it appears that the full (valence) CI results for an AO basis containing 50 functions (12 valence electrons) can be estimated to within a few hundredths of an eV by means of this technique.

Of particular interest is the observation that once sufficient reference configurations are included to give a value for $\sum_i^{ref} c_i^2$ in excess of 0.90, the quantity $E_{MRD-CI} + K_2$ (denoted as the full CI correction in Fig. 3) holds very nearly constant with respect to further expansion of the reference set, which is to say that the optimal $\lambda_2$ factor appears to approach unity under these conditions.

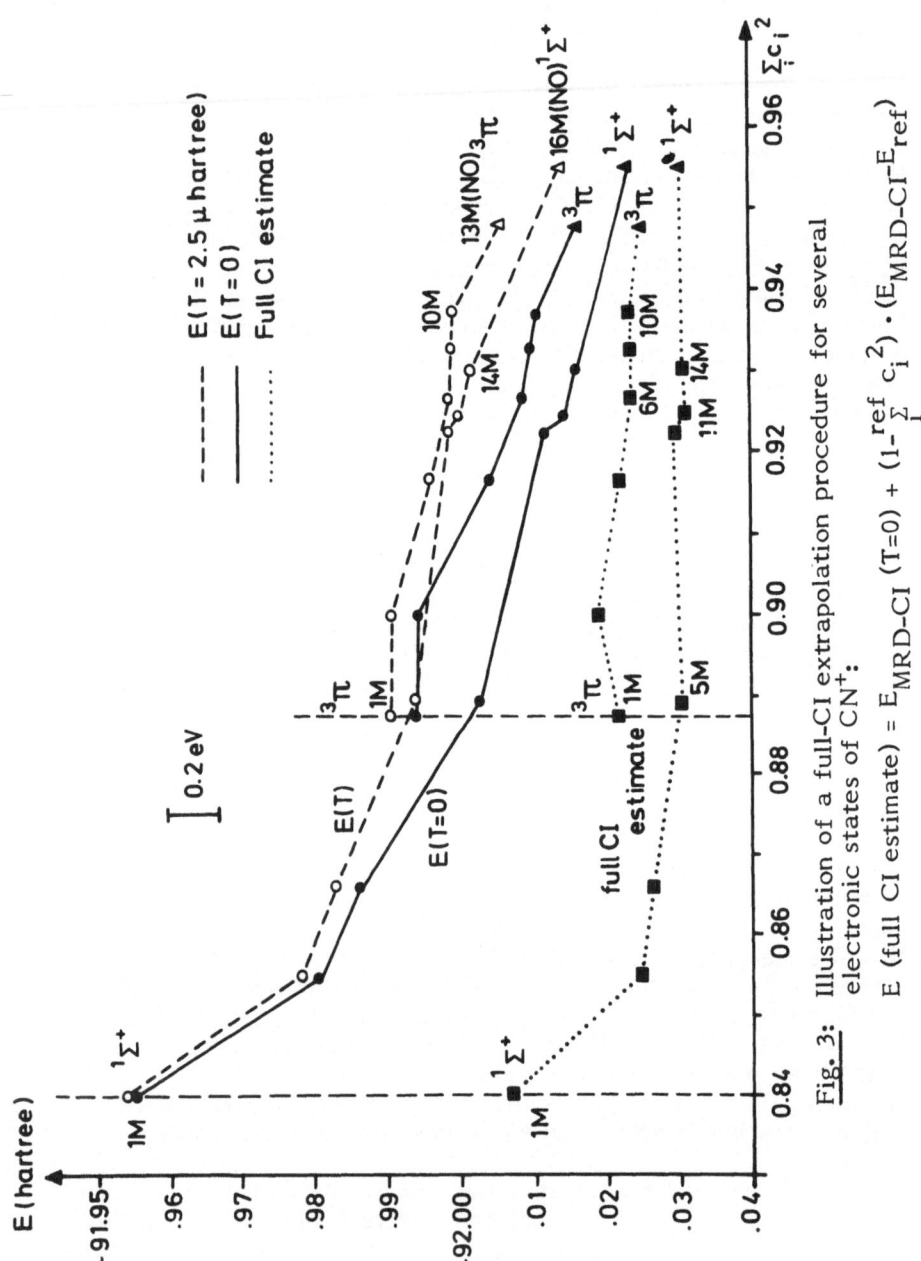

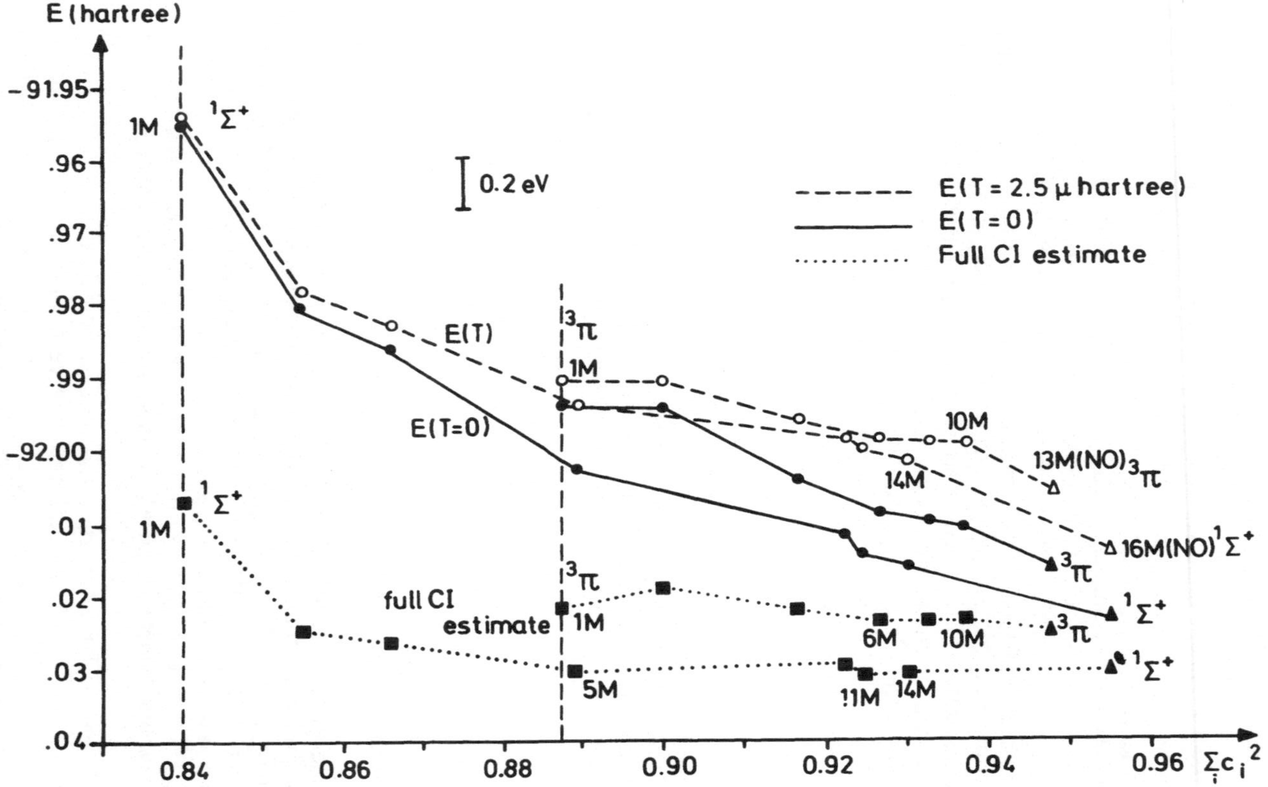


Fig. 3: Illustration of a full-CI extrapolation procedure for several electronic states of $CN^+$:

$$E\ (\text{full CI estimate}) = E_{MRD\text{-}CI}\ (T=0) + (1-\sum_{I}^{ref} c_I^2) \cdot (E_{MRD\text{-}CI} - E_{ref})$$

It would be premature to conclude that this value for $\lambda_2$ is typical, especially since one knows that $\lambda_2=0$ for any two-electron system, but at least it appears optimistic that a plotting technique such as that employed in Fig. 3 can obtain a realistic estimate for this factor if relatively large values of $\sum^{ref} c_i^2$ can be achieved in the MRD-CI calculations. If no clear secondary configurations present themselves in a given case it is difficult to obtain a sufficiently large range for $\sum^{ref} c_i^2$ to allow $\lambda_2$ to be determined very accurately, and thus in practice there is less freedom in applying this second type of extrapolation as in the first instance (for which the threshold T can always be varied over a wide range).

On the other hand experience with this full CI extrapolation technique is rather limited in comparison to the MRD-CI extrapolation procedure for non-zero selection thereshold mentioned first, and thus it is still too soon to judge the extent of the former's usefulness. It is important to recall, however, that such estimates are valid for any type of electronic state and that they thus enjoy a wide range of applicability, whereby most other perturbation techniques (18-20) designed to provide similarly accurate predictions of the full CI energy are only reliable for electronic states which are well described in Hartree-Fock calculations, i.e. which are representable in terms a single reference or dominant electronic configuration.

Thus far in the discussion nothing has been said about estimating AO basis deficiencies and thus it is interesting to consider this point before closing this section. The basic idea behind the MRD-CI and full CI extrapolation procedures can be summarized as one of self-correcting perturbation theory. Explicit diagonalizations are carried out and estimates $K_i$ are made for the energy contributions of excluded terms; then additional diagonalizations for somewhat larger CI spaces are performed in order to test the accuracy of the first estimates, wherupon a new value of $\lambda_i K_i$ is predicted for the corresponding $F_i$ value. If a similar approach is to be developed for the AO basis error the first requirement would thus seem to be the identification of a preliminary correction term $K_3$, analogous to $\Sigma\, {}^1\Delta E_i$ and $(1-\sum_i^{ref} c_i^2)\cdot(E_{MRD-CI}-E_{ref})$ in the first two cases.

One way of at least approaching this goal would be to obtain the full CI energy with the help of the previous extrapolation procedures for a series of AO basis sets and to assume that the contributions of the various one-electron functions are additive. The latter condition implies that all the one-electron functions considered are mutually orthonormal, and thus a first step in this procedure would necessarily involve an orthonormal transformation of the full AO basis to be considered. A standard subset of functions would then be chosen and the full CI energy thereof could be estimated as accurately as possible. The process would then be repeated for somewhat larger AO basis sets, in which groups of new functions are systematically added to the standard basis, whereupon energy lowerings would be noted for each basis augmentation, the sum of which would thus serve as $K_3$. By then carrying out still larger calculations in which several such sets of additional basis functions are added simultaneously, it would be possible to check the additivity of these quantities and thereby elicit an estimate for $\lambda_3$, from which an accurate prediction of the full CI result for the total AO space could be obtained.

Since the latter is still not complete this extrapolation would still fall short of allowing a correction for the total AO basis deficiency $F_3$, but it would go a long way toward achieving this goal. Computationally there is a potentially great advantage to such a procedure in that it would enable accurate CI treatments to be carried out for very large AO basis sets ($m \gtrsim 200$) without requiring a full electron repulsion integral transformation to be executed; the largest full transformation would be for the m' functions in the largest one-electron basis treated basis treated in the actual diagonalizations, although these functions would necessarily be expanded in terms of the m species in the total AO basis because of the orthonormalization requirement. The feasibility of such a third type of extrapolation procedure is something that remains to be probed in future work.

## V. COMPUTATIONAL DETAILS OF THE CI CALCULATIONS

The main conclusion from the error analysis of the previous sections is that it is far better to carry out a multi-reference double-excitation CI calculation in a flexible AO basis than a full CI of comparable computational expense, since the latter would necessarily involve the use of a much more restricted one-electron basis set. In addition there is strong evidence that the use of selection techniques and an accompanying energy extrapolation procedure enhances the overall applicability of such MRD-CI treatments. In the process of developing computer program systems to implement the MRD-CI method, one is faced with essentially two alternatives for constructing the necessary Hamiltonian matrices and carrying out the necessary diagonalizations, namely configuration-driven methods on the one hand or integral- or loop-driven techniques on the other. The distinction between these two types of algorithms is drawn on the basis of whether the major loop structures thereof involve successive comparisons of pairs of configurations to determine what electron repulsion integrals are needed for the associated Hamiltonian matrix elements (configuration-driven), or if instead they proceed via a stepwise analysis of groups of electron repulsion integrals in order to compute to which Hamiltonian matrix elements they contribute (integral- or loop-driven).

When threshold selection techniques are contemplated it is clear that configuration-driven algorithms have a fundamental advantage because they assume no particular relationship among the elements of a given CI space; a throughly random selection is thus possible, with the computational times decreasing with the square of the number of selected species in a truncated MRD-CI treatment. A definite relationship among configurations is essential to the operation of integral- or loop-driven techniques such as the direct CI (4,21) or GUGA (4,8,22) methods, however, since this information is critcal for an efficient identification of those Hamiltonian matrix elements which contain contributions from a given electron repulsion integral. Moreover, because of this distinction the actual implementation of the threshold selection procedure itself is invariably achieved with a configuration-driven algorithm.

In making an evaluation of the various computational methods as yet suggested for carrying out CI calculations it is therefore important to consider the question of whether it is advantageous to employ a threshold selection within the general MRD-CI framework. Since selection introduces an additional type of error into the theoretical treatment ($F_1$ in Fig. 1) it might be argued, for example, that it is better to deal with the T=0 MRD-CI space directly (or at least with the Hartree-Fock interacting subspace thereof), in which case the need for a threshold selection would be eliminated. When one recalls, however, the experience that a full CI is generally inferior to an MRD-CI calculation of comparable expense by reason of the fact that the latter can employ a significantly larger AO basis thereby, it is clear that a conclusion of this nature may be premature.

As a case in point it is instructive to consider the following results for the first two $^1A'$ states of non-symmetric $SO_2$. The Hartree-Fock interacting space for three reference configurations corresponds to a secular equation of 23613 in a double-zeta AO basis and thus requires a considerable computational expenditure when it is treated directly via the GUGA method (8b). A calculation of this nature is substantially more effective for the lowest $^1A'$ state of $SO_2$ than for $2^1A'$, as seen by the fact that three additional secondary configurations not included in the above reference set appear with relatively large coefficients in the final CI expansion for this state. Including these three species in the reference set leads to a tripling of the corresponding MRD-CI space and thus makes a straightforward diagonalization of even the associated Hartree-Fock interacting space quite a large undertaking.

With the use of a threshold selection (T=25 μhartree) and the energy extrapolation procedure discussed in the previous section, however, an MRD-CI treatment for the combined total of six reference configurations can be conveniently carried through (in about a third of the time for the first $SO_2$ treatment mentioned above), and on this basis it is found that the $2^1A'-^1A'$ transition energy is some 0.9 eV smaller in this treatment than in the original calculation with only three reference species. This result is in accord with the qualitative observation that the three new reference species are considerably more important for the representation of the upper state, and the clear indication is that the second calculation with the higher degree of configuration selection but with the larger set of reference configurations is more accurate than the first. In other words, the error introduced by the selection procedure is more than balanced by the fact that through the use of a larger reference set the discrepancy ($F_2$) between the MRD-CI and corresponding full CI results is substantially reduced.

The clear conclusion from the $SO_2$ calculations is that provided the threshold selection technique is accompanied by a suitable increase in the size of the reference set in the MRD-CI treatment, the overall accuracy of the calculations can be improved by such methods rather than worsened thereby. Application of the energy extrapolation technique can hold the error associated with configuration selection to a minimum, while still enabling the treatment of larger reference and AO basis sets than can be handled in a comparable amount of computation time when the MRD-CI space (or Hartree-Fock interacting subspace thereof) is left untruncated.

Another aspect of this discussion which also should not be lost from view is that the ability to carry out a threshold selection represents an additional degree of flexibility to any CI program system, which nonetheless does not exclude the possibility of omitting its use should the circumstances require this action. The key point is that the range of applicability of the overall theoretical method is greatly expanded when the opportunity exists to vary not only the sizes of the AO basis and the set of reference configurations in MRD-CI calculations, but also the value of the selection threshold in defining the configuration space for which explicit diagonalization procedures are to be implemented. By varying all three of these quantities an optimum balance between accuracy and computational expense can be sought out and the essential generality of the CI method can be used to its maximum extent. The variety of problems to which this approach can be successfully applied can be judged from several review articles on this topic which have recently appeared (24,25).

As emphasized at the beginning of this section once it is decided that a threshold selection should be included in a CI program system, there are strong arguments that the type of computational algorithm employed therein should be of the configuration-driven type. Proponents of the integral- or group-driven methods have recently made claims (6,8) that for the treatment of MRD-CI spaces at zero threshold, or alternatively their Hartree-Fock interacting subspaces (9), this type of algorithm is far superior to its configuration-driven counterpart, but this assertion is apparently based on experience with a rather inefficient type of configuration-driven algorithm (26-27). A recent implementation of the Table CI method (23), which falls in the latter category and is directly applicable for threshold selection procedures, including the selection process itself, has led to comparable computer timings for untruncated MRD-CI spaces, for example, as in the best GUGA-CI program system as yet reported (8b). The Table CI technique appears to be 2-3 times slower for the simplest case of a single closed-shell reference species, but the above comparison shows that it runs at the same speed when two-open-shell reference configurations are chosen, and there is a strong indication that as the composition of the reference set becomes more complicated (more and different types of open-shell structures) the Table CI configuration-driven algorithm is superior.

Hereto should be mentioned that the GUGA loop-driven method is more difficult to apply to general reference sets in MRD-CI treatments, whereby the Table CI technique simply requires a list of reference configurations for its use and employs exactly the same program structure for all types of open-shell species. Another advantage of the Table CI procedure is that it is directly applicable to the treatment of spin-orbit and spin-spin-effects in atomic or molecular Hamiltonians, whereas GUGA-CI is based on a strictly spinless representation of the non-relativistic interactions. Hence the fact that a threshold selection is much more effectively carried out with a configuration-driven algorithm does not appear to lead to disadvantages with regard to other aspects of the overall CI program system, which would otherwise detract from the increased flexibility such procedures are designed to provide.

One final point about the use of selection techniques which should be examined is their applicability to the calculation of potential surfaces. It might well be expected that if a different configuration selection is made for each nuclear conformation, problems with the continuity of the resulting energy and property surfaces would develop. This objection can be easily removed, however, by requiring that the same selection be made throughout, but that the perturbative energy lowerings needed in the MRD-CI extrapolation procedure for the unselected species be recomputed at each geometrical point. Similarly in using the full CI extrapolation procedure for the $F_2$ energy difference in Fig. 1, it is also well to require that the same reference set be used in the MRD-CI treatment for the entire surface. None of the various CI techniques discussed in this work is in principle less satisfactory with regard to continuity requirements than is the SCF method, for example, while at the same time from the standpoint of accuracy they are far superior because they allow one to approach the goal of a full CI in a large AO basis much more closely than can be done in a single-configuration treatment. Moreover, the steps which are taken to insure continuity in the MRD-CI calculations are inevitably bound up with computational savings since they eliminate a number of operations from being carried out at each point of the surface, so such measures are seen to be desirable from a variety of standpoints.

## VI. RELATIONSHIP OF MC-SCF TREATMENTS TO THE CI ERROR ANALYSIS

In the foregoing discussion of CI methods and their error analysis very little has been said about what to employ as one-electron functions in constructing the various configurations. In virtually all calculations to date in which the MRD-CI technique has been applied, either SCF orbitals of some leading configuration have been chosen for this purpose, or in cases for which several dominant configurations exist some set of approximate natural orbitals obtained by diagonalization of the first-order density matrix (28) have been used. Such one-electron sets do not in general provide the minimum possible energies for a given CI space, however, and thus the question arises whether care should be taken to optimize such species directly for the CI treatment, i.e. whether a multi-configurational SCF should be carried out. If such orbital optimizations are to be realized essentially two alternatives present themselves as to the nature of the CI space to be treated thereby. The first of these involves what has been referred to in Sect. II as a reference CI with a relatively small number of configuratio (29), oftentimes with all additional singly excited configurations thereto, while the second type employs a much larger configuration space coming under the definition of an MRD-CI calculation.

In the case of a relatively small reference CI it is known that the MC-SCF procedure can effect a considerable energy lowering relative to the use of some conventional SCF-MO set of one-electron functions, but the total energies obtained thereby are still much higher in general that those which result from a large MRD-CI calculation employing the same AO basis but with an unoptimized one-electron function set. Furthermore since the MC-SCF procedure is iterative in nature it requires that at least a partial transformation of the electron repulsion integrals be repeated from 5-10 times in a single treatment; as a result, it is invariably found that the overall computation time expended thereby greatly exceeds what is required for a conventional MRD-CI treatment. In addition there is the danger in the MC-SCF technique that a converged solution cannot be obtained at all or at least that it is attained only with considerable difficulty.

More recently a workable procedure has been developed (30) which allows the MC-SCF procedure to be applied to MRD-CI wavefunctions themselves. The key point to be recognized in this connection, however, is that a typical MRD-CI calculation already affords a good approximation to the corresponding full CI data. Consequently the eigenvalue and eigenvector results which are derived therefrom are relatively insensitive to the choice of one-electron basis. In a true full CI the MC-SCF solution is not even uniquely defined (31), since any non-singular transformation of the AO basis set fulfills the energy minimization criterion.

The latter situation has several consequences, which should be considered in evaluating the advisability of employing the MC-SCF procedure for such large MRD-CI wavefunctions. First of all, the energy lowering achieved by the optimization is likely to be very small. Comparison of MRD-CI results for natural orbitals and related SCF MO's (32), for example, indicates that energy lowerings of no more than a few thousandths of a hartree will generally result from application of the MC-SCF technique for large MRD-CI wavefunctions. Since the main interest in such calculations centers on energy differences either between different states or different nuclear geometries, it is very doubtful that effects of this magnitude would be of practical significance in actual applications.

Furthermore, the fact that at this level of treatment the CI results are only slightly dependent on the choice of one-electron basis may actually be detrimental to the attainment of convergence in the MC-SCF procedure itself. It is well known from conventional SCF calculations, for example, that difficulties arise when two or more different charge distributions of nearly equal energy exist, and thus in the case of MC-SCF treatments in the framework of a large CI, for which a multitude of different nearly optimal basis transformations can be found, it might be anticipated that the attainment of satisfactory convergence can be problematical. The fact that each MC-SCF iteration must be paid for at the minimum by a time-consuming (partial) integral transformation makes such calculations significantly less economical than for a conventional MRD-CI treatment, which point again needs to be weighed against the advantages one hopes to gain from orbital optimizations in the first place.

## VI. CONCLUSION

An error analysis of various types of CI methods shows that best results are generally obtained by employing a somewhat restrictive definition of the configuration space, but while using a very flexible AO basis set. In essence it is found that narrrowing the list of configurations to be treated explicitly with diagonalization techniques can ultimately produce increased overall accuracy in the theoretical treatment, because the gain in efficiency achieved thereby opens up the possibility of employing larger AO basis sets than would otherwise be possible. Thus a multi-reference double-excitation CI in a relatively large AO basis is invariably more effective than a full CI requiring the same computational expenditure, i.e. with a necessarily smaller AO basis.

By the same token it is found to be better to employ selection techniques in processing such MRD-CI spaces, even though reducing the sizes of the associated secular equations clearly leads to diminished accuracy when the choice of AO basis and reference configurations is held constant. Especially if a perturbative technique is employed to extrapolate the MRD-CI energy results back to zero selection threshold, it is found that larger reference configuration and AO basis sets can be employed for a given amount of computation time with this technique, so that the total error in comparably sized calculations is significantly smaller when truncation of the CI space is introduced than when it is not.

These results taken together suggest that the best computational strategy in CI calculations involves a program system which has essentially three input variables: an AO basis, a reference set of configurations from which the CI spaces are generated by including all singly and doubly excited species related thereto, and a selection threshold on which the final choice of configurations to be included directly in the diagonalization procedure is based. An optimum balance between accuracy and amount of computation time expended can be obtained with such a program system by simply varying these three input parameters. The basic technique employed is to carry out a truncated MRD-CI calculation (or a series thereof) and then through the successive application of various energy extrapolation procedures to obtain good approximations first to the zero-threshold MRD-CI, then to the corresponding full CI, and finally to the exact non-relativistic results for a given atomic or molecular system. The extrapolation techniques themselves can best be thought of as a form of self-correcting perturbation theory in which estimated energy values obtained at different levels of the calculations are compared with one another to give an indication of the results which would be obtained if the full diagonalizations were explicitly carried out.

The key feature of such a computational system is that it provides a direct relationship between the accuracy achieved in the calculations and the computational time required to perform them; lowering the selection threshold or expanding either the AO basis or the reference set for the MRD-CI treatment inevitably increases the expense of the calculations, but by varying these three input parameters over an effectively continuous range one exercises a high degree of control over the reliability which can attached to the computed results. Moreover, because the CI method itself is ultimately based on a matrix solution of the Schrödinger equation, it is seen that its range of applicability has no essential restrictions as to type of electronic state or nuclear conformation to be dealt with. Indeed, the latter feature distinguishes this type of calculation from virtually every other computational scheme as yet devised for the treatment of the electronic structure of atoms and molecules.

## ACKNOWLEDGEMENT

The authors are indebted to numerous coworkers of the Universities of Bonn, Wuppertal and Nebraska, especially to Drs. S.-K. Shih and J. Römelt, for numerous enlightening discussions and for help in obtaining the numerical results on which the arguments of this paper are based. The computer time and services made available by the University of Bonn Computing Center (RHRZ) are hereby also gratefully acknowledged.

## REFERENCES

1. R.J. Buenker and S.D. Peyerimhoff, in "Excited States in Quantum Chemistry", NATO ASI Series C46, eds. C.A. Nicolaides and D.R. Beck, D. Reidel Publ. Co., Dordrecht, Holland 1979

2. R.J. Buenker, Gaz. Chim. Ital. 108, 245 (1978)

3. M. Tinkham, "Group Theory and Quantum Mechanics", McGraw-Hill Book Co., New York 1963

4. B.O. Roos, Chem. Phys. Letters 15, 153 (1972)

5. R.J. Buenker and S.D. Peyerimhoff, Theor. Chim. Acta 35, 33 (1974)

6. R.C. Raffenetti, K. Hsu and I. Shavitt, Theor. Chim. Acta 45, 33 (1977)

7. L.E. McMurchie and E.R. Davidson, J. Chem. Phys. 66, 2959 (1977)

8a. B.R. Brooks and H.F. Schaefer III, J. Chem. Phys. 70, 5092 (1979)
b. B.R. Brooks, W.D. Laidig, P. Saxe, N.C. Handy and H.F. Schaefer III, Physica Scripta 21, 312 (1980)

9. A. Bunge, J. Chem. Phys. 53, 20 (1970)

10. T.-J. Tseng and F. Grein, J. Chem. Phys. 59, 6563 (1973)

11. E.W. Thulstrup and Y. Öhrn, J. Chem. Phys. 57, 3716 (1972)

12. R.J. Buenker and S.D. Peyerimhoff, Chem. Phys. 8, 56 (1975)

13. R.J. Buenker and S.D. Peyerimhoff, Chem. Phys. 8, 324 (1975)

14. R.J. Buenker and S.D. Peyerimhoff, Theor. Chim. Acta 39, 217 (1975)

15. E.R. Davidson in "The World of Quantum Chemistry", eds. R. Daudel and B. Pullmann, p. 17, D. Reidel Publ. Co., Dordrecht, Holland 1974

16. G. Hirsch, P.J. Bruna, S.D. Peyerimhoff and R.J. Buenker, Chem. Phys. Letters 52, 442 (1977);
W. Butscher, S.-K. Shih, R.J. Buenker and S.D. Peyerimhoff, Chem. Phys. Letters 52, 457 (1977)

17. P.J. Bruna, R.J. Buenker and S.D. Peyerimhoff, Chem. Phys. Letters 72, 278 (1980)

18. W. Meyer, J. Chem. Phys. 58, 1017 (1972);
R. Ahlrichs, H. Lischka, V. Staemmler and W. Kutzelnigg, J. Chem. Phys. 62, 1225 (1974)

19. L.S. Cederbaum, Theoret. Chim. Acta 31, 329 (1973);
W. v. Niessen, G.H.F. Diercksen, L.S. Cederbaum and W. Domcke, Chem. Phys. 18, 469 (1976)

20. R.J. Bartlett and C.D. Purvis, Int. J. Quantum Chem. 14, 561 (1978)
M.A. Robb, D. Hegarty and S. Prime, in "Excited States in Quantum Chemistry", NATO ASI Series C46, eds. C.A. Nicolaides and D.R. Beck, D. Reidel Publ. Co., Dordrecht, Holland 1979

21. C. Zirz and R. Ahlrichs, Proc. of Daresbury Study Weekend, 1980; preprint communicated to the authors

22. B.O. Roos and P.E.M. Siegbahn, Int. J. Quantum Chem. 17, 485 (1980)

23. R.J. Buenker, in Proc. of the Workshop on Quantum Chemistry and Molecular Physics in Wollongong, Australia, February 1980

24. S.D. Peyerimhoff and R.J. Buenker, Proc. of the IBM Meeting on Computational Methods in Chemistry, Bad Neuenahr, September 1979

25. S.D. Peyerimhoff and R.J. Buenker, in Proc. of the Workshop on Quantum Chemistry and Molecular Physics in Wollongong, Australia, February 1980

26. I. Shavitt, in "Modern Theoretical Chemistry", Vol. 3 (Methods of Electronic Structure Theory), ed. H.F. Schaefer III, Plenum Press, N.Y. 1977, p. 189

27. B.R. Brooks and H.F. Schaefer III, Int. J. Quantum Chemistry 14, 603 (1978)

28. C.F. Bender and E.R. Davidson, J. Phys. Chem. 70, 2675 (1966)

29. B.O. Roos, P.R. Taylor and P.E. Siegbahn, Chem. Phys. 48, 157 (1980)

30. B.R. Brooks, W.D. Laidig, P. Saxe, J.D. Goddard and H.F. Schaefer III, preprint communicated to the authors

31. P.O. Löwdin, Phys. Rev. 97, 1474 (1955);
P.O. Löwdin and H. Shull, Phys. Rev. 101, 1730 (1955)

32. K.H. Thunemann, J. Römelt, S.D. Peyerimhoff and R.J. Buenker, Int. J. Quantum Chem. 11, 743 (1977)

# THE CONFIGURATION INTERACTION METHOD

B.NAGY Otto

Université Catholique de Louvain
Laboratoire de Chimie Générale et Organique
Place L. PASTEUR,1,Louvain-la-Neuve, BELGIUM.

Abstract. The classical configuration interaction method is presented in a straightforward manner. The emphasis is on conceptual and computational details in order to give the beginner a better physical insight and a working knowledge of the method. After having discussed how the configuration interaction method handles the electron correlation problem it is shown how to construct symmetry-and spin-adapted multi-electron wave functions which are the most appropriate to carry out actual calculations. By using various theorems and rules it is explained how to construct matrix elements between electron configurations of different degree of excitation. Finally a few shortcomings of the method are mentioned together with recent developments improving the computational methodology.

## 1. Introduction.

One of the most powerful methods of calculating molecular electronic structure and properties is the HARTREE-FOCK Self-Consistent Field Molecular Orbital Method (HF SCF MO) (1). In its basic form it implies a certain number of approximations. First, the Hamiltonian operator used is a non-relativistic one neglecting therefore spin-spin and spin-orbit interactions. If this approximation is quite appropriate to light elements, it is rather bad for heavier ones. Second, since finite basis functions are used the total electronic energy obtained ($E_{SCF}$) is much higher than the exact



*I. G. Csizmadia and R. Daudel (eds.), Computational Theoretical Organic Chemistry, 77–100.*

non-relativistic energy ($E_{NR}$). This energy difference (basis error) can be reduced by using greater and greater basis sets. In this way the energy is steadily converging to its HARTREE-FOCK limit value ($E_{HFL}$). As a matter of fact the HF SCF MO method yields only approximate solutions of the SCHRÖDINGER equation since $E_{HFL}$ is still higher than $E_{NR}$. The difference is called the correlation energy ($E_C$) (2). $E_C$ satisfies the Virial Theorem and analysis shows that the HARTREE-FOCK potential energy ($V_{HF}$) of the electrons is higher than their exact non-relativistic potential energy ($V_{NR}$) This shows the tendency of electrons to occupy the same space element in the HF SCF MO method i.e. the electronic movements are not well correlated. The correlation between electrons of same spin is better (FERMI correlation ) since the SLATER determinant (see below) vanishes when two electrons have the same spatial coordinates. In other words the PAULI exclusion principle creates a FERMI hole around each electron where the probability of finding an other electron of same spin is vanishingly small. On the other hand electrons of different spin are badly correlated (COULOMB correlation) since the SLATER determinant is not zero when the two electrons occupy the same point in space. Consequently "ionic" structures are over-estimated.

In general the electronic movements are statistically correlated originating from the antisymmetry(permutation) and the symmetry (spin and space) of electronic wave functions and from the COULOMB repulsion (3). The rigorous mathematical treatment of the correlation problem necessitates the use of reduced density operators , second quantization etc. and will not be given here (4,5,6,7).

The correlation problem is extremely important since the physico-chemical properties to be calculated (reactivity, conformation, spectroscopy etc) are of the same order of magnitude as the errors due to the HF SCF MO method. Obviously there is a need for better wave functions which describe more accurately the various electronic systems.

Before going any further it is useful to recall some concepts concerning the multi-electron wave function $\Phi$ used to describe a multi-electron system. Let us consider a closed-shell, n-electron with only non-degenerate molecular orbitals (MO) $\psi_i$: $\psi_1$, $\psi_2$, .......$\psi_n$. The assignment of electrons to the various MO's gives an electron configuration. A spatial MO, $\psi_i$, can be occupied by zero (empty or virtual orbital), 1 (open shell, spin $\alpha$ or $\beta$) or 2 (closed shell, spin $\alpha$ and $\beta$) electrons. On the other hand spin-orbitals i.e. functions having both spatial and spin part, $\psi_i\alpha$ or $\psi_i\beta$, can only be empty or singly occupied. This is due to the PAULI principle. Since the HF SCF MO method handles electrons independently of each other(independent system approximation, see above), a many-electron spin-orbital can be written as a product of one-electron spin-orbitals. If n=2 we have for example $\phi' = \psi_1\alpha(1)\psi_1\beta(2)$. The dash indicates that $\phi'$ satisfies the PAULI principle but it is not antisymmetric. A many-electron wave function must be however antisymmetric to electron exchange i.e. must be eigenfunction of the following eigen-value equation

$$\bar{P}_{ij}\Phi = -\Phi$$

where $\bar{P}_{ij}$ is the permutation operator interchanging the coordinates of any two electrons i and j.
Obviously $\bar{P}_{12}\phi' \neq -\phi'$ but the linear combination $\phi = \Phi - \bar{P}_{12}\Phi'$ will be appropriate as can easily be verified. After normalization we obtain

$$\Phi = \frac{1}{\sqrt{2}}\left[\psi_1\alpha(1)\psi_1\beta(2) - \psi_1\alpha(2)\psi_1\beta(1)\right].$$

All these criteria (PAULI principle, antisymmetry, normalization) can be included in one single way of writing the so-called SLATER determinant :

$$\Phi = \frac{1}{\sqrt{2}}\begin{vmatrix} \psi_1\alpha(1) & \psi_1\beta(1) \\ \psi_1\alpha(2) & \psi_1\beta(2) \end{vmatrix}$$

For a n-electron system in the ground state we have

$$\Phi = \frac{1}{\sqrt{n!}}\begin{vmatrix} \psi_1\alpha(1) & \psi_1\beta(1) & \psi_2\alpha(1)\ldots & \psi_{n/2}\alpha(1) & \psi_{n/2}\beta(1) \\ \psi_1\alpha(2) & \psi_1\beta(2)\ldots & & \cdot & \cdot \\ \vdots & \vdots & & \vdots & \vdots \\ \psi_1\alpha(n) & \psi_1\beta(n)\ldots & & \psi_{n/2}\alpha(n) & \psi_{n/2}\beta(n) \end{vmatrix}$$

A more economical way of writing uses only the main diagonal elements (normalization factor neglected) whereby the permutation procedure is implicitly understood :

$$\Phi = |\psi_1\alpha(1)\psi_1\beta(2)\psi_2\alpha(3)\psi_2\beta(4)\ldots\psi_{n/2}\alpha(n-1)\psi_{n/2}\beta(n)| \equiv |\Phi_d|$$

or in general $\Phi = |\ i\alpha j\beta k\alpha\ldots\ldots\ldots|$

where the spin-orbital $i\alpha$ is occupied by electron 1, $j\beta$ by electron 2, $k\alpha$ by electron 3 etc. i.e. the spin-orbitals are ordered following increasing electron numbering.
Sometimes it is useful to write

$$\Phi = \frac{1}{\sqrt{n!}} \sum_{ij} (-1)^P \overline{P}_{ij} \Phi_d$$

$$\equiv \overline{A} \Phi_D$$

where $\overline{A}$ is a normalized antisymmetryzation operator and P represents the number of electron-pair permutations needed to regenerate $\Phi_d$ from $\overline{P}_{ij}\Phi_d$.
The use of single-configuration wave function in HF SCF MO calculations leads to the correlation problem. Therefore this wave function is not a good approximation since the very concept of independent one-electron spin-orbitals is not valid. There are several ways of improving the situation (perturbation method, alternant molecular orbital method, spin-polarized and unrestricted Hartree-Fock methods, second quantization, many body problem, use of correlated wave functions etc.).
One of the most powerful and most widespread one is the Configuration Interaction (CI) method. In what follows we are going to describe the classical CI method which can be easily developed from simple quantum-chemical principles in a straightforward manner (1,8).

## 2. The CI method.

The exact wave function $\Psi$ is developed in series of SLATER determinants each of which is representing a different electron configuration :

$$\Psi = \sum_r^{\infty} C_{ri} \Phi_r$$

In theory the series is infinite since it is constructed on infinite basis set. For obvious practical reasons the series must be truncated and therefore the CI function $\Psi$ will represent only an approximation of the true wave function. The longer the series the better the approximation. The improved energy and wave function are obtained from the solution of the SCHRÖDINGER equation :

$$\overline{H}|\Psi\rangle = E|\Psi\rangle$$

Using the variation theorem we have

$$E = \frac{\langle\Psi|\overline{H}|\Psi\rangle}{\langle\Psi|\Psi\rangle} = \frac{\sum_{rs} C^*_{ri} C_{si} \langle\phi_r|H|\phi_s\rangle}{\sum_{rs} C^*_{ri} C_{si} \langle\phi_r|\phi_s\rangle} = \frac{\sum_{rs} C^*_{ri} C_{si} H_{rs}}{\sum_{rs} C^*_{ri} C_{si} \Delta_{rs}}$$

where $H_{rs}$ is the matrix element of the HAMILTONIAN operator and $\Delta_{rs}$ is the overlap integral. Rearranging we obtain the coupled secular equations

$$\sum_r C_r (H_{rs} - \Delta_{rs} E) = 0 \quad \text{with} \quad s=1,2,3......n$$

According to standard procedure the solution of the secular determinant

$$|H_{rs} - \Delta_{rs} E| = 0$$

yields the different energy eigenvalues the substitution of which into the secular equations leads to the various $C_r$ coefficients i.e. to the required wave function $\Psi$. It can be seen that the CI method is just a simple exercice in the classical variational procedures. However two very crucial difficulties arise from the practical point of view : the choice of basis set and the choice of electron configurations (5). It can easily be seen that even if the different electron configurations ($\Phi_r$) are constructed from a relatively limited basis set the series expansions of $\Psi$ will contain a very large number of terms whose handling is quite unpractical (9). Therefore the number of electron configurations has to be reduced artificially and a number of procedures has been developed for this purpose (see BUNKER's paper in this book). A natural limitation of electron configurations comes from the space and spin symmetry of the SLATER determinants (see below).

CI is especially important for excited states because of the energetic vicinity of the interacting configurations. For the ground state CI is less important since BRILLOUIN's theorem (see below) is operating and higher energy configurations do not mix well with the ground state configuration.

In order to get a better feeling of what CI method is really doing let us consider the case of the hydrogen molecule. According to HF SCF MO theory its ground state is represented by the single determinantal wave function $\Phi$ :

$$\Phi = \frac{1}{\sqrt{2!}} \begin{vmatrix} \psi_1\alpha(1) & \psi_1\beta(1) \\ \psi_1\alpha(2) & \psi_1\beta(2) \end{vmatrix} \equi | \psi_1\alpha(1)\psi_1\beta(2) |$$

where $\psi_1 = \frac{1}{\sqrt{2}} (\phi_1+\phi_2)$, the doubly occupied bonding molecular orbital. Substituting this LCAO form into the determinant and by developing it we obtain :

$$\Phi=\frac{1}{2\sqrt{2}}\left(\begin{vmatrix}\phi_1\alpha(1) & \phi_1\beta(1)\\ \phi_1\alpha(2) & \phi_1\beta(2)\end{vmatrix}+\begin{vmatrix}\phi_2\alpha(1) & \phi_2\beta(1)\\ \phi_2\alpha(2) & \phi_2\beta(2)\end{vmatrix}+\begin{vmatrix}\phi_1\alpha(1) & \phi_2\beta(1)\\ \phi_1\alpha(2) & \phi_2\beta(2)\end{vmatrix}+\begin{vmatrix}\phi_2\alpha(1) & \phi_1\beta(1)\\ \phi_2\alpha(2) & \phi_1\beta(2)\end{vmatrix}\right)$$

The first two determinants represent "ionic" configurations where both electrons are occupying the same atomic orbital (either $\phi_1$ or $\phi_2$). The last two determinants represent "covalent" configurations since the two electrons occupy different atomic orbitals. One can see that this single determinantal wave function $\Phi$ overestimates the importance of "ionic" configurations in giving them the same weight as to the "covalent" configurations. Clearly electrons of opposite spins are not well correlated since they are allowed to pack together on the same atom very close to each other . This has a serious practical consequence : the dissociation energy is not correctly accounted for since even at infinite interatomic distances the two electrons still have the same probability to be located on only one atom. Let us improve now $\Phi$ by CI i.e. by mixing in an other electronic configuration built on for instance on the virtual molecular orbital $\psi_2=\frac{1}{\sqrt{2}}(\phi_1-\phi_2)$ :

$$\Phi' = \frac{1}{\sqrt{2!}}\begin{vmatrix}\psi_2\alpha(1) & \psi_2\beta(1)\\ \psi_2\alpha(2) & \psi_2\beta(2)\end{vmatrix} \equiv \left|\psi_2\alpha(1)\ \ \psi_2\beta(2)\right|$$

We have the trial CI function

$$\Phi_{CI}= \Phi - a\Phi'$$

since calculation shows that the mixing coefficient a is negative. By replacing in $\Phi$ and $\Phi'$each MO by its LCAO form and by developing the determinant we obtain

$$\Phi_{CI} = \tfrac{1}{2}[(1-a)(\phi_1\alpha(1)\phi_1\beta(2) + \phi_2\alpha(1)\phi_2\beta(2)) +$$

$$+(1+a)(\phi_1\alpha(1)\phi_2\beta(2) + \phi_2\alpha(1)\phi_1\beta(2))]$$ (not normalized!)

It is clear that since $|a|>0$ the CI method decreases the importance of "ionic" terms favouring the "covalent" ones. With increasing interatomic distance $\alpha \longrightarrow 1$; at infinite separation a = 1 and only the "covalent" terms survive and a correct dissociation energy is obtained. This example shows clearly how CI handles the correlation problem. A more detailed picture can be obtained by extending slightly the basis set using the following MO's (10,11) :

$$\sigma_g=\frac{1}{\sqrt{2}}(1s_1+1s_2);\ \sigma_u=\frac{1}{\sqrt{2}}(1s_1-1s_2);\pi_u=\frac{1}{\sqrt{2}}(2p_{x1}+2p_{x2}).$$

The wave function is (CI)

$$\Phi_{CI}=C_1|\sigma_g\alpha\sigma'_g\beta|+C_2|\sigma_u\alpha\sigma'_u\beta|+C_3|\pi_u\alpha\pi_u\beta|$$

The first term introduces in-out or radial correlation if different exponents are used in $\sigma_g$ and $\sigma'_g$ : when one electron is close to the core the other is farther away.
The second term introduces right-left correlation since $\sigma_u$ has a nodal plane perpendicular to the molecular axis obliging the two electrons to sit on different atoms.
The third term accounts for an angular correlation since the nodal plane of $\pi_u$ containing the molecular axes obliges electrons to occupy positions above and below it.

## 3. Construction of symmetry-and spin-adapted single-configuration wave functions.

As it was pointed out previously the HAMILTONIAN operator used is non-relativistic, therefore it does not contain spin coordinates but only space coordinates (kinetic and potential energies). This means that $\overline{H}$ commutes with all electron-spin operators $\overline{S}^2$ and $\overline{S}_z$. On the other hand $\overline{H}$ is symmetry invariant i.e. the symmetry operators $\overline{R}$ commute with $\overline{H}$. In other words any eigenfunction of the HAMILTONIAN operator ($\overline{H}$) is also simultaneously eigenfunction of spin($\overline{S}$) and symmetry ($\overline{R}$) operators.
For any operator $\overline{\Omega}$ commuting with $\overline{H}$ we can write the matrix element

$$\langle\Phi_i|\overline{H\Omega}\ |\Phi_j\rangle = \langle\Phi_i|\overline{\Omega H}\ |\Phi_j\rangle$$

$$\langle\Phi_i|\overline{H\Omega}-\overline{\Omega H}|\Phi_j\rangle=0$$

With the eigenvalue equation $\overline{\Omega}|\Phi_i\rangle=w_i|\Phi_i\rangle$ we obtain

$$(w_j - w_i)\langle\Phi_i|\overline{H}|\Phi_j\rangle = 0$$

Therefore the matrix element $H_{ij}$ is non-zero if and only if $w_i = w_j$ i.e. if $\Phi_i$ and $\Phi_j$ have the same spin eigenvalues and the same space symmetry. The CI matrix will be considerably simplified since no matrix element can survive between wave functions having different spin eigenvalues and different symmetries. One can take advantage of this natural truncation of the size of the CI problem by using at the outset symmetry-and spin-adapted wave functions to construct the various electron configurations.

a) Spin-adapted wave functions.

We are looking for wave functions which are simultaneously eigenfunctions of the total spin operators $\bar{S}^2$ and $\bar{S}_z$. In determining the total spin S only electrons occupying different space orbitals should be considered (non-equivalent electrons) since the contribution of closed shells is zero. For the SLATER determinant

$$\Phi = |\psi_1\alpha\psi_1\beta\psi_2\alpha\psi_2\beta\psi_3\alpha\psi_4\alpha\psi_5\alpha|$$

we have S=3/2 in atomic units. The independence of orbital and spin parts in a spin-orbital makes it possible to use the following practical abbreviation :

$$\Phi = |\alpha\alpha\alpha| \equiv |\alpha\alpha\alpha\rangle$$

We have of course $\bar{S}^2|\alpha\alpha\alpha\rangle = [3/2(3/2+1)]^{1/2}|\alpha\alpha\alpha\rangle$ and

$\bar{S}_z|\alpha\alpha\alpha\rangle = 3/2|\alpha\alpha\alpha\rangle$ i.e. $\Phi$ is a good spin-adapted function.

There are many methods of determining eigen-functions of $\bar{S}^2$ (12) : diagonalization of the $\bar{S}^2$ matrix (8), group theory (9), application of raising $\bar{S}_+$ and lowering $\bar{S}_-$ operators (13) by projection operators (4,14) not to mention but a few.

The most elegant and straightforward method is the projection operator technique and we shall use it in what follows. The procedure can be summarized in the following way :

| | | | | |
|---|---|---|---|---|
| Projection operator of required spin quantum number | x | Trial function in general spin space | ⟶ | Pure spin state function of required multiplicity (not normalized) |

$$\bar{O}_{Si}\Phi' \longrightarrow \Phi_{Si}$$

Thus the projection operator (which is idempotent) annihilates all undesired multiplicities and projects out only the desired one (Si).

The spin-projection operator may be written in various forms :

$$\bar{O}_{Si} = \prod_{j\neq i}[\bar{S}^2 - S_j(S_j+1)]$$

$$= \prod_{j\neq i}[\bar{S}_z^2 + \bar{S}_z + \bar{S}_-\bar{S}_+ - S_j(S_j+1)]$$

$$= \prod_{j\neq i}[\bar{S}_z^2 - \bar{S}_z + \bar{S}_+\bar{S}_- - S_j(S_j+1)]$$

or in a practically useful analytical form (15)

$$\bar{O}_{Si} = \prod_{j \neq i} \left\{ \Sigma \bar{P}^S_{pq} + \frac{1}{4}[(n_\alpha - n_\beta)^2 + 2n] - S_j(S_j+1) \right\}$$

where $\bar{P}^S_{pq}$ is the spin-permutation operator permuting α and β spins, $n_\alpha$ and $n_\beta$ are the number of non-equivalent α and β spins respectively and n is the total number of non-equivalent electrons. In order to know which spin states to project out one has to use the following formula (16) which gives the number (Q) of independent spin states of a given multiplicity (S) what a n non-equivalent electron system may have :

$$Q(n,S) = \binom{n}{n/2-S} - \binom{n}{n/2-S-1} =$$

$$= \frac{n!}{(\frac{n}{2}-S)!(\frac{n}{2}+S)!} - \frac{n!}{(\frac{n}{2}-S-1)!(\frac{n}{2}+S+1)!} =$$

$$= \frac{(2S+1)n!}{(\frac{n}{2}+S+1)!(\frac{n}{2}-S)!}$$

Q(n,S) is in fact the dimension of an irreducible representation of the corresponding symmetric group.
Let us note by passing that the possible outcomes of this formula are usually represented by a branching diagram (8).
Altogether there are

$$2^n = \sum_{a=o}^{n} \binom{n}{a} = \sum_{a=o}^{n} \frac{n!}{a!(n-a)!}$$

different many-electron spin functions (determinants) where a is the number of α spins ($a \leqslant n$).
As an example let us examine the case when n = 3.
The simple product spin eigenfunctions of $\bar{S}_z$ may be directly constructed. They may include a= 0,1,2,3 α spins. Therefore we have

$\binom{3}{3}$= 1 function with 3α spins : $f_1 = |\alpha\alpha\alpha\rangle$; $S_z = \frac{3}{2}$.

$\binom{3}{2}$= 3 functions with 2α spins: $f_2 = |\alpha\alpha\beta\rangle$,
$f_3 = |\alpha\beta\alpha\rangle$ and $f_4 = |\beta\alpha\alpha\rangle$; $S_z = \frac{1}{2}$ .

$\binom{3}{1}$= 3 functions with 1α spin : $f_5 = |\alpha\beta\beta\rangle$; $f_6 = |\beta\alpha\beta\rangle$
and $f_7 = |\beta\beta\alpha\rangle$; $S_z = -\frac{1}{2}$ .

$\binom{3}{0} = 1$ function with no $\alpha$ spin : $f_8 = |\beta\beta\beta\rangle ; S_z = -\frac{3}{2}$.
There are altogether $2^3=8$ functions.
These functions will give rise to Q (3,3/2) = 1 quartet and Q(3,1/2)= 2 doublet spin states. The corresponding spin- adapted wave functions will be projected out of these product functions $f_i$ using the appropriate projection operators:

$$\bar{O}_{3/2} = \left\{ \Sigma\bar{P}^S_{pq} + \frac{1}{4}[(n_\alpha - n_\beta)^2 + 2.3] - \frac{1}{2}(\frac{1}{2}+1) \right\}$$ for the quartet and

$$\bar{O}_{1/2} = \left\{ \Sigma\bar{P}^S_{pq} + \frac{1}{4}[(n_\alpha - n_\beta)^2 + 2.3] - \frac{3}{2}(\frac{3}{2}+1) \right\}$$

for the doublets. Of course, $\Sigma\bar{P}^S_{pq} = \bar{P}^S_{12} + \bar{P}^S_{13} + \bar{P}^S_{23}$ and

$\bar{P}^S_{pq}=0$ if spins p and q are both $\alpha$ or both $\beta$.

Thus

$$\bar{O}_{3/2}|\alpha\alpha\beta\rangle = \left\{ \bar{P}^S_{13} + \bar{P}^S_{23} + \frac{1}{4}[(2-1)^2 + 2.3] - \frac{1}{2}(\frac{1}{2}+1) \right\} |\alpha\alpha\beta\rangle =$$
$$= |\beta\alpha\alpha\rangle + |\alpha\beta\alpha\rangle + |\alpha\alpha\beta\rangle$$

After normalization we have $\frac{1}{\sqrt{3}}(|\beta\alpha\alpha\rangle + |\alpha\beta\alpha\rangle + |\alpha\alpha\beta\rangle)$.

By the same

$$\bar{O}_{1/2}|\alpha\alpha\beta\rangle = \left\{ \bar{P}^S_{13} + \bar{P}^S_{23} + \frac{1}{4}[(2-1)^2 + 2.3] - \frac{3}{2}(\frac{3}{2}+1) \right\} |\alpha\alpha\beta\rangle =$$
$$= |\beta\alpha\alpha\rangle + |\alpha\beta\alpha\rangle - 2|\alpha\alpha\beta\rangle$$

and after normalizing : $\frac{1}{\sqrt{6}}$ ( $\beta\alpha\alpha\rangle + |\alpha\beta\alpha\rangle - 2|\alpha\alpha\beta\rangle$).

The operation with $\bar{O}_{3/2}$ on the various product functions yields the four components of the quartet :
$f_1$ gives $\Phi_{3/2,3/2} = |\alpha\alpha\alpha\rangle$ i.e. $f_1$ is already eigenfunction of $\bar{S}^2$. $f_2, f_3$ and $f_4$ give $\Phi_{3/2,1/2} = \frac{1}{\sqrt{3}}(|\alpha\alpha\beta\rangle + |\beta\alpha\alpha\rangle + |\alpha\beta\alpha\rangle)$.
$f_5, f_6$ and $f_7$ give $\Phi_{3/2,-1/2} = \frac{1}{\sqrt{3}}(|\alpha\beta\beta\rangle + |\beta\alpha\beta\rangle + |\beta\alpha\alpha\rangle)$. Finally $f_8$ leads to $\Phi_{3/2,-3/2} = |\beta\beta\beta\rangle$ (see the case of $f_1$).

All these functions are orthonormal as they should be.
The use of $\bar{O}_{1/2}$ is a little more complicated.

$\bar{O}_{1/2}|\alpha\alpha\alpha\rangle = \bar{O}_{1/2}|\beta\beta\beta\rangle = 0$ i.e. functions $f_1$ and $f_8$ have no doublet character. With the other functions one obtains:

$$\bar{O}_{1/2}|\alpha\alpha\beta\rangle \longrightarrow \Phi^{a}_{1/2,1/2}=\frac{1}{\sqrt{6}}(|\alpha\beta\alpha\rangle+|\beta\alpha\alpha\rangle-2|\alpha\alpha\beta\rangle)$$

$$\bar{O}_{1/2}|\alpha\beta\alpha\rangle \longrightarrow \Phi^{b}_{1/2,1/2}=\frac{1}{\sqrt{6}}(|\beta\alpha\alpha\rangle+|\alpha\alpha\beta\rangle-2|\alpha\beta\alpha\rangle)$$

$$\bar{O}_{1/2}|\beta\alpha\alpha\rangle \longrightarrow \Phi^{c}_{1/2,1/2}=\frac{1}{\sqrt{6}}(|\alpha\alpha\beta\rangle+|\alpha\beta\alpha\rangle-2|\beta\alpha\alpha\rangle)$$

$$\bar{O}_{1/2}|\alpha\beta\beta\rangle \longrightarrow \Phi^{d}_{1/2,-1/2}=\frac{1}{\sqrt{6}}(|\beta\alpha\beta\rangle+|\beta\beta\alpha\rangle-2|\alpha\beta\beta\rangle)$$

$$\bar{O}_{1/2}|\beta\alpha\beta\rangle \longrightarrow \Phi^{e}_{1/2,-1/2}=\frac{1}{\sqrt{6}}(|\alpha\beta\beta\rangle+|\beta\beta\alpha\rangle-2|\beta\alpha\beta\rangle)$$

$$\bar{O}_{1/2}|\beta\beta\alpha\rangle \longrightarrow \Phi^{f}_{1/2,-1/2}=\frac{1}{\sqrt{6}}(|\beta\alpha\beta\rangle+|\alpha\beta\beta\rangle-2|\beta\beta\alpha\rangle)$$

Obviously there are too many functions for the two doublets (only four functions are needed!) . Furthermore these functions e.g. $\Phi^a, \Phi^b, \Phi^c$ are neither orthogonal (i.e. $\langle\Phi^a|\Phi^b\rangle\neq 0$) nor linearly independent (i.e. $\sum_i C_i\Phi^i=0$ for $C_i\neq 0$). The situation becomes normal if we orthogonalize these functions for instance by taking their appropriate linear combinations. We have finally the required two doublet functions :

$$\Phi^{a}_{1/2,1/2} = \Phi_{1/2,1/2} = \frac{1}{\sqrt{6}}(|\alpha\beta\alpha\rangle+|\beta\alpha\alpha\rangle-2|\alpha\alpha\beta\rangle)$$

$$\Phi^{d}_{1/2,-1/2}= \Phi_{1/2,-1/2}= \frac{1}{\sqrt{6}}(|\beta\alpha\beta\rangle+|\beta\beta\alpha\rangle-2|\alpha\beta\beta\rangle)$$

and

$$\Phi^{b}_{1/2,1/2} - \Phi^{c}_{1/2,1/2} \longrightarrow \Phi_{1/2,1/2} = \frac{1}{\sqrt{2}}(|\beta\alpha\alpha\rangle-|\alpha\beta\alpha\rangle)$$

$$\Phi^{e}_{1/2,-1/2} - \Phi^{f}_{1/2,-1/2} \longrightarrow \Phi_{1/2,-1/2}= \frac{1}{\sqrt{2}}(|\beta\beta\alpha\rangle-|\beta\alpha\beta\rangle).$$

b) Symmetry-and spin-adapted wave functions.

In order to take advantage of the space symmetry of a system, the various Slater determinants have to be constructed with symmetry-adapted functions. These latter can easily be obtained by using the appropriate symmetry projection operator (17) :

$$\bar{O}_{\Gamma i}= \sum_j \chi_j^{\Gamma i} R_j$$

where $\Gamma i$ is the required symmetry to be projected,

$\chi_j^{\Gamma i}$ is the corresponding character which belongs to the symmetry operation $R_j$. When the functions are symmetry-adapted they have also to be spin-adapted.
The whole procedure goes as follows :

1) establishment of SLATER determinants which are eigenfunctions of $\bar{S}_Z$ and the MO part of which is symmetry-adapted by projection;

2) projection of eigenfunctions of $\bar{S}^2$ out of these functions ;

3) projection of symmetry of the obtained eigenfunctions of $\bar{S}_Z$ and $\bar{S}^2$ to get the final symmetry- and spin-adapted wave function (this is a verification).

In summary :

SLATER determinants, eigenfunctions of $\bar{S}_Z$;
symmetry-adapted molecular orbitals.

$\downarrow$ spin projection $\bar{O}_{Si}$

Spin-adapted functions, eigenfunctions of $\bar{S}_Z$ and of $\bar{S}^2$.

$\downarrow$ symmetry projection $\bar{O}_{\Gamma i}$

Spin-and symmetry-adapted single-configuration wave functions, eigenfunctions of $\bar{S}_Z$ and $\bar{S}^2$ .

Let us examine the case of hydrogen molecule using the subminimal base $\phi_i = |1S_i>$. This molecule belongs to the symmetry point group $C_i$ the character table of which is

| $C_i$ | E | i |
|---|---|---|
| $\Sigma_g$ | 1 | 1 |
| $\Sigma_u$ | 1 | -1 |
| $\Gamma_{red}$ | 2 | 0 |

The character of the reducible representation ($\Gamma_{red}$) is also included . It can easily be obtained by standard group-theoretical techniques (17). It can also be shown by reduction of the reducible representation that

$$\Gamma_{red} = \Sigma_g \oplus \Sigma_u$$

i.e. $\Gamma_{red}$ containes the two irreducible representations $\Sigma_g$ and $\Sigma_u$. This means that the MO's must have these symmetries. The symmetry projection operators are therefore

$$\bar{O}_{\Sigma g} = E + i$$

and $$\bar{O}_{\Sigma u} = E - i$$

1) The symmetry-adapted molecular orbitals will be

$$\bar{O}_{\Sigma g}\phi_1 = (E + i)\phi_1 = \phi_1 + \phi_2 \longrightarrow \psi_b = \frac{1}{\sqrt{2}}(\phi_1 + \phi_2)$$

$$\bar{O}_{\Sigma u}\phi_1 = (E - i)\phi_1 = \phi_1 - \phi_2 \longrightarrow \psi_v = \frac{1}{\sqrt{2}}(\phi_1 - \phi_2)$$

where $\psi_b$ isthe bonding and $\psi_v$ is the virtual MO.

With these MO's the following SLATER determinants can be constructed

$\Phi_1 = |b\alpha b\beta|$, $S_z = 0$ ;

$\Phi_2 = |b\alpha v\alpha|$, $S_z = 1$ ;

$\Phi_3 = |b\alpha v\beta|$ and $\Phi_4 = |b\beta v\alpha|$, for both $S_z = 0$;

$\Phi_5 = |b\beta v\beta|$, $S_z = -1$ ;

$\Phi_6 = |v\alpha v\beta|$, $S_z = 0$ ;

of which $\Phi_1$ represents an unexcited, $\Phi_2$, $\Phi_3$, $\Phi_4$ and $\Phi_5$ represent singly excited and $\Phi_6$ stands for a doubly excited electron configuration (see below) . They are all eigenfunctions of $\bar{S}_z$.

2) With two non-equivalent electrons (n=2) one has one triplet and one singlet function, i.e.

$Q(2,1) = 1$ and $Q(2,0) = 1$.

Therefore, the total number of spin functions is $2^2=4$. The required spin-projection operators are

$$\bar{O}_1 = \left\{ \bar{P}^S_{12} + \tfrac{1}{4}[(n_\alpha - n_\beta)^2 + 2.2] \right\}$$

and
$$\bar{O}_o = \left\{ \bar{P}^S_{12} + \tfrac{1}{4}[(n_\alpha - n_\beta)^2 + 2.2] - 1(1 + 1) \right\}$$

The projection goes as follows :

$$\bar{O}_1\Phi_1 = \left\{ \bar{P}_{12} + \tfrac{1}{4}[(1-1)^2 + 2.2] \right\}|b\alpha b\beta| = |b\beta b\alpha| + |b\alpha b\beta| = 0$$

since the two determinants differ only by electron permutation as can be verified by developing the determinants.

$$\bar{O}_1\Phi_2=\left\{0+\frac{1}{4}\left[(2-0)^2+2.2\right]\right\}|b\alpha v\alpha|=2|b\alpha v\alpha| \longrightarrow \Phi_2$$

$$\bar{O}_1\Phi_3=\left\{\bar{P}^S_{12}+\frac{1}{4}\left[(1-1)^2+2.2\right]\right\}|b\alpha v\beta|=|b\beta v\alpha|+|b\alpha v\beta| \longrightarrow \frac{1}{\sqrt{2}}(\Phi_3+\Phi_4)$$

since the two determinants differ by spin permutation.

Furthermore

$$\bar{O}_1\Phi_4 \longrightarrow \frac{1}{\sqrt{2}}(\Phi_3+\Phi_4);\ \bar{O}_1\Phi_5 \rightarrow \Phi_5 \text{ and } \bar{O}_1\Phi_6 = 0.$$

Therefore the triplet wave functions are

$$\Phi_{1,1} = \Phi_2\ ;\ \Phi_{1,0} = \frac{1}{\sqrt{2}}(\Phi_3+\Phi_4)\ ;\ \Phi_{1,-1} = \Phi_5$$

The singlet wave function is calculated with $\bar{O}_o$ in a similar way and has the form :

$$\Phi_{o,o} = \frac{1}{\sqrt{2}}(\Phi_3-\Phi_4).\ \text{Here}\ \bar{O}_o\Phi_2 = \bar{O}_o\Phi_5 = 0.$$

Let us note that there are two more singlet functions originating from the closed-shell configuration $\Phi_1$ and $\Phi_6$.

3) The final wave functions are obtained by symmetry projection.

$$\bar{O}_{\Sigma u}\Phi_{1,0} \longrightarrow \bar{O}_{\Sigma u}(\Phi_3+\Phi_4)=(E-i)(|b\alpha v\beta|+|b\beta v\alpha|) =$$

$$=(E-i)|b\alpha v\beta| + (E-i)|b\beta v\alpha|=$$
$$=|Eb\alpha Ev\beta|-|ib\alpha iv\beta|+|Eb\beta Ev\alpha|-|ib\beta iv\alpha|=$$
$$=|b\alpha v\beta|+|b\alpha v\beta|+|b\beta v\alpha|+|b\beta v\alpha|=2(\Phi_3+\Phi_4)$$

Therefore $\Phi_{1,o}$ transforms like $\Sigma_u$. Of course $\bar{O}_{\Sigma g}\Phi_{1,o}=0$.

After having projected all functions one obtains the following wave functions :

$\Phi_1({}^1\Sigma_g)$, $\Phi_6({}^1\Sigma_g)$ and $\Phi_{o,o}=\frac{1}{\sqrt{2}}(\Phi_3-\Phi_4)$ $({}^1\Sigma_u)$ singlet functions and $\Phi_{1,1}=\Phi_2({}^3\Sigma_u)$, $\Phi_{1,o}=\frac{1}{\sqrt{2}}(\Phi_3+\Phi_4)$ $({}^3\Sigma_u)$, $\Phi_{1,-1}=\Phi_5({}^3\Sigma_u)$ triplet functions.

One can see that this last symmetry projection appears superfluous since it does not modify the function obtained after spin projection. This is because in the system at hand the MO's are non-degenerate and completely filled i.e. there are no partially filled MO's. In the latter case the full projecting procedure must be used however (for proof and example of benzene see ref. 15).

4. Calculation of CI matrix elements.

a) Construction of electron configurations.

In the independent model approximation the solution of the multi-electron SCHRÖDINGER equation

$$\bar{H}|\Phi\rangle = E|\Phi\rangle$$

where $\bar{H}$ is the multi-electron HAMILTONIAN operator and $\Phi$ is a SLATER determinant implies in fact the solution of a one-electron SCHRÖDINGER equation, the HARTREE-FOCK equation

$$\bar{F}|\psi_j\rangle = E_j|\psi_j\rangle$$

where $\bar{F}$ is a "one-electron" HAMILTONIAN called the HARTREE-FOCK operator and $\psi_j$ is the jth molecular orbital having energy $E_j(1)$. More explicitly

$$\bar{F} = \bar{h} + \bar{G} \;;\; \bar{h} = \bar{T} + \bar{V} \;;\; \bar{G} = 2\bar{J} - \bar{K}$$

$\bar{h}$ is a one-electron operator including the operators $\bar{T}$ and $\bar{V}$ which are the kinetic energy and potential energy operators respectively. $\bar{G}$ is a two-electron operator including the well-known COULOMB operator

$$\bar{J}(1) = \sum_i \langle\psi_i(2)|r_{12}^{-1}|\psi_i(2)\rangle$$

and the exchange operator

$$\bar{K}(1) = \sum_i \langle\psi_i(2)|r_{12}^{-1}|\psi_j(2)\rangle$$

both in atomic units.

The SCF solution of the HF equation in the LCAO approximation (i.e. $\psi_j = \sum_r C_{rj}\phi_r$) leads to the desired MO's and their energies. For instance if the bonding MO's are $\psi_g, \psi_e$ and $\psi_a$ with the corresponding energies $E_g<E_e<E_a$ the multi-electron wave function representing the ground state will be given by the following SLATER determinant

$$^1\Phi_o = |g\alpha g\beta e\alpha e\beta a\alpha a\beta| = {}^1|0\rangle$$

This represents a singlet electron configuration (unexcited) with all bonding MO's singly occupied . Other electron configurations can be constructed by removing one or more electrons from the bonding MO's and by putting them in the virtual ones ($\psi_b, \psi_c$ and $\psi_d$ with $E_b < E_c < E_d$). This "excitation" leads to mono-, bi-or higher excited configurations depending on the number of electrons involved. Thus typical mono-excited configurations will be for instance

$$^{1,3}\Phi_a^b = {}^{1,3}|_a^b> = \frac{1}{\sqrt{2}}(|j\alpha j\beta a\alpha b\beta| \pm |j\alpha j\beta b\alpha a\beta|)$$

$$^{1,3}\Phi_a^c = {}^{1,3}|_a^c> = \frac{1}{\sqrt{2}}(|j\alpha j\beta a\alpha c\beta| \pm |j\alpha j\beta c\beta a\beta|)$$

$$^{1,3}\Phi_e^b = {}^{1,3}|_e^b> = \frac{1}{\sqrt{2}}(|e\alpha b\beta a\alpha a\beta| \pm |b\alpha e\beta a\alpha a\beta|)$$

For practical reasons only the most important spinorbitals are given in the SLATER determinants and deeper lying closed shells are neglected. Of course, they must be taken into account when the matrix elements are calculated.

By the same bi-excited configurations might be e.g.

$$^{1}\Phi_{aa}^{cc} = {}^{1}|_{aa}^{cc}> = |e\alpha e\beta c\alpha c\beta|$$

$$^{1,3}\Phi_{ae}^{bb} = {}^{1,3}|_{ae}^{bb}> =$$

$$= \frac{1}{\sqrt{2}}(|e\alpha b\beta b\alpha a\beta| \pm |b\alpha e\beta a\alpha b\beta|)$$

$$^{1,3}\Phi_{ee}^{cd} = {}^{1,3}|_{ee}^{cd}> \frac{1}{\sqrt{2}} \; (|c\alpha d\beta a\alpha a\beta| \pm |d\alpha c\beta a\alpha a\beta|)$$

In the state symbols lower letters designate bonding orbitals while the upper ones represent virtual orbitals. We limit our discussion to singly and doubly excited configurations. In constructing excited configurations it is very important to bear in mind the following rules: a) virtual orbitals should be written in the place of bonding orbitals which are vacated by the excited electrons; b) the spin of spin-orbitals must be preserved.

b) Calculation of CI matrix elements.

The construction of CI secular determinant requires the calculation of matrix elements such as

$$<0|\bar{H}|0>,\ <0|\bar{H}|_a^b>,\ <0|\bar{H}|_{aa}^{cc}>,\ <_a^b|\bar{H}|_a^b>,$$

$$<_a^b|\bar{H}|_a^c>,\ <_a^b|\bar{H}|_{ae}^{bb}>,\ <_{ae}^{bb}|\bar{H}|_{ae}^{bb}>,\ <_{ae}^{bb}|\bar{H}|_{gi}^{bb}> \text{ etc.}$$

In order to get a deeper insight into the algebraic techniques we shall carry out the calculations in a direct manner. The generalizations of the results in form of useful rules will be given later.

$$<{}^{b}_{a}|\bar{H}|0>=\frac{1}{\sqrt{2}}<(|j\alpha j\beta a\alpha b\beta|+|j\alpha j\beta b\alpha a\beta|)\,|\bar{H}|j\alpha j\beta a\alpha a\beta>=$$

$$=\frac{1}{\sqrt{2}}(<j\alpha j\beta a\alpha b\beta|\bar{H}|j\alpha j\beta a\alpha a\beta>+$$

$$<j\alpha j\beta b\alpha a\beta|\bar{H}|j\alpha j\beta a\alpha a\beta>)$$

Taking $\bar{H}$ explicitly we have for the first term

$$<j\alpha j\beta a\alpha b\beta|\sum_i\bar{h}_i + \sum_{i<k}\sum r_{ik}^{-1}|j\alpha j\beta a\alpha a\beta>=$$

$$=<b|\bar{h}_4|\ a><\beta|\beta><j\alpha|j\alpha><j\beta|j\beta><a\alpha|a\alpha>+$$

$$+<j\alpha b\beta|r_{14}^{-1}|j\alpha a\beta><j\beta|j\beta><a\alpha|a\alpha>+$$

$$+<j\beta b\beta|r_{24}^{-1}|j\beta a\beta><j\alpha|j\alpha><a\alpha|a\alpha>+$$

$$+<a\alpha b\beta|r_{34}^{-1}|a\alpha a\beta><j\alpha|j\alpha><j\alpha|j\beta>$$

These are the only elements which survive because of the orthonormality of spinorbitals. As a matter of fact the electrons on which the HAMILTONIAN do not operate must occupy the same spin-orbitals in both bra and ket otherwise the corresponding integral is zero. Consequently the spinorbitals which differentiate bra and ket (i.e. the two SLATER determinants) must always be included in the integral . Thus for example the following arrangements are all zero :

$$<j|h_1|j><\alpha|\alpha><j\beta|j\beta><a\alpha|a\alpha><b\beta|a\beta>=0$$

$$<j\alpha a\alpha|r_{13}^{-1}|j\alpha a\alpha\ ><j\beta|j\beta><b\beta|a\beta>=0$$

According to classical HF SCF MO theory a bielectronic integral gives rise to a COULOMB integral

$$<j\alpha k\alpha|r_{12}^{-1}|j\alpha k\alpha>\ \equiv J_{jk}\equiv<jj|kk>$$

and an exchange integral

$$<j\alpha k\alpha|r_{12}^{-1}|k\alpha j\alpha>\ \equiv K_{jk}\equiv<jk|kj>$$

if the spins of the implied spinorbitals are identical; only a COULOMB integral is obtained if the spins are different (In $<jk|kj>$ the bra and ket are representing electron 1 and 2 respectively) . It should be noted that if the sign of $J_{jk}$ is positive that of $K_{jk}$ is negative since the latter contains an additional permutation in the ket. With this in mind we have therefore for the first term :

$h_{ba} + \langle jj|ba\rangle + \langle jj|ba\rangle - \langle ja|bj\rangle + \langle aa|ba\rangle$

The second term yields an identical expression and we have

$\langle {}^{b}_{a}|\bar{H}|0\rangle = \frac{1}{\sqrt{2}}(2h_{ba}+2\langle aa|ba\rangle+4\langle jj|ba\rangle-2\langle ja|bj\rangle)$

The spinorbital j may be any spinorbital of the bonding orbital manyfold. Therefore one has to sum over j . Finally we obtain

$$\langle {}^{b}_{a}|\bar{H}|0\rangle = \sqrt{2}[h_{ba}+\sum_{j}(2\langle ba|jj\rangle-\langle bj|ja\rangle)]=\sqrt{2}\, F_{ba}$$

since when j=a, $2\langle ba|aa\rangle-\langle ba|aa\rangle = \langle ba|aa\rangle$.
However, since SCF spinorbitals are used, the HF matrix is diagonal in this base i.e. $F_{ij} = 0$ if $i \neq j$.
Therefore $\langle {}^{b}_{a}|\bar{H}|0\rangle = 0$ and we obtain BRILLOUIN's theorem:

There is no CI matrix element between the ground state wave function and mono-excited wave functions.

Of course, when non SCF spinorbitals are used to construct SLATER determinants this theorem does not apply.

In calculating CI matrix elements one has to be sure that both bra and ket have the maximum number of identical spin-orbitals on which the HAMILTONIAN does not operate. This can be achieved by reaching maximum coincidence of bra and ket by appropriate permutations in the latter. This is illustrated in the following calculation.

$\langle {}^{b}_{a}|\bar{H}|{}^{b}_{e}\rangle =$

$$= \frac{1}{2}\langle (|e\alpha e\beta a\alpha b\beta| + |e\alpha e\beta b\alpha a\beta|)|\bar{H}|(|e\alpha b\beta a\alpha a\beta| + |b\alpha e\beta a\alpha a\beta|)\rangle =$$

$$= \frac{1}{2}(\langle e\alpha e\beta a\alpha b\beta|\bar{H}|e\alpha b\beta a\alpha a\beta\rangle - \langle e\alpha e\beta b\alpha a\beta|\bar{H}|e\alpha b\beta a\alpha a\beta\rangle +$$

$$+\langle e\alpha e\beta a\alpha b\beta|\bar{H}|b\alpha e\beta a\alpha a\beta\rangle + \langle e\alpha e\beta b\alpha a\beta|\bar{H}|b\alpha e\beta a\alpha a\beta\rangle)$$

Let us calculate the first term :

$$\langle e\alpha e\beta a\alpha b\beta|\bar{H}|e\alpha b\beta a\alpha a\beta\rangle = \langle e\beta b\beta|r_{24}^{-1}|b\beta a\beta\rangle -$$

$$-\langle e\beta b\beta|r_{24}^{-1}|a\beta b\beta\rangle = \langle eb|ab\rangle - \langle ea|bb\rangle$$

Let us permute electrons 2 and 4 in ket : $\bar{P}_{24}|\rangle = -|\rangle$.

One obtains :

$$-\langle e\alpha e\beta a\alpha b\beta|\bar{H}|e\alpha a\beta a\alpha b\beta\rangle = -\langle e\beta|\bar{h}_2|a\beta\rangle -$$

$$-\langle e\alpha e\beta|r_{12}^{-1}|e\alpha a\beta\rangle - \langle e\beta a\alpha|r_{23}^{-1}|a\beta a\alpha\rangle -$$

$$-\langle e\beta b\beta|r_{24}^{-1}|a\beta b\beta\rangle + \langle e\beta b\beta|r_{24}^{-1}|b\beta a\beta\rangle =$$

$$= -h_{ea} - \langle ee|ea\rangle - \langle ea|aa\rangle - \langle ea|bb\rangle + \langle eb|ba\rangle$$

It is clear that the maximum coincidence principle ensures a higher value for the integrals.
By proceeding in the same way for the other terms we have

$$\langle {}^{b}_{a}|\bar{H}|{}^{b}_{e}\rangle = -h_{ea} - \langle ea|ee\rangle - \langle ea|aa\rangle - \langle ea|bb\rangle + 2\langle eb|ba\rangle.$$

Again orbitals e and a may be any j bonding orbital therefore we have to add a contribution equal to $-\sum_j (2\langle ea|jj\rangle - \langle ej|ja\rangle)$. Thus

$$\langle {}^{b}_{a}|\bar{H}|{}^{b}_{e}\rangle = -h_{ea} - \sum_j (2\langle ea|jj\rangle - \langle ej|ja\rangle) - \langle ea|bb\rangle + 2\langle eb|ba\rangle$$

$$= -F_{ea} - \langle ea|bb\rangle + 2\langle eb|ba\rangle$$

since, when j=a, $2\langle ea|aa\rangle - \langle ea|aa\rangle = \langle ea|aa\rangle$ and

when j=e, $2\langle ea|ee\rangle - \langle ea|ee\rangle = \langle ea|ee\rangle$.
Introducing the LCAO approximation

$$\psi_a = \sum_p C_{pa}\phi_p,\quad \psi_e = \sum_t C_{te}\phi_t,\quad \psi_b = \sum_r C_{rb}\phi_r = \sum_s C_{sb}\phi_s$$

where $\phi_i$'s constitute an orthonormal basis set one obtains

$$\langle {}^{b}_{a}|\bar{H}|{}^{b}_{e}\rangle = -\sum_p\sum_t C_{pa}C_{te}F_{pt} - \sum_t\sum_r C_{te}C_{ta}C_{rb}^2\langle tt|rr\rangle +$$

$$+2\sum_t\sum_r C_{te}C_{tb}C_{ra}C_{rb}\langle tt|rr\rangle$$

When BRILLOUIN's theorem applies, $F_{ea} = 0$ and we have

$$\langle {}^{b}_{a}|\bar{H}|{}^{b}_{e}\rangle = 2\langle eb|ba\rangle - \langle ea|bb\rangle$$

Bi-excited configurations mix well with ground state configuration:

$$\langle {}^{cc}_{aa}|\bar{H}|0\rangle = \langle e\alpha e\beta c\alpha c\beta|\bar{H}|e\alpha e\beta a\alpha a\beta\rangle = \langle ca|ca\rangle$$

There are also monoexcited-biexcited and biexcited-biexcited matrix elements. Their calculation is straightforward according to the principles presented above.
For triplet states there is a phase ambiguity. The phase must be chosen in such a way that each triplet component

gives the same result. For instance,

$^{3}\Phi_{a}^{b}(S_Z=0)=\frac{1}{\sqrt{2}}(|a\alpha b\beta|-|b\alpha a\beta|)$ is the good choice since it gives the same result as $^{3}\Phi_{a}^{b}(S_Z=-1)=|b\beta a\beta|$. Therefore possibility $^{3}\Phi_{a}^{b}(S_Z=0)=\frac{1}{\sqrt{2}}(|b\alpha a\beta|-|a\alpha b\beta|)$ should not be envisaged.

In the following lines a few general formulas are presented for practical purpose :

$$^{1}\langle 0|\bar{H}|0\rangle=E_o,\quad ^{1}\langle 0|\bar{H}|{}^{s}_{r}\rangle=\sqrt{2}\,F_{rs},$$

$$^{1}\langle 0|\bar{H}|{}^{kk}_{ii}\rangle=\langle ki|ki\rangle,\quad ^{1}\langle 0|\bar{H}|{}^{kl}_{ii}\rangle=\sqrt{2}\langle ki|li\rangle,$$

$$^{1}\langle 0|\bar{H}|{}^{kk}_{ij}\rangle=\sqrt{2}\langle ki|kj\rangle,$$

$$^{1,3}\langle{}^{s}_{r}|\bar{H}|{}^{l}_{k}\rangle=\delta_{rk}\delta_{sl}E_o+\delta_{rk}F_{sl}-\delta_{sl}F_{rk}-\langle rk|sl\rangle+\langle sr|kl\rangle\pm\langle sr|kl\rangle$$

$$^{1}\langle{}^{s}_{r}|\bar{H}|{}^{kk}_{ii}\rangle=\sqrt{2}[\delta_{ri}\delta_{sk}F_{ik}-\delta_{sk}\langle ir|ik\rangle+\delta_{ri}\langle ik|sk\rangle]$$

$$^{1,3}\langle{}^{s}_{r}|\bar{H}|{}^{kk}_{ij}\rangle=\delta_{sk}\delta_{ri}F_{jk}+\delta_{ri}\langle jk|sk\rangle-\delta_{sk}[\langle jk|ir\rangle\pm\langle ik|jr\rangle]$$

$$^{1,3}\langle{}^{s}_{r}|\bar{H}|{}^{kl}_{ii}\rangle=\pm\delta_{ri}\delta_{sk}F_{il}\mp\delta_{sk}\langle il|ir\rangle+\delta_{ri}[\langle ik|sl\rangle\pm\langle il|sk\rangle]$$

$$^{1}\langle{}^{kk}_{ii}|\bar{H}|{}^{tt}_{rr}\rangle=\delta_{kt}\delta_{ir}[E_o+2F_{kk}-2F_{ii}-4\langle ii|kk\rangle+2\langle ik|ik\rangle]+$$
$$+\delta_{kt}\langle ir|ir\rangle+\delta_{ir}\langle kt|kt\rangle$$

$$^{1}\langle{}^{kk}_{ii}|\bar{H}|{}^{tt}_{rs}\rangle=\sqrt{2}[\delta_{kt}\delta_{ir}(-F_{is}-2\langle is|kk\rangle+\langle ik|ks\rangle)+\delta_{kt}\langle ir|is\rangle]$$

$$^{1}\langle{}^{kk}_{ii}|\bar{H}|{}^{tu}_{rr}\rangle=\sqrt{2}[\delta_{kt}\delta_{ir}(F_{ku}-2\langle ku|ii\rangle+\langle ki|ui\rangle)+\delta_{ir}\langle kt|ku\rangle]$$

$$^{1,3}\langle{}^{kl}_{ii}|\bar{H}|{}^{tt}_{rs}\rangle=\delta_{kt}\delta_{ir}[\pm\langle st|li\rangle-\langle si|lt\rangle\mp\langle si|lt\rangle]$$

$$^{1,3}\langle{}^{kk}_{ij}|\bar{H}|{}^{tt}_{rs}\rangle=\delta_{kt}\delta_{ir}[-F_{js}-2\langle js|kk\rangle+\langle jk|ks\rangle]+\delta_{ir}\delta_{js}\langle kt|kt\rangle+$$

$+\delta_{kt}[<ir|js>\pm<is|jr>]+\delta_{kt}\delta_{ir}\delta_{js}[E_o+2F_{kk}-F_{ii}-2<ii|kk>+$

$<ik|ik>]$

$^{1,3}<^{kl}_{ii}|\bar{H}|^{tu}_{rr}>=\delta_{ir}\delta_{kt}\delta_{lu}[E_o+F_{kk}+F_{ll}-2F_{ii}-2<ii|kk>+<ik|ki>-$

$-2<ii|ll>+<il|li>+<ii|ii>+<kk|ll>\pm<kl|lk>]$

Instead of the direct calculation of matrix elements one may apply a faster method using SLATER's rules:

1) The matrix element between determinants differing by more than two spin-orbitals is zero.
2) If two determinants differ by one spin-orbital ($\psi_i \neq \psi_{i'}$) the matrix element will contain a mono-electronic term ($h_{ii'}$) and a sum of three-center bi-electronic integrals : $\sum_{j\neq i}(<ii'|jj>-<ij|ji'>)$.

   $<ij|ji'>$ appears only when j has the same spin as i and i'.
3) If two determinants differ by two spin-orbitals ($\psi_i \neq \psi_{i'}$ and $\psi_j \neq \psi_{j'}$) the matrix element will contain a sum of four-center bi-electronic integrals: $<ii'|jj'>-<ij'|ji'>$ the latter appearing only if j and j' have the same spin as i and i'.
4) If two determinants are identical the matrix element will be composed of a sum of mono-electric terms $\sum_i h_{ii}$ and a sum of two-center bi-electronic integrals:

   $\sum\sum_{i<j}(<ii|jj>-<ij|ji>)$. $<ij|ji>$ intervenes only when i and j have the same spin.

For actual calculation the SÁNDORFY-DAUDEL rules (18) are more appropriate since they give useful tricks for electron book-keeping :

1) The determinants are represented by two parentheses containing the spinorbitals in natural order; the first contains spinorbitals with $\alpha$ spins and the second contains spinorbitals with $\beta$ spins.

E.g.: $|1\alpha 1\beta 2\alpha 2\beta 3\alpha 4\beta|$ will be written as (123)(124).

2) Calculation of diagonal elements $H_{pp}$. The two identical determinants must be written above each other so that the $\alpha$ and $\beta$ parentheses coincide :

   (...ij)(ij)
   (...ij)(ij)

Any orbital pair chosen "vertically" (e.g.ii) in a given parenthesis (i.e. the two spin-orbitals must have the same spin!) will give rise to $h_{ii}$ terms. The interaction with other orbital pairs (e.g.jj) produces COULOMB integrals $<ii|jj>$ and exchange integrals $<ij|ji>$. These latter appear only when i and j have same spin. Thus, in the present case, we have

$$H_{pp}=2h_{ii}+2h_{jj}+4<ii|jj>-2<ij|ji>+<ii|ii>+<jj|jj>$$

$$\longrightarrow 2\sum_j h_{jj}+\sum_i\sum_j (2J_{ij}-K_{ij})$$

E.g. : $\left.\begin{matrix}(124)(34)\\(124)(34)\end{matrix}\right\}\longrightarrow h_{11}+h_{22}+h_{33}+2h_{44}+<44|44>+$

$+<11|22>+<11|33>+2<11|44>+<22|33>+2<22|44>+$

$+2<33|44>-<12|12>-<14|14>-<24|24>-<34|34>.$

3) Calculation of off-diagonal elements $H_{pq}$.

a) one spin-orbital difference between bra(p) and ket(q):

$$\left.\begin{matrix}(\dots i\dots)(\dots)\\(\dots k\dots)(\dots)\end{matrix}\right\}\longrightarrow H_{pq}=(-1)^n[h_{ik}+\sum_{r\neq i,k}<ik|rr>-\sum_{s\neq i,k}<is|ks>]$$

n=number of permutations necessary to put bra and ket into coincidence;

r=any spin-orbital; the integral $<ik|rr>$ will appear twice if bra and ket contain both $r\alpha$ and $r\beta$;

s=any occupied orbital having the same spin as i.

E.g. : $\left.\begin{matrix}(125)(34)\\(135)(34)\end{matrix}\right\}\longrightarrow$

$$(-1)^{\circ}[h_{23}+<23|11>+<23|55>+<23|33>+<23|44>-<21|13>-<25|35>]$$

b) two spin-orbitals difference between bra (p) and ket (q):

$$\left.\begin{matrix}(\dots ij)(\dots)\\(\dots kl)(\dots)\end{matrix}\right. \text{ or } \left.\begin{matrix}(\dots i)(\dots j)\\(\dots k)(\dots l)\end{matrix}\right. \rightarrow H_{pq}=(-1)^n[<ik|jl>+\tau<il|jk>]$$

$\tau=-1$ if i and j have same spin and $\tau=0$ otherwise.

E.g.: $\begin{matrix}(123)(3)\\(134)(1)\end{matrix}\xrightarrow[\text{for coincidence}]{\text{n=1 permutation}}\begin{matrix}(123)(3)\\(143)(1)\end{matrix}\rightarrow H_{pq}=-<24|31>$

Let us add that when higher order excitations are to be included their matrix elements may be easily calculated by graphical methods (19). The CI method in its classical form has a very serious disadvantage stemming from the use of virtual orbitals in the establishment of excited electronic configurations. Virtual orbitals are not really solutions of a variational problem but they are simply the by-products of the SCF method. Consequently they are too diffuse and the CI problem will have a slow convergence. This means that a great many configurations must be included in the calculation in order to get a significant lowering of energy. Therefore it is not astonishing that several new methods have been proposed to avoid this drawback: use of natural orbitals as basis set (20), multi-configurational SCF (MC SCF where the variation principle is applied simultaneously to all orbitals used and to the coefficients of CI development)(21), CI from molecular integrals (CIMI)(22), pseudo natural orbitals CI (PNO-CI)(23), self-consistent electron pair method (SCEP)(24), unitary group CI (25) etc. In these more advanced methods not only the simple orbital picture of electron excitation disappears as it happened already in classical CI method but also the integral occupation number of orbitals loses its significance.

Let us conclude by reemphasizing the fact that the configuration interaction method in its various form remains one of the most powerful means of accounting for electron correlation.

References.

(1) Csizmadia, I.G.:"Theory and Practice of MO Calculations on Organic Molecules", Elsevier, Amsterdam, 1976.
(2) Wigner, E.P.: 1934, Phys. Rev. 46, pp. 1002.
(3) Kutzelnigg, W.: 1973, Fortschr. Chem. Forsch.41, pp. 31.
(4) Von Neumann, J.:"Mathematical Foundations of Quantum Mechanics" Princeton University Press, 1955.
(5) Shavitt, I.: in "Modern Theoretical Chemistry" (H.F. Schaefer, Ed.), Vol. 3, Plenum Press 1977.
(6) Peat, F.D.: in "Physical Chemistry, an Advanced Treatise" (D. Henderson, Ed.), Vol. XIA, Academic Press 1975.
(7) Ruelle, J.L.: Ph.D.Thesis, U.C.L. Louvain-la-Neuve, 1980.
(8) Daudel, R., Lefebvre, R. and Moser, C.:"Quantum Chemistry; Methods and Applications", Interscience Publishers, New-York, 1959.
(9) McWeeny, R. and Sutcliffe, B.T.:"Methods of Molecular Quantum Mechanics", Academic Press, New-York, 1969.
(10) Davidson, E.R. and Jones, L.L.:1962, J. Chem. Phys. 37, pp. 721.
(11) Coffey, P. and Jug, K.:1974, J. Chem. Ed. 51, pp.252.

(12) Pauntz, R.: "Spin Eigenfunctions " Plenum Press, New York, 1979.

(13) WERTZ, J.E. and BOLTON, J.R.:"Electron Spin Resonance ", McGraw -Hill, New York,1972.

(14) Löwdin, P.O.: 1956, Phys. Rev.97, pp. 1509.

(15) McGlynn, S.P., Vanquickenborne, L.G.,Kinoshita, M. and Carroll, D.G.: " Introduction to Applied Quantum Chemistry", Holt, Rinehart and Winston, Inc., New-York, 1972.

(16) Wigner, E.P. : "Group Theory and its Applications to the Quantum Mechanics of Atomic Spectra",Academic Press, 1959.

(17) Heine, V.: "Group Theory in Quantum Mechanics", Pergamon Press, Oxford, 1960.

(18) Sándorfy, C.:"Les Spectres Electroniques en Chimie Théorique", Revue d'Optique, Paris, 1959.

(19) Cizek, J. : 1969, Adv. Chem. Phys. 14, pp. 35

(20) Shull, H. and Löwdin, P.O.:1959, J. Chem.Phys.30, pp. 617.

(21) Wahl, A.C. and Das, G.:in "Modern Theoretical Chemistry" (Ed. Schaefer, H.F.), Vol.3, Plenum Press 1977.

(22) Roos, B.O. and Siegbahn,E.M.: in "Modern Theoretical Chemistry "(Ed.Schaefer, H.F.) Vol. 3, Plenum Press, 1977.

(23) Meyer, W.: in "Modern Theoretical Chemistry"(Ed. Schaefer, H.F.), Vol.3, Plenum Press, 1977.

(24) Meyer, W.: 1976, J. Chem. Phys. 64, pp. 2901.

(25) Hegarty, D. and Robb, M.A.: 1979, Mol. Phys. 38, pp. 1795..

# OPTIMIZATION AND ANALYSIS OF ENERGY HYPERSURFACES

Paul G. Mezey

Department of Chemistry and Chemical Engineering, University of Saskatchewan, Saskatoon, Saskatchewan, CANADA S7N OWO

ABSTRACT

Continuous and discrete optimization problems of computational theoretical chemistry are reviewed, with particular emphasis on potential energy hypersurfaces.

## 1. OPTIMIZATION METHODS IN THEORETICAL CHEMISTRY

One of the central problems of computational theoretical chemistry is to find optimum wavefunctions $\psi(\underline{r})$ that minimize the energy expectation value functional

$$E(\underline{r}) = \langle\psi(\underline{r})|\hat{H}(\underline{r})|\psi(\underline{r})\rangle \tag{1.1}$$

for various chemical systems, subject to the constraints

$$\langle\psi(\underline{r})|\psi(\underline{r})\rangle = 1 \ . \tag{1.2}$$

If an approximate quantum chemical model is specified, the limitations of the model usually lead to constraints on the actual representation of the Hamiltonian $\hat{H}(\underline{r})$. Similarly, we usually place some constraints on the wavefunction $\psi(\underline{r})$ in addition to requiring that it must be a well behaved function and normalized according to (1.2): e.g. we restrict our choice of $\psi(\underline{r})$ to a subspace of the Hilbert space by specifying a finite basis set,

$$\{\sigma_i\}_{i=1}^{n_b}$$

where $n_b$ is the dimension of the basis. Most frequently the $\sigma_i$



*I. G. Csizmadia and R. Daudel (eds.), Computational Theoretical Organic Chemistry, 101–128.*

basis functions are gaussian type atomic orbitals and the molecular wavefunction is generated as a Slater determinant (1) or a linear combination of such Slater determinants (2)

$$\psi = \sum_t C_t \psi_t \tag{1.3}$$

where

$$\psi_t = \det|\phi_{t_1} \phi_{t_2} \dots \phi_{t_{n_b}}| . \tag{1.3a}$$

The $\phi_{t_k}$ orbitals are linear combination of the AO basis functions

$$\phi_{t_k} = \sum_{j=1}^{n_b} c_{t_k j} \sigma_j \tag{1.3b}$$

The restrictions, inherent in the actual formulation of the problem, are usually made before any computation, and optimization of $\psi$ is carried out within a restricted model.

The application of various optimization techniques represents an important aspect of computational theoretical chemistry, in fact, a large part of computational effort and computer time involved in calculating molecular wavefunctions is spent on optimization.

Optimization problems are usually formulated in terms of an objective function F(p), defined over a set D, where elements p of D correspond to possible solutions. A $p \in D$ solution is optimum if F(p) is minimum (in some cases it is customary to define the optimum as the maximum of function F' = -F). The properties of set D and function F fully characterize the optimization problem. In theoretical chemistry the objective function is usually the energy, F = E, and in principle D contains all well behaved functions satisfying the boundary conditions of the given chemical problem. In practice, however, restrictions such as (1.3a) and (1.3b) lead to a simpler set D; e.g. in the LCAO single determinental formalism D is a domain of abstract vectors $\underline{p}$ containing the $\alpha_j$ orbital exponents and the $c_{kj}$ LCAO coefficients as components, fulfilling the orthonormality constraints for the $\phi_k$ orbitals. In this example the $\underline{p}$ elements of D may be regarded as continuous vector variables and the objective function $F(\underline{p}) = E(\underline{p})$ corresponds to a continuous optimization problem. However, considering e.g. the economic aspects of such calculations, one may define an objective function, that depends on both the calculated energy expectation value and the computer time T required for its calculation, F' = F'(E,T). Such an objective function could reflect the optimum compromise between accuracy and computer costs. The optimum of F' clearly depends on the actual restrictions placed on the approximate wavefunction, e.g. on the integer number $n_b$, characterizing the size of the basis set. In this example D becomes a union of $D^{(n_b)}$ sets, corresponding to various discrete choices of $n_b$,

$$D = \bigcup D^{(n_b)} \quad .$$

There is a discrete $D^{(n_b)}$ set that contains a vector $\underline{p}^{(n_b)}$ which optimizes the actual objective function F', and the selection of this optimum $D^{(n_b)}$ set (i.e. the selection of the optimum $n_b$ number) corresponds to a discrete optimization problem.

A more general construction of D may correspond to a union of feasible $D^{(m_i)}$ domains, representing the $m_i$ elements of a set of discrete mathematical models, $\{m_i\}$,

$$D = \bigcup D^{(m_i)} \quad .$$

Chemical intuition is more closely related to the choice of the mathematical model than to an actual numerical solution, whereas most recent improvements of the mathematical optimization techniques are related to finding a given type of solution more rapidly. Usually it is possible to classify the above two types of problems, i.e. finding the best model and finding the best numerical solution, as discrete and continuous optimization problems, respectively.

### 1.1. Discrete optimization problems

Most discrete optimization problems may be formulated in terms of an objective function $F(\ell)$ that is defined for every element of a countable set $\{\ell_i\}$. The selection of an optimum element $\ell_{opt}$ corresponds to finding the minimum of $F(\ell)$, $F(\ell_{opt})$ over the discrete set $\{\ell_i\}$. Although the objective function $F(\ell)$ is nondifferentiable, extensions of methods used for continuous optimization, e.g. the Lagrange multiplier technique, may be applied in searching for $\ell_{opt}$. Matters become much more complicated, if elements $\ell_i$ themselves depend on a set of continuous parameters, $\underline{r}$:

$$\ell_i = \ell_i(\underline{r}) \qquad (1.4)$$

Unfortunately most discrete optimization problems of computational theoretical chemistry fall into this class, consequently, these problems are difficult to cast in a rigorous mathematical form that may be converted into programmable algorithms.

Our fundamental problem is, how to choose a quantum chemical

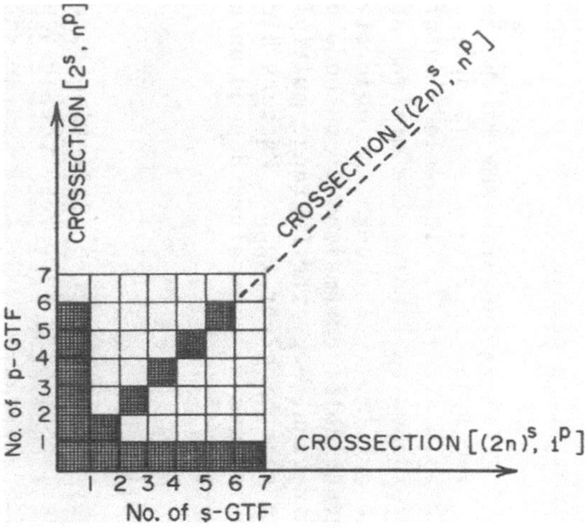


Figure 1. Schematic representation of various Gaussian bases.

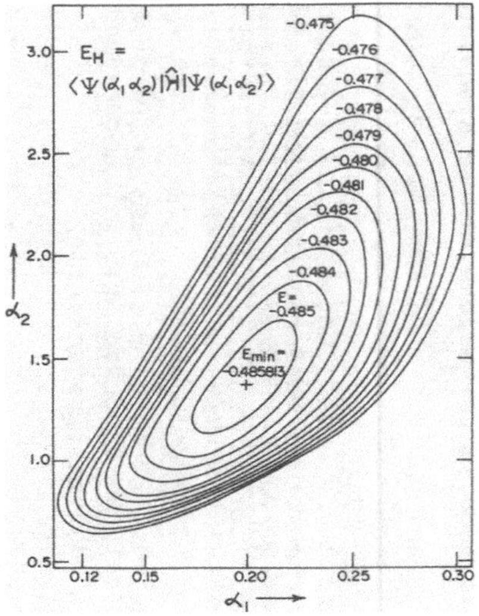


Figure 2. Energy contours of the $E(\alpha_1, \alpha_2)$ function for the H atom.

model that adequately represents our chemical problem and is also solvable by currently available computational techniques. There are several aspects one must consider in choosing a quantum chemical model. As an example, if one studies the migration of triple bond in a series of alkyn-1-ols, only relatively short alkyne chains may be considered. However, a minimum size is required to be able to study the effect at all. If the model compounds are chosen, an appropriate quantum chemical technique has to be selected. Clearly, the triple bond migration is a time dependent phenomenon, however, at present the solution of the time dependent Schrodinger equation for such molecules represents an insurmountable problem. The same is valid for a relativistic solution. It is likely that one has to settle for the separation of nuclear and electronic motion and calculate wavefunctions within the Born-Oppenheimer approximation. All the above choices between discrete theoretical models are more or less limited and very often determined by computational possibilities. However, there are further choices to be made between various feasible models. Shall we use a small basis set for an SCF calculation and carry out a limited CI treatment involving few configurations, or shall we rather use an extended basis set and no CI techniques? If the limitations on the basis set size are given (e.g. by the dimensions of the program or by the available computer time) then what is the best choice for the dimensions of the subspaces of s, p, d etc. type LCAO basis functions? Shall we use polarization functions, and if yes, how many such functions and on what nuclei?

These latter questions are usually answered on the basis of previous experience, although for the given chemical problem and computational constraints they all have a discrete solution, or best choice.

In Figure 1 some of the possible choices for a Gaussian basis set for the fluorine atom are shown schematically. Series $[2^s, n^p]$ and $[(2n)^s, 1^p]$ are poor choices, since

TABLE 1. Orbital exponent variations for the Fluorine atom

| | $2^p$ subset from $[2^s, 2^p]$ | $2^p$ subset from $[4^s, 2^p]$ |
|---|---|---|
| $\alpha_1$ | 4.092113 | 4.163853 |
| $\alpha_2$ | 0.696367 | 0.707438 |

either the s-subspace or the p-subspace is under-represented. Optimum representations are those along or near the $[(2n)^s, n^p]$ series, since for these bases the dimensions of the s and p subspaces correspond to the best compromise if the total dimension of the $\{\sigma_i\}$ basis is fixed. (3)

The selection of the best ratio of s and p type functions in a basis set of a fixed overall dimension is a well defined discrete optimization problem for the given molecule. Nevertheless, this discrete optimization problem is intimately connected with a continuous optimization problem: i.e. with the search for the best orbital exponents $\{\alpha_i\}$ that must be selected from a continuous set of real numbers. This interrelation between the discrete and continuous problems complicates the solution, and this is why very few rigorous optimizations have been carried out on molecules, and usually we rely on a somewhat arbitrary combination of experience and intuition.

This interrelation is clearly shown in Table I, where optimum $2^p$ subsets of two fully optimized fluorine basis sets are compared. These subsets correspond to two discrete models, one is $[2^s, 2^p]$ basis, the other is a $[4^s, 2^p]$ basis set. Although the s and p subsets of these bases are orthogonal to each other by symmetry, the optimum p-orbital exponents depend on the actual representation of the s subspace, since different s functions generate a different potential in the fluorine atom. The optimum choice of the continuous parameters (orbital exponents) $\alpha_1$ and $\alpha_2$ correspond to a continuous optimization problem.

Naturally, similar discrete optimum problems arise in studying reaction paths on potential energy surfaces. Since a complete analysis of the entire hypersurface even for a relatively small molecule is seldom possible, an initial choice between various reaction mechanisms must be made, and only a few of the many possibilities may be followed up by detailed calculations (4). That is, there is a choice from a set of discrete reaction mechanisms that constitutes a discrete optimum problem, and the actual properties of the minimum energy path, or more generally the properties of a reaction domain along this path, (5) correspond to a continuous optimization problem.

Stereochemistry shows very clearly this duality of our optimization problems: there is a discrete point group that corresponds to the most stable arrangements of the given nuclei - a discrete optimum problem, whereas the best bond length, bond angle parameters represent the optimum of a set of continuous parameters - a continuous optimization problem. The discrete and continuous optimum problems are interrelated. If a molecule may have different conformations $A^{(1)}, \ldots A^{(i)}$, belonging to point groups $G^{(1)}, \ldots G^{(i)}$, then the optimum point group cannot be selected without knowing the optimum bond length values, since bond length variations may change the order of relative stabilities. On the other hand, if a bond length is optimum for a given point group, it may be different from the overall optimum, as long as there is a chance that the molecule favors a different point group.

It is usually possible to formulate the discrete optimization problem as a choice between neighbourhoods of various local minima of an objective function, whereas the continuous optimization problem corresponds to minimization of the function in the given neighbourhood of a local minimum. The discrete and continuous problems are then related to the global and local properties of this objective function, respectively.

Some discrete optimization problems of theoretical chemistry may be formulated very elegantly in graph theoretical terms. Graph theoretical methods are particularly suitable for π-electron systems in semiempirical formalism (see e.g. refs 6-9), however, the extension of these techniques to more sophisticated quantum chemical treatments is a complicated task. Another area where graph theoretical techniques are very promising is the analysis of multidimensional potential energy surfaces (10) and the determination of possible reaction mechanisms.

1.2. Continuous optimization problems

It is somewhat ironic that quantum chemistry, the science of discrete, quantized eigenstates of molecular systems, is so heavily dependent on the mathematical apparatus of continuous functions. The approach of using hermitian operators on well behaved functions in order to obtain a set of discrete eigenvalues appears, at best, very indirect. The mathematical theory of Hilbert space and well behaved functions, however, is highly developed, and continuity (and continuous optimization methods) allow for direct analogies between classical and quantum mechanics. The continuity of wavefunction $\psi$, as a function of some generalized coordinates and additional parameters, makes the application of standard optimization methods like the Ritz variational technique possible.

The most commonly used objective function is the energy expectation value (1.1) which is almost everywhere a continuous and differentiatiable function of both the parameters of the approximate wavefunction and the coordinates used for the specification of the molecular system. These coordinates are usually cartesian coordinates of the nuclei or a set of internal coordinates, and no electronic coordinates are explicitly considered in optimization problems, since the integration in the expectation value expression (1.1) involves all the electronic coordinates. By collecting all these variables in vector $\underline{r}$, one may define the energy gradient

$\underline{g}(\underline{r})$ at point $\underline{r}$ by

$$g_i(\underline{r}) = \frac{\partial E(\underline{r})}{\partial r_i} \tag{1.5}$$

A point $\underline{r}_c$ where the gradient vanishes

$$\underline{g}(\underline{r}_c) = \underline{0} \tag{1.6}$$

is a critical point. At every critical point where $E(\underline{r})$ is twice differentiable the Hessian matrix $H(\underline{r}_c)$ is defined by

$$H_{ij}(\underline{r}_c) = \frac{\partial^2 E(\underline{r}_c)}{\partial r_i \partial r_j} \tag{1.7}$$

We usually require that in our model $E(\underline{r})$ is twice continuously differentiable.

The set of critical points, $\{\underline{r}_c\}$ may be classified according to their Hessian matrices $H(\underline{r}_c)$ (11). If the Hessian matrix is singular,

$$\det|H(\underline{r}_c)| = 0 \tag{1.8}$$

then $\underline{r}_c$ is a degenerate critical point. The index $\lambda(\underline{r}_c)$ is equal to the number of negative eigenvalues of the Hessian. If

$$\lambda(\underline{r}_c) = n \tag{1.9}$$

where n is the dimension of vector $\underline{r}$, then $\underline{r}_c$ is a maximum, whereas if a non-degenerate critical point $\underline{r}_c$ has a zero index,

$$\lambda(\underline{r}_c) = 0 \tag{1.10}$$

then $\underline{r}_c$ is a minimum. If $\underline{r}_c$ is a non-degenerate critical point and

$$0 < \lambda(\underline{r}_c) < n \tag{1.11}$$

then $\underline{r}_c$ is a saddle point.

For a degenerate critical point $\underline{r}_c$ the Hessian matrix $H(\underline{r}_c)$ formally has at least one zero eigenvalue. A necessary condition for a degenerate critical point to be a minimum is that the orders of the first non-vanishing derivatives along each $r_i$ variables are even numbers, and all these derivatives are positive. If a degenerate critical point $\underline{r}_c$ with index $\lambda(\underline{r}_c) = 0$ does not satisfy the above condition then $\underline{r}_c$ is called a shoulder.

In computational practice minima $\lambda(\underline{m}) = 0$ and saddle points with index $\lambda(\underline{r}_c) = 1$ are of the greatest importance. If an approximate wavefunction $\psi(\underline{r})$ depends on parameters $\underline{p}$ and nuclear coordinates $\underline{x}$, then vector $\underline{r}$ may be represented as the direct sum

$$\underline{r} = \underline{p} \oplus \underline{x} \tag{1.12}$$

i.e.

$$r_i = p_i \quad (i \leq k) \tag{1.12a}$$

$$r_{k+i} = x_i \tag{1.12b}$$

if the dimension of the parameter vector $\underline{p}$ is k.

If we are interested in optimizing an approximate wavefunction for a fixed nuclear configuration, then $\underline{x}$ is fixed and may be omitted from $\underline{r}$, i.e. n = k, and

$$\underline{r} = \underline{p} \tag{1.13}$$

The simplest examples for such wavefunctions are approximate atomic wavefunctions where no $\underline{x}$ vector is involved. In Figure 2 the $E(\underline{r})$ surface of the hydrogen atom ground state is shown as calculated with a basis set of two Gaussian functions. For this surface

$$r_i = p_i = \alpha_i, \quad i = 1, 2,$$

that is, the variables are the two orbital exponents $\alpha_1$ and $\alpha_2$ of the ($2^s$) Gaussian basis $\{\sigma_1, \sigma_2\}$.

For geometry optimization problems of molecules we may assume that the optimum $\underline{p}$ parameters are depending on the nuclear coordinates,

$$\underline{p} = \underline{p}\ (\underline{x}) \tag{1.14}$$

and by considering the optimum parameter values for each nuclear arrangement, the wavefunction may be written as

$$\psi = \psi(\underline{x}) \tag{1.15}$$

and vector $\underline{r}$ as

$$\underline{r} = \underline{x}\ . \tag{1.16}$$

If the optimization is restricted to the $\underline{p}$ parameters, eq. (1.13), then only minima of the $E(\underline{r})$ functional are of interest and critical points $\underline{r}_c$ with index $\lambda(\underline{r}_c) > 0$ may have significance only as false "solutions" to our optimization problem. On the other hand, in the case of geometry optimization (eq. (1.16) both minima $\underline{m}$ with index

HYPERBOLIC PARABOLOID

$$\left(\frac{x}{a}+\frac{y}{b}\right)\left(\frac{x}{a}-\frac{y}{b}\right)=cz.$$

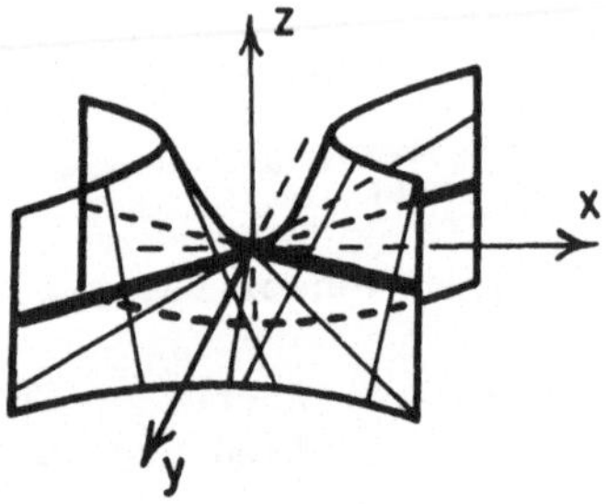


systems of lines

$$\frac{x}{a}+\frac{y}{b}=mcz, \quad \frac{x}{a}-\frac{y}{b}=\frac{1}{m}z,$$

$$\frac{x}{a}+\frac{y}{b}=\frac{1}{n}z, \quad \frac{x}{a}-\frac{y}{b}=ncz,$$

any member of which lies wholly on the surface and is a generator of the paraboloid.

Figure 3. Hyperbolic paraboloid.

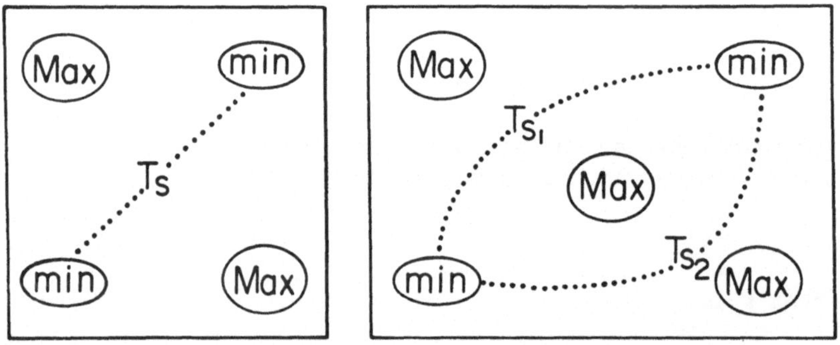


Figure 4. Schematic representation of reaction paths.

$$\lambda(\underline{m}) = 0 \tag{1.17}$$

and saddle points $\underline{r}_c$ with index

$$\lambda(\underline{r}_c) = 1 \tag{1.18}$$

are of importance. Minima correspond to "equilibrium geometries" and the latter type of critical points correspond to "transition state structures" on a potential energy hypersurface

$$E(\underline{r}) = E(\underline{x}) \ . \tag{1.19}$$

The objective of most geometry optimizations, carried out within the Born-Oppenheimer approximation (12), is finding minima of the total energy hypersurface. However, the properties of saddle points (13) or the location of critical points of energy component hypersurfaces (14), corresponding to various partitionings of the total energy, are also important in the interpretation of chemical processes. Whereas numerous minimization techniques have been applied for the determination of energy minima (e.g. the very efficient gradient methods, (15-25)), there are relatively few techniques that are directly applicable to more general critical points, e.g. to saddle points. In the next chapter we shall review some of these techniques.

## 2. OPTIMIZATION OF TRANSITION STATE STRUCTURES

Theoretical determination of transition state geometries of molecules involved in conformational changes or chemical reactions is of considerable importance, since experimental determination of structural features of such species is extremely difficult and often impossible. The concept of transition state plays a central role in the interpretation and prediction of chemical processes. Although in the rigorous quantum mechanical sense the transition state as represented by a given geometric arrangement of the atoms is only an approximation and in reality no chemical reaction follows a definite geometric course along a potential energy hypersurface, the geometries associated with such points on an $E(\underline{r})$ energy hypersurface may be used for the classification of chemical processes.

### 2.1. The saddle point problem

Within the framework of the Born-Oppenheimer (12) and energy hypersurface approximations transition states are associated with such saddle points $\underline{s}$ of the hypersurface, which are maxima with respect to one direction (13); that is, the Hessian matrix $H(\underline{s})$ has exactly one negative eigenvalue, $\lambda(\underline{s}) = 1$ (11). Saddle points with larger $\lambda$ index, $\lambda(\underline{s}) > 1$ are of little chemical importance, since they are maxima along at least two linearly independent directions, and such "hill tops" are certainly avoided by minimum energy reaction paths.

In Figure 3 the simplest saddle surface, the hyperbolic paraboloid

$$\frac{X^2}{a^2} - \frac{Y^2}{b^2} = cZ$$

is shown. The saddle point is a minimum along coordinate X and a maximum along coordinate Y. The hyperbolic paraboloid has a very attractive feature: it is quadratic in both X and Y. Energy hypersurfaces of molecular systems are, unfortunately, far from quadratic in most regions of the internal coordinates; nevertheless, close enough to a (non-degenerate) saddle point the error of a quadratic approximation is small. Since it is always easier to find a critical point of a quadratic function than that of more general functions, for transition state determination a good initial guess is particularly important.

In addition to the problems associated with non-quadratic hypersurfaces there are further problems related to the curvature properties at the saddle point. If the curvatures along the maximum and minimum cross sections are very different, i.e. if the positive and negative eigenvalues of the Hessian matrix $H(\underline{s})$ are very different in magnitude, then similar numerical problems may arise as in the case of minima of long and narrow valleys. The extreme case with zero curvature along one direction corresponds to a singular Hessian matrix, i.e. to a degenerate critical point. The two main types of near-degenerate saddle points of index $\lambda(\underline{s}) = 1$ are sometimes referred to as Don Quixote and Sancho Panza type saddle points (11,14); the Hessian of the former has a large negative and at least one near-zero positive eigenvalue, whereas the Hessian of the latter has a near-zero negative and large positive eigenvalues.

## 2.2. Algorithms for saddle point determination

The determination of saddle points of index $\lambda(\underline{s}) = 1$ is a considerably more complicated task than finding minima of a multidimensional hypersurface. There are very efficient methods that can be applied to various minimization problems, e.g. various gradient methods: the conjugate gradient methods (15-17), variable metric methods (17-23) and supermemory gradient methods (24,25); however, their direct application to transition state optimization is possible only if the direction $\underline{e}_1$ of the canonical coordinate along which saddle point $\underline{s}$ ($\lambda(\underline{s}) = 1$) is a maximum, is known. In such a case a series of minimizations may be carried out in the subspace of the internal coordinates orthogonal to $\underline{e}_1$ at various values of the $r^1$ component along coordinate $\underline{e}_1$ and a one dimensional maximum search along $\underline{e}_1$ leads to the transition state. In chemical practice the $\underline{e}_1$ direction is often known from the symmetry of the molecules, and frequently the location of the maximum along $\underline{e}_1$ is also known, as is the case for the planar transition state of the $NH_3$ pyramidal inversion process. In such cases a direct energy minimization within the symmetry of the transition state (in the case of $NH_3$ an in-plane optimization) directly leads to the transition state and no one dimensional maximization is needed.

It is possible, however, to apply unconstrained minimization methods to the scalar quantity

$$\sigma(\underline{r}) = \underline{g}(\underline{r})\cdot\underline{g}(\underline{r}) \tag{2.1}$$

i.e. to the scalar square of the gradient norm.

Function $\sigma(\underline{r})$ is zero at every critical point of the energy hypersurface $E(\underline{r})$

$$\sigma(\underline{r}) = 0 \text{ if } \underline{g}(\underline{r}) = \underline{0}, \tag{2.2}$$

that is, for saddle point $\underline{s}$

$$\sigma(\underline{s}) = 0 \ . \tag{2.3}$$

However, if a solution $\underline{r}_c$ is found for which

$$\sigma(\underline{r}_c) = 0 \tag{2.4}$$

than $\underline{r}_c$ may be any critical point with any index $\lambda$

$$0 \leq \lambda(\underline{r}_c) \leq n \ . \tag{2.5}$$

Consequently further tests are required to decide whether $\underline{r}_c$ is indeed a transition state, even if the minimization procedure has started with a reasonable guess near the saddle point $\underline{s}$, $\lambda(\underline{s}) = 1$. Such tests involve the evaluation of the Hessian matrix $H(\underline{r}_c)$ at point $\underline{r}_c$. In addition, function $\sigma(\underline{r})$ may have non-zero minimum points, i.e. points where condition (2.4) is not satisfied, and these minima do not correspond to critical points of the hypersurface $E(\underline{r})$. Nevertheless, an unconstrained minimization of $\sigma(\underline{r})$ may lead to such local minima.

The first systematic procedure for transition state determination based on the minimization of the gradient norm $\sigma(\underline{r})$ has been proposed by McIver and Komornicki (26). In an iterative procedure $\sigma(\underline{r})$ is reduced to zero and for solution $\underline{r}_c$ the Hessian matrix $H(\underline{r}_c)$ is evaluated and condition (1.18) is checked. In a modification due to Poppinger (27), the Hessian matrix with its eigenvalues and eigenvectors is evaluated repeatedly, and if convergence occurs, then it is to a saddle point $\underline{s}$, $\lambda(\underline{s}) = 1$, i.e. to a transition state. Variants of the gradient norm minimization method have been used with semiempirical methods (26,28,29) and with ab initio single determinant methods (27,30, 31) for various molecular rearrangements and reactions. An adaptation of the gradient norm method to multiconfigurational SCF wavefunctions has been reported recently (32).

Another procedure, the "synchronous transit method" has been proposed for transition state determination by Halgren and Lipscomb (33). The method is based on the assumption that the structural variations of a molecular system along a reaction path are not very different from geometry variations generated by quadratic interpolation between "reactant" and "product" geometries.

A series of alternating linear and quadratic interpolations are carried out between the "reactant" and "product" minimum points. A maximum point obtained in a linear interpolation defines a relative distance in the 3N-6 dimensional space from the two minima, and in the following quadratic interpolation step the energy is minimized on a hypersurface where the relative distance is constant. This simple search strategy works well for relatively simple hypersurfaces and converges to a transition state; however, it may also generate non-convergent loops far from the saddle point (34). Alternative strategies that may improve the convergence properties are currently under testing (ref. 44 in (34)).

A method, proposed by Muller and Brown (35) and Muller (34), is based on the determination of ascending valley points on a series of hyperspheres, centered on the reaction path. The valley points converge to the saddle point on both sides. For each hypersphere an n-1 dimensional minimization problem has to be solved in order to find the next valley point. This method requires the previous determination of the two minimum points that are on the two sides of the saddle.

The so-called X-method (36,37) does not require previous determination of minimum points nor the evaluation of the gradient at the search point. The X-method is based on a property of quadratic saddle surfaces that may be best visualized on the example of the hyperbolic paraboloid (Fig. 3): there is one and only one pair of horizontal straight lines that lie wholly on the saddle surface. In the case of the

$$\frac{X^2}{a^2} - \frac{Y^2}{b^2} = cZ \qquad (2.6)$$

hyperbolic paraboloid the equations of these lines are

$$\frac{X}{a} = \frac{Y}{b} \qquad (Z = 0) \tag{2.7a}$$

$$\frac{X}{a} = -\frac{Y}{b} \qquad (Z = 0) \tag{2.7b}$$

The crossing point of these lines is the saddle point (0, 0, 0).

This feature may be utilized to locate saddle points of surfaces that are approximately quadratic in the neighbourhood of the critical point. If one starts with a reasonably good guess, then this condition is usually met. For quadratic surfaces the crossing point of the best fitting pair of horizontal line segments is the saddle point. Consequently, the determination of the saddle point is equivalent to the determination of the best fitting horizontal line pair. The method may be termed the "method of horizontal lines" or, since the line segments form an X, "X-method".

This idea may be generalized for multidimensional hypersurfaces (37). After suitable translation of the coordinate frame any quadratic hypersurface E over coordinates $\{x_i\}_1{}^n$ may be written as

$$\underline{r}'\underline{Hr} = E(\underline{r}) \tag{2.8}$$

where H is the Hessian matrix expressed at critical point $\underline{r} = \underline{0}$. For non-degenerate first order saddle points H has exactly n-1 positive and 1 negative eigenvalues.

The generalization for multidimensional non-degenerate saddle surfaces is as follows. Let us assume that $0 < \lambda < n$ and that the Hessian H is diagonal, what always can be achieved by a unitary transformation. We may also assume that the first set of $n-\lambda$ coordinates are those with positive and the second set of $\lambda$ coordinates are those with negative eigenvalues. Since the saddle point is non-degenerate, the Hessian has no zero eigenvalue. Choosing one coordinate from the first and one from the second set of coordinates and keeping all other coordinates zero, one obtains a two dimensional subspace over which the corresponding cross-section of the hypersurface is necessarily a hyperbolic paraboloid, as the one shown in Fig. 3. There are $(n-\lambda)\lambda$ such coordinate pairs and one may construct $(n-\lambda)\lambda$ such hyperbolic paraboloids. Within any of these subspaces the corresponding two dimensional Hessian is non-singular, it has one positive and one negative eigenvalue. That is, none of the two horizontal lines on this hyperbolic paraboloid may coincide with any of the coordinate directions, since this would imply a zero eigenvalue for the Hessian. Consequently, both horizontal lines must have components along both of the coordinate directions. It follows then, that no two such hyperbolic paraboloids in two different subspaces can have common horizontal lines. Consequently, there are at least $2(n-\lambda)\lambda$ different horizontal lines passing through the saddle point and lying wholly on the hypersurface. Since each of the coordinate directions can be expressed as a linear combination of two of these lines, the lines form a generator set for the n-dimensional coordinate space. Consequently, one always can select n linearly independent horizontal lines that lie on the surface, with a common point at the saddle point.

On the other hand, if n linearly independent horizontal lines pass through a point of a quadratic hypersurface over the n-dimensional coordinate space, and all of them lie on the surface, then the gradient at this point must be zero. Consequently, n linearly independent horizontal lines, lying on the hypersurface, can have a common point only at the critical point.

The strategy for the determination of saddle points is clear then: one should fit a set of n linearly independent, horizontal lines with one common point onto the hypersurface. This set of lines may fit on the surface only if their common point coincides with the saddle point.

Test calculations with the multidimensional X-method indicate that the method converges to the saddle point even if the initial guess was near to an energy minimum. However, in the case of near-degenerate saddle points, such as Don Quixote and Sancho Panza type surfaces (11,14) the convergence properties of the method are poor.

## 3. OPTIMIZATION OF REACTION PATHS

### 3.1. The reaction path concept

The notion of reaction path is associated with the classical image of the geometric rearrangement of various nuclei during the chemical process. For a given "reactant" - "product" pair of nuclear configurations all those geometric changes that lead from the geometry of the reactant to the geometry of the product, may be regarded as reaction paths in a multidimensional space R, spanned by the coordinates used to specify the geometry of the nuclei. It is advantageous to describe geometric variations of the molecular system by using the formalism of Riemannian geometry (38), since it is possible then to obtain results in a form that is independent of the actual choice of the internal coordinates.

In order to interpret chemical reactions with the aid of

reaction paths on abstract energy hypersurfaces one usually relies on the Born-Oppenheimer (12) or an equivalent approximation. For the given electronic state of the molecular system the energy may be represented as a functional of the nuclear positions. This approximation involves the separation of nuclear and electronic motion, or a similar separation of "rapidly changing" and "slowly changing" variables (39), and is not strictly valid. In addition even if one uses an approximate Born-Oppenheimer model, the energy functional is not an analytic function of the nuclear coordinates at every point $\underline{r} \in R$, and points where $E(\underline{r})$ is non-differentiable may occur, typically at intersections of energy hypersurfaces (40,41), and several examples are known for the occurrence of instable or erratic regions of both RHF and UHF solutions (e.g. 42,43). Furthermore, most standard ab initio methods cannot handle the neighbourhoods of Coulomb singularities properly, i.e. the vicinity of those points where two or more nuclei have coincident positions. (Such processes correspond to nuclear reactions). Such points, however, may be excluded from any analysis that requires continuous and smooth hypersurfaces, by defining a domain (5)

$$D_{excl} = \bigcup_{\gamma} G(\underline{r}_{\gamma}, \rho). \tag{3.1}$$

Here $\underline{r}_{\gamma}$ is either a point where the coordinates of two or more of the nuclei become identical or any point where $E(\underline{r})$ is not twice continuously differentiable. The $\rho$ radius of the $G(\underline{r}_{\gamma}, \rho)$ open balls is a suitably chosen small positive value (5).

Another problem may arise in regions of the coordinate space where the potential energy hypersurface is flat along some of the coordinates (44). On such surfaces small energy variations may be associated with large geometry changes and small errors in the optimization may lead to large errors in the calculated geometry. Although, in addition to the associated vibrational problem, all essential features of a complete minimum energy path may be treated using the large amplitude motion formalism (45) originally developed for spectroscopic applications, this method is not ideally suited for the analysis of conformational problems or reactions of larger molecules.

A rigorous mathematical formulation of the reaction coordinate concept has been given by Hofacker (46), Marcus (47,48), Polanyi and Schreiber (49) and for special "multivalued" representations by Miller and Light (50). This formulation is designed for dynamical studies, and is ideally suited only for small (e.g. triatomic) chemical systems, although for low energy processes it may be extended for larger molecules (51). More suitable for larger organic molecules is a simplified concept: the concept of minimum energy reaction path. A precise definition of minimum energy reaction path, called "intrinsic reaction coordinate", has been given by Fukui (52) and later refined by Tachibana and Fukui (53,54). This reaction path may be visualized as a path that interconnects the reactant and product minima on an energy hypersurface "relief map" (5), passes through a transition state and is orthogonal to any equipotential contour hypersurface crossed.

Using the formalism of Riemannian geometry (38), a vector $\underline{r}$, specifying the geometry of the actual molecular system, may be given in terms of contravariant components $r^i$ or covariant components $r_i$ that are related by the $g_{ij}$ Riemannian metric tensor:

$$r_i = g_{ij} r^j \quad . \tag{3.2}$$

Here the usual summation convention is used: if an index occurs twice in the same term, that implies summation. In terms of the metric tensor infinitesimal distance ds is defined as

$$ds^2 = g_{ij}\, dr^i\, dr^j \ . \tag{3.3}$$

The scalar product $\underline{r}\cdot\underline{p}$, norm $|\underline{r}|$ and angle $\alpha$ between vectors $\underline{r}$ and $\underline{p}$ are defined as

$$\underline{r}\cdot\underline{p} = g_{ij}\, r^i\, p^j = r^i p_i \ , \tag{3.4}$$

$$|\underline{r}| = (r^i r_i)^{1/2}, \tag{3.5}$$

$$\cos\alpha = \frac{\underline{r}\cdot\underline{p}}{|\underline{r}|\,|\underline{p}|} \ , \tag{3.6}$$

respectively. This formulation allows a very general description of various surface properties, and orthogonality relations, gradient and curvature properties may be studied in terms of coordinate invariant quantities (38). A review of the relevant concepts of differential geometry has been given by Tachibana and Fukui (53).

It is convenient to regard the $E(\underline{r})$ energy value as component $r^{n+1}$ of an n+1 dimensional vector ${}^{n+1}\underline{r}$ with the first n components identical to those of vector $\underline{r}$. Vector ${}^{n+1}\underline{r}$ belongs to an n+1 dimensional space ${}^{n+1}R$, ${}^{n+1}\underline{r} \in {}^{n+1}\overline{R}$, embedding the energy hypersurface. In space ${}^{n+1}R$ the hypersurface may be visualized as a relief map where the energy corresponds to elevation, and a reaction process may be imagined as following a <u>relief path</u> ${}^{n+1}P$ on this relief hypersurface. Changes along this relief map involve variations in the energy, i.e. in component ${}^{n+1}r^{n+1}$. By contrast, the projection of such a relief path ${}^{n+1}P$ onto the coordinate space R corresponds to geometry variations only, with no direct reference to energy. This projected path P may be referred to as the reaction path (5).

If the relief path ${}^{n+1}P$, interconnecting two minimum points on the hypersurface, is given, then one may easily define a numerical measure of how far the reaction has progressed along ${}^{n+1}P$. Such a measure is the ${}^{n+1}s$ arc length, measured from the minimum of the reactants:

$$ {}^{n+1}s = \int_{{}^{n+1}P} {}^{n+1}ds \, . \tag{3.7} $$

Here ${}^{n+1}ds$ is defined similarly to eq. (3.3) as

$$ {}^{n+1}ds^2 = {}^{n+1}g_{ij} \; {}^{n+1}dr^i \; {}^{n+1}dr^j \tag{3.8} $$

The ${}^{n+1}s$ arc length is the relief path coordinate along relief path ${}^{n+1}P$. An analogous measure may be defined for reaction path P, as the reaction coordinate,

$$ s = \int_P ds \, . \tag{3.9} $$

If path P is a steepest descent path, i.e. one that follows the negative gradient $-\underline{g}(\underline{r})$ at every non-critical point of the path, $\underline{r} \in P$, then the path vector $\underline{a}(\underline{r})$ is defined as the normalised gradient vector:

$$ \underline{a}(\underline{r}) = \tilde{\underline{g}}(\underline{r}) = \frac{\underline{g}(\underline{r})}{|\underline{g}(\underline{r})|} \, . \tag{3.10} $$

The extremity of path P is usually (but not always) a critical point $\underline{r}_c, \underline{g}(\underline{r}_c)=\underline{0}$, and in such a case one may define a path vector $\underline{a}(\underline{r}_c)$ as the limit of path vectors along P, leading to point $\underline{r}_c$ (5). If P leads into a domain where $E(\underline{r})$ is non-differentiable i.e.

$$ P \cap D_{excl} \neq \emptyset \tag{3.11} $$

then in this domain no path vector is defined. Using the path vector concept and the curvature properties of the hypersurface it is possible to determine which are the most important coordinate domains $D_\mu$ of the hypersurface and in what domains are steepest descent paths stable with respect to small vibrational perturbations (5). Such domains are those with index $\mu = 0$.

The minimum energy reaction path concept is frequently used for qualitative description of chemical reactions. A minimum energy path corresponds to an idealized, infinitely slow rearrangement of the nuclei, in which no vibration occurs and energy is continuously provided or removed as the rearrangement progresses. Such an ideal model is clearly a gross oversimplification of real reaction processes, where vibrational motion "across" the reaction path is always present and the energy redistribution between various vibrational modes, as well as the rotational transitions, cannot be described by a classical mechanical model. Nevertheless, the minimum energy path concept is useful to describe the crude features of conformational and reaction processes, and distinct reaction mechanisms are usually associated with distinct minimum energy paths (see Figure 4). Such an assignment of minimum energy paths to reaction mechanisms is appropriate, if there are no instability domains $D_\mu$, $\mu\geq 1$ along the path P. However, if P enters an instability domain, $D_\mu$, $\mu\geq 1$, then small vibrational perturbations may alter the outcome of the reaction and the assignment of a unique reaction mechanism to the given minimum energy path P is not justified (5).

3.2. Optimization of minimum energy paths

Ideally, a method for minimum energy path determination could follow the definition given by Fukui (52) and it could start with the determination of the transition state, i.e. a saddle point $\underline{s}$ with index $\lambda(\underline{s})=1$. The eigenvector of the Hessian $H(\underline{s})$ that corresponds to the single negative eigenvalue defines the path vector $\underline{a}(\underline{s})$ at this point (5) and simple relaxation methods could be used to determine the two steepest descent path portions of the minimum energy path, leading to the two minima. By locating all saddle points $\{\underline{s}^{(i)}\}$, $\lambda(\underline{s}^{(i)}) = 1$, in a given region D of the hypersurface, all minimum energy paths of D may be determined. Such a technique, however, is seldom used, since the determination of transition states is usually much more cumbersome than determination of minimum energy configurations. That is, in most actual calculations one starts at the "wrong ends", i.e., by optimizing two minimum energy geometries at $\underline{m}^{(1)}$ and $\underline{m}^{(2)}$ and various algorithms are employed to find a minimum energy path interconnecting the two minima. Such a technique may fail for the simple reason that there is no direct minimum energy path between the two minima $\underline{m}^{(1)}$ and $\underline{m}^{(2)}$. This is the case whenever there is a stable intermediate (or several such intermediates) that corresponds to another minimum, $\underline{m}_3$ (or a series of such minima, resp.). In the simplest case the complete reaction may be regarded as two consecutive minimum energy processes, $\underline{m}^{(1)}\rightarrow\underline{m}^{(3)}$ followed by $m^{(3)}\rightarrow\underline{m}^{(2)}$. Even more problematic are those cases where the search for a minimum energy path, starting from a given minimum $\underline{m}^{(1)}$, runs into "blind alleys".

For the simplest problems one may rely on chemical intuition in selecting an approximate reaction path and one may carry out simple minimizations in all directions orthogonal to this path at a series of points. Ideally, such an optimization leads to a relaxed path, which is the minimum energy path if the initial approximate path was a good one (55,56). However, the method is known to be very sensitive to the initial choice of the guessed reaction path, and it often fails to converge to the minimum energy path (56,26). Recently an improved version of this method and an analysis of its convergence properties have been published (32).

The "synchronous transit" method of Halgren and Lipscomb, if it converges to a true transition state, also gives points along the minimum energy path (33). In general, this technique is more reliable than the simple path relaxation method. However, failures of the method have been reported (34).

A technique, called "reference coordinate method", is based on the initial determination of an equipotential contour line that stretches from the vicinity of one minimum to that of the other (57). For a set of points along the equipotential curve the energy is minimized along all directions orthogonal to the curve. For equipotential curves with small curvatures the calculated minimum points give a good approximation to the minimum energy path. However, if the curvature is large, then the number of points along the equipotential curve must be increased considerably. The method may fail to locate the most important middle section of the minimum energy path unless a large number of extra energy optimizations are performed (34). In addition, the determination of an appropriate line segment on a multidimensional contour surface is a rather complicated task, particularly if one needs a line that approaches both the reactant and product minima "close enough." Without knowning the topological features of the hypersurface at least approximately, it is difficult to determine such a line segment, even with the aid of contour-following algorithms.

The method of ascending valley points (34,35), proposed recently, is based on a series of energy minimizations on hyperspheres of various radii. The series of minima eventually leads to a transition state and on simple energy hypersurfaces where the entire minimum energy path is contained in a $D_0$ domain (5) these minimum points give a good approximation to the minimum energy path. However, the method fails to locate parts of the path if it enters an instability domain $D_\mu$, $\mu \geq 0$ (5).

None of the direct methods proposed for the determination of reaction paths is as reliable as the method based on the initial determination of the transition state and the subsequent determination of the relaxation (steepest descent) path. Transition state determination, however, is itself a complex task, and one may expect that the search for efficient methods of minimum energy path determination will continue.

## 4. FITTING OF APPROXIMATE SURFACES

In computational practice conformational energy surfaces of larger organic molecules are generated in a coordinate frame where the origin is fixed to a given nucleus or a point specified by some local symmetry of a molecular moiety. If single determinant HF or MCSCF CI wavefunctions and their energy expectation values are used to generate portions of the energy hypersurface, then the conformational motion or reaction is usually approximated by assigning essentially rigid molecular models to each point of the reaction path. The cartesian axes are then defined with respect to such a rigid molecular model, e.g. by using the Z-matrix technique, employed in the Gaussian 70 program of Pople and coworkers (58). The 3N variables are reduced to 3N-6 as six cartesian components are fixed at zero in <u>any conformation</u>, the usual choice being

$$X_1 = Y_1 = Z_1 = X_2 = Y_2 = Y_3 = 0 , \qquad (4.1)$$

i.e. the origin is at nucleus 1, the positive Z axis is along the vector pointing from nucleus 1 to nucleus 2 and the XZ plane contains nucleus 3. This choice does not define the axis system uniquely for every nuclear arrangement, nevertheless, the $3N \rightarrow 3N-6$ reduction remains valid even if atoms 1, 2, and 3 become collinear. To make the cartesian coordinate system unique, one may always define it with reference to a <u>fixed</u> non-collinear arrangement of atoms 1, 2 and 3 (that may correspond to a distorted conformation of a molecule with a linear equilibrium structure). By requiring $X_3 > 0$ in <u>this</u> reference conformation and taking axes X, Y and Z as a right hand system, the coordinate system becomes unique. Any nuclear configuration of the molecular system may be specified in terms of an $n = 3N-6$ dimensional vector $\underline{r}$ with n components in the above cartesian coordinate system.

These coordinates may be used directly to describe conformational motions or chemical reactions if no general dynamical studies are involved. However, unless specific symmetry preserving motions are considered only, this coordinate specification is not ideally suited for dynamic studies since corrections for the motion of the center of mass of the molecule are required. The image of the molecular motion on the hypersurface in such a coordinate system may show unexpected features, inconsistent with the simple three dimensional model of a particle moving in a cartesian frame. Whereas visualization of chemical processes on multidimensional hypersurfaces may be important in order to recognize the topological features and simple empirical rules, such simplified images may be very misleading (59).

For some simple problems, e.g. double minimum potential problems, it is relatively easy to find analytic functions that fit well to the corresponding crossection of the hypersurface, e.g. by using gaussian functions and polynomials (60-64). However, constructing a faithful analytic representation of larger domains of the hypersurface is a complicated problem (65,11). In some cases the fitted surface may show extra critical points, this being the case for various polynomial approximations or spline fitting schemes (66). The optimization of the fitting functions represent an additional optimization step that is carried out after an approximate optimization of the quantum chemical objective function (67). Since for a many dimensional problem the direct calculation of a dense grid of points is often impractical the analysis of potential energy hypersurfaces in terms of critical points, reaction paths and reactive domains critically depends on the choice of the analytic representation.

For many physical problems the favored analytic representation is a polynomial. Polynomials are the natural choice for a variety of models, e.g. quadratic functions for harmonic potentials, power series expansions for Coulomb potentials in crystal fields, etc. On a closed interval [a, b] polynomials have the attractive property that they may approximate any continuous function to any desired accuracy (the set of polynomials P[a, b] is everywhere dense in the set of continuous functions C[a, b]). However, polynomials that fit well in an interval containing only few of their roots often show large, "irregular" oscillations in the vicinity of their remaining roots. The asymptotic behaviour of polynomials is often a disadvantage as they diverge to $+\infty$ or $-\infty$. The degree of polynomials sufficiently accurate in a given interval [a, b] may be very high and in quantum chemical applications it is often possible to generate approximate hypersurfaces more economically by a combination of gaussian and trigonometric functions. Nevertheless, the conceptual simplicity of polynomial fitting and that of the corresponding computer algorithms are distinct advantages.

An exact fit to a finite set of points may be obtained by a Lagrange interpolation polynomial. In the one dimensional case the set of $\{f_i\}_1^n$ energy values at points $\{x_i\}_1^n$ are exactly reproduced by

$$E(x) = \sum_{j=1}^{n} f_j \frac{\prod_{i\neq j}^{n} (x_i - x)}{\prod_{i\neq j}^{n} (x_i - x_j)} \tag{4.2}$$

which is a polynomial of degree n-1. This formula may be generalized to the m dimensional case of $\{f_i\}_1^n$ function values at points $\{\underline{x}_i\}_1^n$ by replacing the parentheses by vector norms,

$$E'(\underline{x}) = \sum_{j=1}^{n} f_j \frac{\prod_{i\neq j}^{n} |\underline{x}_i - \underline{x}|}{\prod_{i\neq j}^{n} |\underline{x}_i - \underline{x}_j|} \tag{4.3}$$

However, the calculation of the norm is often inconvenient and it is more advantageous to construct separate products for each vector component:

$$E(\underline{x}) = \sum_{j=1}^{n} f_j \frac{\sum_{k=1}^{m} \prod_{i\neq j}^{n} (x_{ik} - x_k)}{\sum_{k=1}^{m} \prod_{i\neq j}^{n} (x_{ik} - x_{jk})} \tag{4.4}$$

The overall degree of this polynomial does not exceed n-1. Both (4.2) and (4.4) reproduce the $f_i$ function values at every specified point.

With the ready calculation of energy gradients by the more advanced <u>ab initio</u> programs (68,69) methods for fitting both energy values and their derivatives become particularly important. An exact fit to a finite set of energy values and the corresponding derivatives may be obtained by the Hermite interpolation polynomial. In the one dimensional case of $\{f_i\}_1^n$ energy values and $\{g_i\}_1^n$ derivatives at points $\{x_i\}_1^n$ both energies and their derivatives are exactly reproduced by

$$E(x) = \sum_{j=1}^{n} f_j \frac{\prod_{i\neq j}^{n} (x_i - x)^2}{\prod_{i\neq j}^{n} (x_i - x_j)^2} + \sum_{j=1}^{n} g_j G_j(x) \tag{4.5}$$

The first sum reproduces the $f_j$ function values at each $x_j$ point and it does not contribute to the derivative at any of the $x_j$ points. The $G_j(x)$ polynomial must fulfill the following conditions:

$$G_j(x_i) = 0 \quad \text{for every } i,\, j \tag{4.6}$$

i.e. it does not contribute to the energy at any of the $x_j$ points, and

$$\frac{dG_j(x_i)}{dx} = \delta_{ij}\,, \tag{4.7}$$

i.e. its derivative vanishes at every $x_i$ point except $x_j$. A suitable choice is

$$G_j(x) = -(x_j - x)\frac{\prod_{i\neq j}^{n}(x_i - x)^2}{\prod_{i\neq j}^{n}(x_i - x_j)^2} \tag{4.8}$$

and (4.5) may be re-written in a more concise form as

$$E(x) = \sum_{j=1}^{n} (f_j - g_j(x_j - x))\frac{\prod_{i\neq j}^{n}(x_i - x)^2}{\prod_{i\neq j}^{n}(x_i - x_j)^2} \tag{4.9}$$

The degree of the Hermite interpolation polynomial is 2n-1.

An m-dimensional generalization of the Hermite interpolation polynomial may be given as

$$E(\underline{x}) = \sum_{j=1}^{n} f_j \frac{\sum_{k=1}^{m}\prod_{i\neq j}^{n}(x_{ik} - x_k)^2}{\sum_{k=1}^{m}\prod_{i\neq j}^{n}(x_{ik} - x_{jk})^2} + \sum_{j=1}^{n}\sum_{k=1}^{m} g_{jk}G_{jk}(\underline{x}) \tag{4.10}$$

which polynomial exactly reproduces both the $\{f_i\}_1^n$ function values and the $\{\underline{g}_i\}_1^n$ gradient vectors of each of the specified points $\{\underline{x}_i\}_1^n$. As in the one dimensional case, the first sum generates the function values at the $\underline{x}_i$ points and does not contribute to the gradient vectors there, whereas the second sum generates the gradients and does not influence the function value at the $\underline{x}_i$ points. Polynomial $G_{jk}(\underline{x})$ must fulfill the conditions

$$G_{jk}(\underline{x}_i) = 0 \quad \text{for every i, j and k} \tag{4.11}$$

where k is the component index, and

$$\frac{\partial G_{jk}(\underline{x}_i)}{\partial x_\ell} = \delta_{k\ell}\,\delta_{ij} \tag{4.12}$$

i.e. its derivative vanishes for every component at every $\underline{x}_i$ point except for the k-th component at point $\underline{x}_j$, where it is unity.

The following construction fulfills conditions (4.11) and

and (4.12):

$$G_{jk}(\underline{x}) = -(x_{jk}-x_k) \frac{\prod_{i\neq j}^{n} (x_{ik}-x_k)^2}{\prod_{i\neq j}^{n} (x_{ik}-x_{jk})^2} \tag{4.13}$$

With the above $G_{jk}(\underline{x})$ the overall degree of polynomial (4.10) is 2n-1.

An exact fit to function values and derivatives at some points does not necessarily mean that the quality of the fit is good at intermediate points. In fact, polynomials are prone to large oscillations between two specified points $\underline{x}_i$ and $\underline{x}_j$. A Hermite interpolation usually gives smooth results if the number of points, n, is either too small (hence the surface is inaccurate) or exceeedingly large. It is often advantageous to generate approximate least square fits to polynomials of lower degree.

For practical purposes of visualizing surfaces spline fitting methods (66) are very useful, as such approximate surfaces usually do not show oscillations between two neighbouring points $\underline{x}_i$ and $\underline{x}_j$. Cubic splines, i.e. segments of cubic polynomials, joined at the segment boundaries, may satisfy some of the differentiability conditions required for a reaction path analysis (e.g. continuous second derivatives). However, as follows from the theorem of analytical continuation, at the segment boundaries at least one of the non-vanishing derivatives must be discontinuous.

For periodic surfaces (e.g. those where one or more of the coordinates describe bond rotations) trigonometric functions may be used for the fitting, possibly in combination with gaussian functions and polynomials. In such a case exact fitting is seldom possible or practical and least square fitting is used. For double minimum potentials a combination of gaussian functions and polynomials (60) gave excellent agreement between the calculated and observed vibrational frequencies of a series of tetraatomic molecules (61-64).

REFERENCES

1. C.C.J. Roothaan, Rev. Mod. Phys. 23, 69 (1951).
2. A.C. Wahl, G. Das, Adv. Quantum Chem. 5, 261 (1969). J. Hinze, J. Chem. Phys. 59, 6424 (1973), A. Veillard, E. Clementi, Theor. Chim. Acta 7, 133 (1967).
3. R.E. Kari, P.G. Mezey, I.G. Csizmadia. J. Chem. Phys. 64, 632 (1976).

4. "Applications of Electronic Structure Theory" Edited by H.F. Schaefer III, Plenum Press, 1977, New York.
5. P.G. Mezey, Theor. Chim. Acta 54, 95 (1980).
6. I. Gutman, N. Trinajstić, Chem. Phys. Letters, 20, 257 (1973).
7. I. Gutman, Chem. Phys. Letters 46, 169 (1977).
8. M. Randić, J. Am. Chem. Soc. 97, 6609 (1975).
9. M. Randić, Int. J. Quant. Chem., Quant. Biol. Symposia 5, 245 (1978).
10. O. Sinanoğlu, J. Am. Chem. Soc. 97, 2309 (1975).
11. P.G. Mezey, Analysis of Conformational Energy Hypersurfaces, Progr. Theor. Org. Chem. 2, 127 (1977).
12. M. Born, J.R. Oppenheimer, Ann. Phys. 84, 457 (1927).
13. J.N. Murrell, K.J. Laidler, Trans. Faraday Soc. 64, 371 (1968).
14. P.G. Mezey, Chem. Phys. Letters 47, 70 (1977).
15. M.J.D. Powell, Comput. J. 7, 155 (1964).
16. R. Fletcher, C.M. Reeves, Comput. J. 7, 147 (1964).
17. W.I. Zangwill, Comput. J. 10, 293 (1967).
18. R. Fletcher, M.J.D. Powell, Comput. J. 6,163 (1963).
19. N.C. Davidon, A.E.C. Research and Development Report ANL-5990 (1959).
20. J.D. Pearson, Comput. J, 12, 171 (1969).
21. H.Y. Huang, J. Optim. Theor. Appl. 5, 405 (1970).
22. H.Y. Huang, A.V. Levy, J. Optim. Theor. Appl. 6, 269 (1970).
23. N. Adachi, J. Optim. Theor. Appl., 7, 391 (1971).
24. A. Miele, J.W. Cantrell, J. Optim. Theor. Appl. 3, 459 (1969).
25. E.E. Cragg, A.V. Levy, J. Optim. Theor. Appl. 4, 191 (1969).
26. J.W. McIver, A. Komornicki, J. Am. Chem. Soc. 94 , 2625 (1972).
27. D. Poppinger, Chem. Phys. Letters 35, 550 (1975).
28. A. Komornicki, J.W. McIver Jr., J. Am. Chem. Soc., 95, 4512 (1973).
29. A. Komornicki, J.W. McIver Jr., J. Am. Chem. Soc. 96, 5798 (1974), 98, 4553 (1976).
30. A. Komornicki, K. Ishida, K. Morokuma, R. Ditchfield, M. Conrad, Chem. Phys. Letters, 45, 595 (1977).
31. D. Poppinger, L. Radom, J.A. Pople, J. Am. Chem. Soc. 99 7806 (1977).
32. M.J. Rothman, L.L. Lohr Jr., Chem. Phys. Letters 70, 405 (1980).
33. T.A. Halgren, W.N. Lipscomb, Chem. Phys. Letters, 49, 225 (1977).
34. K. Müller, Angewandte Chemie, Internat Ed. 19, 1, (1980).
35. K. Müller, L.D. Brown, Theor. Chim. Acta 53, 75 (1979).
36. P.G. Mezey, M.R. Peterson, I.G. Csizmadia, Can. J. Chem. 55, 2941 (1977).
37. P.G. Mezey, to be published.
38. L.P. Eisenhart, Riemannian Geometry, Princeton, University Press 1963.
39. H. Essén, Int. J. Quant. Chem. 12, 721 (1977).
40. E. Teller, J. Phys. Chem. 41, 109 (1937).
41. E.R. Davidson, J. Am. Chem. Soc., 99, 397 (1977).
42. J. Paldus, A. Veillard, Mol. Phys. 35, 445 (1978).

43. P.G. Mezey, O.P. Strausz, R.K. Gosavi, J. Comput. Chem. 1, 178 (1980).
44. P.G. Mezey, A. Kapur, Can. J. Chem. 58, 559 (1980).
45. P.G. Mezey, C.V.S.R. Rao, J. Chem. Phys., 72, 121 (1980).
46. G.L. Hofacker, Z. Naturforsch., A 18, 607 (1963).
47. R.A. Marcus, J. Chem. Phys. 45, 4493, 4500 (1966).
48. R.A. Marcus, J. Chem. Phys. 49, 2610, 2617 (1968).
49. J.C. Polanyi, J.L. Schreiber, Chem. Phys. Letters 29 319 (1974).
50. C.C. Miller, J.C. Light, J. Chem. Phys. 54, 1635 (1971).
51. M.V. Basilevsky, Chem. Phys. 24, 81 (1977).
52. K. Fukui, J. Phys. Chem. 74, 4161 (1970).
53. A. Tachibana, K. Fukui, Theor. Chim. Acta 49, 321 (1978).
54. A. Tachibana, K. Fukui, Theor. Chim. Acta 51, 189 (1979).
55. P. Empedocles, Int. J. Quant. Chem. 3S, 47 (1969).
56. M.J.S. Dewar, S. Kirschner, J. Am. Chem. Soc. 93, 4290 (1971).
57. R.F. Nalewajski, T.S. Carlton, Acta Phys. Pol. A 53 321 (1978).
58. W.J. Hehre, W.A. Lathan, R. Ditchfield, M.D. Newton, J.A. Pople: QCPE 236, Indiana University, Bloomington, Indiana.
59. X. Chapuisat, Chem. Phys. Letters 63, 389 (1979).
60. J.B. Coon, N.W. Naugle, R.D. McKenzie, J. Mol. Spectrosc. 20, 107 (1966).
61. A. Kapur, R.P. Steer, P.G. Mezey, J. Chem. Phys. 69, 968 (1978).
62. A. Kapur, R.P. Steer, P.G. Mezey, J. Chem. Phys. 70, 745 (1979).
63. A. Kapur, R.P. Steer, P.G. Mezey, J. Chem. Phys. 71, 588 (1979).
64. A. Kapur, P.G. Mezey, R.P. Steer, Chem. Phys. Letters, to be published.
65. G.D. Billing, A. Hunding, Chem. Phys. Letters 44, 30 (1976).
66. J.H. Alberg, E.N. Nielson, J.L. Walsh, The Theory of Splines and their Applications, Academic Press, 1967, New York.
67. D. Garton, B.T. Sutcliffe, Theoretical Chemistry (Specialist Periodical Report) Vol. 1, 34 (1974).
68. P. Pulay, Mol. Phys., 17, 197 (1969), Program FORCE, P. Pulay, W. Meyer.
69. J.A. Pople, private communication.

# AB INITIO ENERGY DERIVATIVES CALCULATED ANALYTICALLY

H. Bernhard Schlegel

Merck Sharp & Dohme Research Laboratories
Rahway, New Jersey 07065

## Abstract

The efficiency and accuracy of analytical differentiation in ab initio calculations are compared with the numerical approach. Formulae for computing integral derivatives are developed and the expression for the first derivative of the SCF energy is presented. The coupled-perturbed Hartree-Fock equations are solved for the first derivative of the molecular orbital coefficients. These are then used to calculate the second derivative of the SCF energy and the first derivatives of two energy schemes involving electron correlation: second order Møller-Plesset and all-doubles configuration interaction. Timing data indicate that as much as an order of magnitude increase in efficiency can be obtained through the use of derivative methods. Examples of geometry optimization and vibrational frequency calculation are given; additional applications to NMR magnetic shielding and electric polarizibility are discussed briefly.

## Introduction

To increase the efficiency of theoretical calculations, two general philosophies can be used:

a) to perform the individual calculations faster, through the development of better algorithms and the use of higher speed computers or

b) to extract more information from each calculation and to use that information effectively.



*I. G. Csizmadia and R. Daudel (eds.), Computational Theoretical Organic Chemistry, 129–159.*

This chapter will be concerned with the latter approach applied to molecular orbital calculations. In particular, attention will be focused on computing energy derivatives analytically and using these energy derivatives in the determination of molecular structure and properties. The contents of this chapter follow closely a series of recent articles[2-5] on the development and implementation of energy derivative methods. For details, the original work should be consulted.

A wide range of problems in quantum chemistry formally involves evaluation of derivatives of the electronic energy with respect to external parameters. Differentiation of the energy with respect to nuclear coordinates corresponds to calculation of forces and force constants determining nuclear motion. These nuclear displacement energy derivatives are important in the exploration of potential surfaces to find stationary points such as equilibrium structures and transition structures. Additional examples are the calculation of electric moments and polarizabilities (energy derivatives with respect to applied electric fields) and the calculation of magnetic properties such as diamagnetic susceptibilities and nuclear magnetic resonance chemical shifts (energy derivatives with respect to external and nuclear magnetic fields). A summary of the relation between these properties and energy derivatives is given in Table 1.

Energy derivatives could be calculated by a finite difference method of numerical differentiation. While it is easy to compute the energy E(x) at two closely spaced points, and approximate the derivative by

$$\frac{\partial E(x)}{\partial x} \simeq \frac{[E(x + \Delta x) - E(x)]}{\Delta x} \qquad (1)$$

it is not very efficient if numerous first or higher derivatives are required. Nor is it accurate unless $\Delta x$ is very small. An analytical differentiation is preferable if efficient algorithms can be devised.

Most of the work on first and second energy derivatives and associated properties has been carried out in conjunction with Hartree-Fock or single-determinant wave functions. The general theory of energy derivatives within this framework has been outlined by several authors. Bratoz[6], Bishop _et al._[7] and Moccia[8] have given analytical expressions for the first and second derivatives of the SCF energy for closed shells. Gerratt and Mills[9] have outlined a perturbed Hartree-Fock theory to calculate one-electron second-order properties and have calculated force constants as the analytical derivative

TABLE 1

Molecular Properties as Energy Derivatives

| | Coordinate | | |
|---|---|---|---|
| x = | atom position | electric field | magnetic field |
| **First Derivative** | | | |
| (dE/dx) | forces on the atoms in a molecule | dipole moment | magnetic moment |
| **Second Derivative** | | | |
| ($d^2E/dxdy$) | | | |
| y = atom position | force constants (vibrational frequencies) | dipole moment derivatives (infra red intensities) | |
| electric field | | dipole polarizibility | |
| magnetic field | | | magnetic susceptibility |
| nuclear spin | | | magnetic shielding (NMR chemical shift) |

of the Hellmann-Feynman force. Thomsen and Swanstrom[12] have analytically calculated the full second-derivativesof the energy for $H_2O$ with a gaussian basis set. However, major applications to the computation of force constants have been with the force method of Pulay.[10,11] In the force method the forces are obtained analytically and these are then differentiated numerically to obtain the force constants. This method has been used to calculate the force fields of a variety of molecules by Pulay and Meyer[13-15] and by Schlegel, Wolfe and Bernardi.[16-18] Ishida, Morokuma and Komornicki[19] have used analytical energy derivatives to study reactions paths. Only a limited amount of work on energy derivatives has been carried out with wavefunctions beyond the Hartree-Fock level, most by the finite difference method,[20-23] but a few with the analytical method.[24]

The following sections outline methods for calculating the integrals and their derivatives, as well as various energies and their derivatives, in the manner that is implemented in the Carnegie-Mellon version of GAUSSIAN 80.[30]

## Wavefunctions

The Hartree-Fock wavefunction for an n-electron molecule can be written as a single determinant,

$$\Psi_{HF} = (n!)^{-1/2} \, |\chi_1 \chi_2 \cdots \chi_n| \tag{2}$$

where $\chi_1$, $\chi_2 \cdots \chi_n$ are a set of orthonormal spinorbitals (one-electron functions of cartesian and spin coordinates). Wave-functions that include electron correlation can be constructed from linear combinations of such determinants. The spinorbitals are written as linear combinations of a set of basis functions $\omega_\mu$,

$$\chi_p = \sum_\mu C_{\mu p} \, \omega_\mu \tag{3}$$

Normally the $\omega_\mu$ will be products of a set of cartesian coordinate basis functions $\phi_1$, $\phi_2$, ... and the spin-functions $\alpha$ and $\beta$. If there are N $\phi$-type functions, there will be 2N spinorbital basis functions $\omega_\mu$, viz. $\phi_1\alpha$, $\phi_1\beta$, $\phi_2\alpha \ldots \phi_N\beta$. In turn, each of the cartesian basis functions is usually expressed as fixed linear combination of normalized guassian type primitives

$$\phi_\mu = \sum_k \xi_{k\mu} \, g_k \tag{4}$$

Each primitive function is written as a polynomial times a spherical gaussian

$$g_a(\vec{\ell}_A \vec{r}_A \alpha) = N_a x_A^{\ell_{Ax}} y_A^{\ell_{Ay}} z_A^{\ell_{Az}} \exp(-\alpha |\vec{r}_A|^2) \tag{5}$$

where $\vec{\ell}_A = (\ell_{Ax}, \ell_{Ay}, \ell_{Az})$, $\vec{r}_A = \vec{r} - \vec{A} = (x_A, y_Z, z_A)$, $\vec{r}$ is the electron cordinate, $N_a$ is the normalization coefficient, and $\vec{A} = (A_x A_y A_z)$, the gaussian center.

The gaussian centers are chosen to follow the nuclei rigidly. Thus, the basis functions depend on the nuclear coordinates directly. In the calculation of properties such as polarizibility or magnetic shielding, there are significant advantages to be obtained by including an explicit dependence on the electric or magnetic field in the basis function.[8,25,26] For example:

$$g_a^E = e^{-k \vec{E} \cdot \vec{r}} g_a \tag{6}$$

$$g_a^M = e^{-i \vec{H} \times \vec{r}_A \cdot \vec{r}} g_a \tag{7}$$

The higher order gaussians can also be expressed as derivatives of the corresponding spherical gaussian

$$g_a(\vec{\ell}_A \vec{r}_A \alpha) = \hat{M}^{\ell_{Ax}} \hat{M}^{\ell_{Ay}} \hat{M}^{\ell_{Az}} g_a(\vec{0}\, \vec{r}_A \alpha) \tag{8}$$

$$\text{where } \hat{M}^{\ell_{Ax}} = 1 \qquad \text{for } \ell_{Ax} = 0 \tag{9}$$

$$= \frac{1}{2}\alpha \frac{\partial}{\partial A_x} \qquad \text{for } \ell_{Ax} = 1 \tag{10}$$

$$\text{and } \hat{M}^{\ell_{Ax}+1} = \frac{1}{2\alpha} \hat{M}^{\ell_{Ax}} \frac{\partial}{\partial A_x} + \frac{\ell_{Ax}}{2\alpha} \hat{M}^{\ell_{Ax}-1} \tag{11}$$

$\hat{M}^{\ell_{Ay}}$ and $\hat{M}^{\ell_{Az}}$ are defined similarly. These relations are useful for the development of integral derivative expressions.

## Integrals

The calculation of total energies for electronic wavefunctions can be divided into two parts,

a) computing the integrals over the basis functions of the necessary one and two electron operators in the Hamiltonian and

b) manipulating these integrals to determine the appropriate linear combination coefficients and energy.

To determine the energy, we must first compute the following terms:

the overlap integrals

$$S_{\mu\nu} = \int \omega_\mu^* \omega_\nu \, d\tau \tag{12}$$

the Core-Hamiltonian integrals

$$H_{\mu\nu} = \int \omega_\mu^* \, \hat{H}^{core} \omega_\nu \, d\tau \tag{13}$$

and the two-electron integrals

$$(\mu\lambda||\nu\sigma) = \iint \omega_\mu^*(1) \; \omega_\lambda^*(2) \; (1/r_{12}) \; [\omega_\nu(1)\,\omega_\sigma(2) - \omega_\sigma(1)\,\omega_\nu(2)] \, d\tau_1 d\tau_2 \tag{14}$$

$$= (\mu\nu|\lambda\sigma) - (\mu\sigma|\lambda\nu)$$

where $\hat{H}^{core}$ is the one-electron core Hamiltonian (kinetic energy + potential energy in the electrostatic field of the nuclei). The two-electron operator in (14) is the interelectronic repulsion energy. Note that $(\mu\lambda||\nu\sigma)$ is antisymmetric in both the pairs $\mu\lambda$ and $\nu\sigma$.

The integrals over the basis functions are composed of integrals over the gaussian primitives. The overlap and kinetic energy integrals are relatively simple[29] and will not be considered here. The nuclear-electron attraction integrals can be treated as a special case of the two electron integrals discussed below.

The evaluation of the two electron integrals is the most time consuming part of the integral step in ab initio computations. Although the use of gaussian functions permits each individual integral to be computed relatively rapidly, there are ca. $N^4$ integrals that must be calculated (where N is the number of primitives in the basis set). Thus, considerable attention has been paid to developing schemes to compute two electron integrals rapidly.

The general two electron integral involving four primitives on different centers can be denoted as:

$$(\vec{\ell}_A\vec{\ell}_B|\vec{\ell}_C\vec{\ell}_D) = \iint g_a(\vec{\ell}_A\vec{r}_A(1)\alpha)\; g_b(\vec{\ell}_B\vec{r}_B(1)\beta)\; \frac{1}{|\vec{r}(1) - \vec{r}(2)|}\; g_c(\vec{\ell}_C\vec{r}_C(2)\gamma)\; g_d(\vec{\ell}_D\vec{r}_D\,(2)\delta)\; dr_1 dr_2 \qquad (15)$$

For efficiency in the evaluation of the integrals, the gaussians can be arranged in shells[28] such that all gaussians within a shell share the same exponent, but differ in the polynomial factor (e.g. and s type and three p type gaussians with the same exponent are used to form the 2sp shell of an STO-3G basis set). The collection of all two electron integrals in a product of four shells is termed a shell block (256 integrals for 4 sp type shells). When these are evaluated together, substantial savings in computation can be achieved since the integrals share a great deal of common information. For 1s spherical gaussians, the general 4 center two electron integral is:

$$(\vec{0},\vec{0}|\vec{0},\vec{0}) = (ss|ss)$$

$$= N_{abcd}\,\Theta F_0(\rho PQ^2)\ \exp\,(-uAB^2)\ \exp\,(-vCD^2) \qquad (16)$$

where $p = \alpha + \beta$ $\qquad q = \gamma + \delta \qquad \rho = pq/(p + q)$

$$\Theta = 2\Pi^{5/2}/(pq\sqrt{p + q}) \qquad u = \alpha\beta/(\alpha + \beta) \qquad v = \gamma\delta/(\gamma + \delta)$$

$$AB^2 = |\vec{A} - \vec{B}|^2 \qquad CD^2 = |\vec{C} - \vec{D}|^2 \qquad PQ^2 = |\vec{P} - \vec{Q}|^2$$

$$\vec{P} = (\alpha\vec{A} + \beta\vec{B})/p \qquad \vec{Q} = (\gamma\vec{C} + \delta\vec{D})/q$$

and

$$N_{abcd} = N_a N_b N_c N_d$$

The function $F_m(t)$ is closely related to the error function, and has the properties

$$F_m(t) = \int_0^1 u^{2m} \exp(-tu^2)\, du$$

$$\frac{dF_m(t)}{dt} = -F_{m+1}(t). \tag{17}$$

Two electron integrals over other types of gaussians can be derived integrating eq. (15) directly. Alternatively, the derivative operators used in eq. (8) to construct higher order gaussians can be applied to the two electron integrals over spherical gaussians,

$$(\vec{\ell}_A \vec{\ell}_B | \vec{\ell}_C \vec{\ell}_D) = \Big( \prod_{\substack{U=ABCD \\ i=xyz}} \hat{M}^{\ell_{Ui}} \Big) (ss|ss)$$

$$= \hat{M}(\omega)\ (ss|ss) \tag{18}$$

There are a number of ways to expand eq. (18) to facilitate computation. For example, an expression can be obtained as a sum of terms containing $F_m(t)$ and angular factors,[29]

$$(\vec{\ell}_A \vec{\ell}_B | \vec{\ell}_C \vec{\ell}_D) = \sum_{m=0}^{m_{max}} C_x^m C_y^m C_z^m \ F_m(t) \tag{19}$$

$$m_{max} = |\vec{\ell}_A| + |\vec{\ell}_B| + |\vec{\ell}_C| + |\vec{\ell}_D|$$

The $C_x^m$'s depend on the exponents and only the x components of the gaussian coordinates and angular momenta; similarly for the $C_y^m$'s and $C_z^m$'s. Rotation to an intermediate, local axis system[28] can force many of the integrals in a shell block to be zero and also simplify the $C^m$'s. The net gain in speed is substantial, as demonstrated by the GAUSSIAN programs[30] which use this scheme to evaluate sp two electron integrals.

An alternate approach recently developed by Dupuis, Rys and King[31] is especially efficient for higher angular momentum basis functions. Exploiting the properties of the Rys polynomials, they have used numerical integration techniques to derive an exact, analytical formula for the two electron integral:

$$(\vec{\ell}_A\vec{\ell}_A|\vec{\ell}_C\vec{\ell}_D) = \sum_{n=1}^{n_{max}} IP_x(t_n)IP_y(t_n)IP_z(t_n)\,\omega_n \tag{20}$$

$$n_{max} > (|\vec{\ell}_A| + |\vec{\ell}_B| + |\vec{\ell}_C| + |\vec{\ell}_D|)/2$$

and

$$IP_x = \exp(\rho PQ_x^2 t_n^2)\cdot(1 - t_n^2)^{-1/2}$$

$$\iint dx(1)dx(2)x_A(1)^{\ell_{Ax}} x_B(1)^{\ell_{Bx}} x_C(2)^{\ell_{Cx}} x_D(2)^{\ell_{Dx}}$$

$$\exp(-\alpha x_A(1)^2 - \beta x_B(1)^2 - \gamma x_C(2)^2 - \delta x_D(2)^2$$

$$- t_n^2(x(1)-x(2))/(1-t_n^2))$$

The $t_n$ and $\omega_n$ are the roots and weights of the Rys polynomials, which have the property

$$F_m(X) = \sum_n t_n^{2m}\,\omega_n \exp(-Xt_n^2)$$

Like the $C_x^m$'s, $IP_x$ depends only on the x components of the gaussian coordinates and the angular momenta. The Rys polynomial approach offers two advantages:[31]

a) the sum in eq. (20) contains only half as many terms as the sum in eq. (19)

b) the small number of IP integrals can be evaluated readily.

For integrals involving d orbitals, the Rys polynomial method becomes the preferred approach.

## Integral Derivatives

To compute energy derivatives analytically, we must first calculate the corresponding derivatives of the one and two electron integrals over the basis functions, and hence over the primitive gaussians. The first derivative with respect to the gaussian center is

$$\frac{dg_a}{dA_x}(\vec{\ell}_A\vec{r}_A\alpha) = 2\ \alpha g_a(\vec{\ell}_A^{+1},\vec{r},\alpha)-\ell_{Ax}g_a(\vec{\ell}_A^{-1},\vec{r},\alpha) \tag{21}$$

where $\vec{\ell}_A^{\pm 1} = (\ell_{Ax} \pm 1,\ \ell_{Ay},\ell_{Az})$

and the second derivative is

$$\frac{d^2 g_a(\vec{\ell}_A\vec{r}_A\alpha)}{dA_x^{\ 2}} = 4\alpha^2 g_a(\vec{\ell}_A^{+2},\vec{r}_A\alpha)-(2\ell_{Ax}+1)g_a(\vec{\ell}_A,\vec{r}_A\alpha)$$

$$+\ell_{Ax}(\ell_{Ax}-1)g_a(\vec{\ell}_A^{-2},\vec{r}_A\alpha) \tag{22}$$

where $\vec{\ell}_A^{\pm 2} = (\ell_{Ax} \pm 2,\ \ell_{Ay},\ell_{Az})$

For gaussians depending explicitly on the electric or magnetic field, derivatives with respect to the field can be expressed in terms of derivatives with respect to the gaussian centers.[25,27] Thus it is sufficient to limit the discussion of integral derivatives to differentiation with respect to the gaussian function centers. Two different approaches can be taken to evaluate two electron integral derivatives.

a) Using Higher Angular Momentum Basis Functions

From eq. (21), the two electron integral derivative with respect to the x component of the gaussian at center A can be written as

$$\frac{\partial}{\partial A_x}(\vec{\ell}_A\vec{\ell}_B|\vec{\ell}_C\vec{\ell}_D) = 2\alpha(\vec{\ell}_A^{+1}\vec{\ell}_B|\vec{\ell}_C\vec{\ell}_D)-\ell_{A_x}(\vec{\ell}_A^{-1}\vec{\ell}_B|\vec{\ell}_C\vec{\ell}_D) \tag{23}$$

Similar expressions can be written for second derivatives. This approach leads to a relatively simple form for the derivatives provided integrals over the higher angular momentum functions are available. The analogues of eq. (23) for one electron integrals may be quite practical since one electron integrals for higher angular momentum gaussians can be calculated easily. However, for two electron integrals, the additional calculations involving the higher angular momentum functions are significantly more difficult to program and more costly to evaluate. Gradient calculations using this method are several times slower than the energy calculation alone, even when special steps are taken to optimize the derivative evaluation.

b) Avoiding Higher Angular Momentum Basis Functions

To compute the first derivative of the Hartree-Fock energy, the individual integral derivatives are not required explicitly, but can be combined with appropriate elements of the density matrix and summed. An algorithm can be constructed that takes advantage of this fact, while avoiding the use of code for higher angular momentum functions.[2]

For the two electron integral over spherical gaussians, eq. (16), the derivatives with respect to a cartesian component of a gaussian center, $R_i$ (R=A,B,C,D; i=x,y,z), can be written as

$$\frac{d(ss|ss)}{dR_i} = -\left(u\frac{d\overline{AB}^2}{dR_i} + v\frac{d\overline{CD}^2}{dR_i}\right) N_{abcd}\ \theta F_0(\rho\overline{PQ}^2)\exp(-u\overline{AB}^2)\exp(-v\overline{CD}^2)$$

$$\rho - \frac{d\overline{PQ}^2}{dR_i} N_{abcd}\ \theta\ F_1(\rho\overline{PQ}^2)\exp(-u\overline{AB}^2)\exp(-v\overline{CD}^2)$$

$$= \Lambda(R_i)\ (ss|ss) + \Lambda(R_i)'(ss|ss)' \qquad (24)$$

where $(ss|ss)'$ is $(ss|ss)$ with $F_n$ replaced by $F_{n+1}$, $\Lambda$ and $\Lambda'$ are linear functions of $\vec{A}$ $\vec{B}$ $\vec{C}$ and $\vec{D}$.

The fact that the expression for a derivative has two similar terms, each of which is closely related to the integral, can be used to considerable advantage in obtaining expressions for higher angular momentum gaussians. As with the integrals themselves, the integral derivatives for higher angular momentum gaussians can be obtained by applying the operators $\hat{M}$ to eq. (24)

$$\frac{d}{dR_i}(\vec{\ell}_A\vec{\ell}_B|\vec{\ell}_C\vec{\ell}_D) = \frac{d}{dR_i}\hat{M}(\omega)\ (ss|ss)$$

$$= \hat{M}(\omega)\ \frac{d}{dR_i}\ (ss|ss)$$

$$= \hat{M}(\omega)\ (R_i)(ss|ss) + \hat{M}(\omega)\ (R_i)'(ss|ss)' \qquad (25)$$

Both parts of this expression for the integral derivative can be expanded in terms of the integral itself and integrals over lower angular momentum functions. Furthermore, much of the code for computing the primed and unprimed quantities is the same as needed for the original integral.

The first derivative of the Hartree-Fock energy does not require the integral derivatives individually (see below). Instead, they can be combined directly with the appropriate density matrix elements, denoted by $\mathbb{P}(\omega)$ in eq. (26), and summed. The special form of eq. (25) permits a great deal of factoring and simplification when the contributions of an entire shell block to all of the derivatives are computed at one time.

$$
\begin{aligned}
&\sum_{\omega}^{\text{block}} \mathbb{P}(\omega)\, d(\vec{\ell}_A\vec{\ell}_B|\vec{\ell}_C\vec{\ell}_D)/dA_x \\
&= \Lambda[\sum_{\omega}\mathbb{P}(\omega)(\vec{\ell}_A\vec{\ell}_B|\vec{\ell}_C\vec{\ell}_D)] + \Lambda_A[\sum_{\omega}\mathbb{P}(\omega)(\vec{\ell}_A^{\,-1}\vec{\ell}_B|\vec{\ell}_C\vec{\ell}_D)] \\
&\quad + \Lambda_B[\sum_{\omega}\mathbb{P}(\omega)(\vec{\ell}_A\vec{\ell}_B^{\,-1}|\vec{\ell}_C\vec{\ell}_D)] + \Lambda'[\sum_{\omega}\mathbb{P}(\omega)(\vec{\ell}_A\vec{\ell}_B|\vec{\ell}_C\vec{\ell}_D)'] \\
&\quad + \Lambda'_A[\sum_{\omega}\mathbb{P}(\omega)(\vec{\ell}_A^{\,-1}\,\vec{\ell}_B|\vec{\ell}_C\vec{\ell}_D)'] + \Lambda'_B[\sum_{\omega}\mathbb{P}(\omega)(\vec{\ell}_A\,\vec{\ell}_B^{\,-1}|\vec{\ell}_C\vec{\ell}_D)'] \\
&\quad + \Lambda'_C[\sum_{\omega}\mathbb{P}(\omega)(\vec{\ell}_A\vec{\ell}_B|\vec{\ell}_C^{\,-1}\vec{\ell}_D)'] + \Lambda'_D[\sum_{\omega}\mathbb{P}(\omega)(\vec{\ell}_A\vec{\ell}_B|\vec{\ell}_C\vec{\ell}_D^{\,-1})'] \qquad (26)
\end{aligned}
$$

where $\Lambda = \Lambda(A_x)$

$$\Lambda_A = -1/2\alpha \quad \partial\Lambda/\partial A_x$$

and $\vec{\ell}_A^{\,-1} = (\ell_{Ax}-1, \ell_{Ay}, \ell_{Az})$, etc.

Such a scheme has been implemented for s and p gaussians and has proven to be one of the fastest algorithms for integral derivatives[2] of this type. This approach has also been applied to exponent derivatives[2] but it has not been practical to extend the method to include d orbitals.

The Rys polynomial method can also be adapted to the computation of first and second derivatives for s, p and d gaussians. Also, there is no particular disadvantage to computing each integral derivative individually. As will be seen below, the first derivatives of the integrals are required explicitly for the differentiation of correlated wavefunctions and the calculation of Hartree-Fock second derivatives. The first derivative is given by

$$\frac{d}{dA_x}(\vec{\ell}_A\vec{\ell}_B|\vec{\ell}_C\vec{\ell}_D) = \sum_{n=1}^{n_{max}} IP'_x(t_n)IP_y(t_n)IP_z(t_n)\omega_n \tag{27}$$

$$n_{max} > (|\vec{\ell}_A| + |\vec{\ell}_B| + |\vec{\ell}_C| + |\vec{\ell}_D| + 1)/2$$

$$IP'_x(t_n) = dIP_x(t_n)/dA_x$$

The $IP'(t_n)$ can be expressed in terms of the $t_n$ and the IP's for only the same and lower angular momenta. Similar formulae can be derived for the second derivatives.

## Energies

In terms of the one and two electron integrals over the basis functions, the Hartree-Fock energy can be written (in the unrestricted formalism) as

$$E_{HF} = \sum_{\mu\nu} P_{\mu\nu}H_{\mu\nu} + \frac{1}{2}\sum_{\mu\nu\lambda\sigma} P_{\mu\nu}P_{\lambda\sigma}(\mu\lambda||\nu\sigma) + V_{nuc} \tag{28}$$

where $P_{\mu\nu}$ is the spinorbital density matrix defined by

$$P_{\mu\nu} = \sum_{i=1}^{n} C^*_{\mu i} C_{\nu i} \tag{29}$$

and $V_{nuc}$ is the nuclear-nuclear repulsion. Note that the sums over greek suffixes in eq. (28) and elsewhere in this section are over the 2N spinorbital basis functions. Thus $P_{\mu\nu}$ are elements of a 2N x 2N matrix. With separate $\alpha$- and $\beta$-type orbitals, this matrix will be blocked with N x N $\alpha$- and $\beta$-parts and no $\alpha\beta$ interaction elements. Orthonormalization of the spinorbitals must be retained and requires the condition

$$\sum_{\mu\nu} C^*_{\mu p} S_{\mu\nu} C_{\nu q} = \delta_{pq} \tag{30}$$

Minimization of the Hartree-Fock energy with respect to the linear coefficients $C_{\mu p}$ (subject to the orthonormality conditions) leads to the Fock type equations

$$\sum_{\nu} (F_{\mu\nu} - \varepsilon_p S_{\mu\nu}) C_{\nu p} = 0 \tag{31}$$

Here $F_{\mu\nu}$ is the 2N x 2N Fock matrix

$$F_{\mu\nu} = H_{\mu\nu} + \sum_{\lambda\sigma} P_{\lambda\sigma}(\mu\lambda||\nu\sigma) \tag{32}$$

and $\varepsilon_p$ is the one-electron energy of the p-th spinorbital. These are the Hartree-Fock self-consistent equations. For UHF theory they separate into two sets of coupled equations for the α- and β-spinorbitals.

Equations (31) will be soluble for 2N possible values of the one-electron energy $\varepsilon_p$, only n of which will correspond to occupied molecular spinorbitals. It is convenient to use suffixes i,j,k (=1,2,...n) for occupied spinorbitals and a,b,c,...(=n+1,...2N) for the remainder (usually described as <u>virtual</u> spinorbitals). We shall continue to use p,q,r,... for the entire set of 2N spinorbitals. It is convenient to define matrix elements with respect to the spin orbitals. The notation for matrix elements is

$$H_{pq} = \int \chi_p^* H^{core} \chi_q \, d\tau = \sum_{\mu\nu} C_{\mu p}^* C_{\nu q} H_{\mu\nu} \tag{33}$$

$$(pq||rs) = \iint \chi_p^*(1)\chi_q^*(2) \left(\frac{1}{r_{12}}\right) [\chi_r(1)\chi_s(2)-\chi_s(1)\chi_r(2)]d\tau_1 d\tau_2$$

$$= \sum_{\mu\nu\lambda\sigma} C_{\mu p}^* C_{\nu q}^* C_{\lambda r} C_{\sigma s} (\mu\nu||\lambda\sigma) \tag{34}$$

Thus, the expression for the Hartree-Fock energy becomes

$$E_{HF} = \sum_{i=1}^{n} H_{ii} + \frac{1}{2} \sum_{i=1}^{n} \sum_{j=1}^{n} (ij||ij) + V_{nuc} \tag{35}$$

To go beyond the Hartree-Fock approximation, the electronic wavefunction can be constructed from a linear combination of determinants. If we limit ourselves to double excitations from the Hartree-Fock reference determinant then the unnormalized wavefunction can be written as

$$\Psi_{corr} = \sum_s a_s \Psi_s = \Psi_{HF} + \frac{1}{4} \sum_{ijab} a_{ij}^{ab} \Psi_{ij}^{ab} \tag{36}$$

where $\Psi_{ij}^{ab}$ is a determinant in which the spinorbitals $\chi_i$ and $\chi_j$ are replaced by spinorbitals $\chi_a$ and $\chi_b$ respectively. The coefficients $a_{ij}^{ab}$ and the energy corresponding to $\Psi_{corr}$ can be computed variationally or by perturbation theory.

Second order Møller-Plesset perturbation theory[32] leads to

$$a_{ij}^{ab} = -(ij||ab)/\ \Delta_{ij}^{ab} \tag{37}$$

and a total energy of

$$E_{MP2} = E_{HF} - \frac{1}{4} \sum_{ij}^{occ} \sum_{ab}^{virt} (ij||ab)^2 (\Delta_{ij}^{ab})^{-1} \tag{38}$$

where

$$\Delta_{ij}^{ab} = \varepsilon_a + \varepsilon_b - \varepsilon_i - \varepsilon_j \tag{39}$$

Once the Hartree-Fock equations are solved for the coefficients $C_{\nu p}$, the matrix elements $(ij||ab)$ are easily obtained and eq. (38) then gives an easily computable expression for the total energy including correlation.

Calculation of $a_{ij}^{ab}$ variationally results in a configuration interaction wavefunction which includes double excitations, $\Psi_{CID}$. The energy $E_{CID}$ is given by the expectation value

$$E_{CID} = \langle\Psi_{CID}|\hat{\mathcal{H}}|\Psi_{CID}\rangle/\langle\Psi_{CID}|\Psi_{CID}\rangle$$

$$= \sum_{st} a_s \mathcal{H}_{st}\ a_t / \sum_s a_s^2 \tag{40}$$

$$= \sum_{abij} a_{ij}^{ab} ((E_{HF} + \Delta_{ij}^{ab})\ a_{ij}^{ab} + w_{ij}^{ab}) / \sum_{abij} (a_{ij}^{ab})^2 \tag{41}$$

The $a_{ij}^{ab}$ are obtained by the itertive solution of:

$$a_{ij}^{ab} = w_{ij}^{ab}\ /\ (E_{CID} - E_{HF} - \Delta_{ij}^{ab}) \tag{42}$$

where

$$w_{ij}^{ab} = (ab||ij) + \frac{1}{2} \sum_{cd} (ab||cd)\ a_{ij}^{cd} + \frac{1}{2} \sum_{kl} (kl||ij)\ a_{kl}^{ab}$$

$$+ \sum_{kc} [-(kb||jc)a_{ik}^{ac} + (ka||jc)a_{ik}^{bc} - (ka||ic)a_{jk}^{bc} + (kb||ic)a_{jk}^{ac}] \tag{43}$$

## Energy Derivatives

In this section, we finally combine the efforts of the previous sections on integrals, integral derivatives and energy expressions to develop the formulae for energy derivatives $E/ x_i$ and $^2E/ x_i x_j$. Direct differentiation of the Hartree-Fock energy eq. (28) with respect to the parameter x gives

$$\partial E_{HF}/\partial x = \sum_{\mu\nu} P_{\mu\nu}(\partial H_{\mu\nu}/\partial x) + \frac{1}{2}\sum_{\mu\nu\lambda\sigma} P_{\mu\nu}P_{\lambda\sigma}(\partial/\partial x)(\mu\lambda||\nu\sigma) + \partial V_{nuc}/\partial x$$

$$+ \sum_{\mu\nu}(\partial P_{\mu\nu}/\partial x)H_{\mu\nu} + \sum_{\mu\nu\lambda\sigma}(\partial P_{\mu\nu}/\partial x)P_{\lambda\sigma}(\mu\lambda||\nu\sigma) \qquad (44)$$

The first three parts on the right-hand side of eq. (44) directly involve the derivatives of the integrals $H_{\mu\nu}$, $(\mu\lambda||\nu\sigma)$ and the nuclear repulsion energy $V_{nuc}$. The remaining terms involve the derivative of the density-matrix and hence of the spinorbital coefficients $C_{\mu i}$. However, explicit evaluation of $\partial P_{\mu\nu}/\partial x$ can be avoided at this point if we note that the final two parts of eq. (44) can be written

$$\sum_{\mu\nu}\sum_{i=1}^{n}(\partial C^*_{\mu i}/\partial x)H_{\mu\nu}C_{\nu i} + \sum_{\nu\mu\lambda\sigma}\sum_{i=1}^{n}(\partial C^*_{\mu i}/\partial x)P_{\lambda\sigma}(\nu\lambda||\nu\sigma)C_{\nu i} + \text{complex conj.}$$

$$= \sum_{\mu\nu}\sum_{i=1}^{n}(\partial C^*_{\mu i}/\partial x)\, F_{\mu\nu}C_{\nu i} + \text{complex conj.}$$

$$= \sum_{\mu\nu}\sum_{i=1}^{n}(\partial C^*_{\mu i}/\partial x)\, \varepsilon_i\, S_{\mu\nu}\, C_{\nu i} + \text{complex conj.} \qquad (45)$$

using eqs. (31) and (32). Further, differentiation of the orthonormality equation (30) leads to (with p = q = i)

$$\sum_{\mu\nu}[(\partial C^*_{\mu i}/\partial x)\, S_{\mu\nu}C_{\nu i} + C^*_{\nu i}\,(\partial S_{\mu\nu}/\partial x)C_{\nu i} + C^*_{\mu i}S_{\mu\nu}(\partial C_{\nu i}/\partial x)] = 0 \qquad (46)$$

Equation (46) may be used to eliminate the coefficient derivatives in eq. (45) and hence obtain the final formula

$$\partial E_{HF}/\partial x = \sum_{\mu\nu} P_{\mu\nu}(\partial H_{\mu\nu}/\partial x) + \frac{1}{2}\sum_{\mu\nu\lambda\sigma} P_{\mu\nu}P_{\lambda\sigma}(\partial/\partial x)(\mu\lambda||\nu\sigma)$$

$$+ \partial V_{nuc}/\partial x - \sum_{\mu\nu} W_{\mu\nu}(\partial S_{\mu\nu}/\partial x) \qquad (47)$$

where $W_{\mu\nu}$ is an 'energy-weighted density matrix'

$$W_{\mu\nu} = \sum_{i=1}^{n} \varepsilon_i C^*_{\mu i} C_{\nu i} \tag{48}$$

We may obtain second derivatives of the Hartree-Fock energy by differenetiating eq. (47) with respect to a second variable y. This leads to

$$\begin{aligned} \partial^2 E_{HF}/\partial x\partial y = & \sum_{\mu\nu} P_{\mu\nu}(\partial^2 H_{\mu\nu}/\partial x\partial y) + \frac{1}{2}\sum_{\mu\nu\lambda\sigma} P_{\mu\nu}P_{\lambda\sigma}(\partial^2/\partial x\partial y)\ (\mu\lambda||\nu\sigma) \\ & + \partial^2 V_{nuc}/\partial x\partial y - \sum_{\mu\nu} W_{\mu\nu}(\partial^2 S_{\mu\nu}/\partial x\partial y) \\ & + \sum_{\mu\nu} (\partial P_{\mu\nu}/\partial y)(\partial H_{\mu\nu}/\partial x) + \sum_{\mu\nu\lambda\sigma} (\partial P_{\mu\nu}/\partial y)P_{\lambda\sigma}(\partial/\partial x)(\mu\lambda||\nu\sigma) \\ & - \sum_{\mu\nu} (\partial W_{\mu\nu}/\partial y)(\partial S_{\mu\nu}/\partial x) \end{aligned} \tag{49}$$

The first four parts of eq. (49) involve the second derivatives of the integrals and $V_{nuc}$. They can be handled in a manner strictly analogous to eq. (47). The remaining terms involve the first derivatives of the density matrices $P_{\mu\nu}$ and $W_{\mu\nu}$; computation of these can no longer be avoided.

Differentiation of the Hartree-Fock wavefunction with respect to a variable y is accomplished by coupled perturbed Hartree-Fock theory (CPHF). The problem is to find solutions of the Fock-type equations (31) for values of y in the vicinity of a value y=0 for which solutions are already available. The necessary equations can be obtained expanding the terms in the Fock equations as powers of y, and insisting that the equations be satisfied for each power of y. Alternately, the same equations can be obtained by formal differentiation of the Fock equation and the orthonormality constraint.

$$\sum_{\nu} \left[\frac{\partial F_{\mu\nu}}{\partial y} - \frac{\partial \varepsilon_p}{\partial y} S_{\mu\nu} - \varepsilon_p \frac{\partial S_{\mu\nu}}{\partial y}\right) C_{\nu p} + (F_{\mu\nu} - \varepsilon_p S_{\mu\nu})\frac{\partial C_{\nu p}}{\partial y}\Big] = 0 \tag{50}$$

and $$\left[\frac{\partial C^*_{\mu q}}{\partial y} S_{\mu\nu} C_{\nu p} + C^*_{\mu q} \frac{\partial S_{\mu\nu}}{\partial y} C_{\nu p} + C^*_{\mu q} S_{\mu\nu} \frac{\partial C_{\nu p}}{\partial y}\right] = 0 \tag{51}$$

These equations can be simplified by converting from the basis function space to the spin orbital space. The derivative of the orbital coefficients can be expressed in terms of the unperturbed coefficients

$$\frac{\partial C_{\nu p}}{\partial y} = \sum_r C_{\nu r} U_{rp} \tag{52}$$

Substituting into eq. (50) and transforming by $C_{\mu q}$ we obtain

$$\sum_{\mu\nu} \Big[C^*_{\mu q}\Big(\frac{\partial F_{\mu\nu}}{\partial y} - \frac{\partial \varepsilon_p}{\partial y} S_{\mu\nu} - \varepsilon_p \frac{\partial S_{\mu\nu}}{\partial y}\Big) C_{\nu p}$$
$$+ \sum_r C_{\mu q}(F_{\mu\nu} - \varepsilon_p S_{\mu\nu})C_{\mu\nu} U_{rp}\Big] = 0 \tag{53}$$

Simplifying, using eq. (30) and eq. (31)

$$\widetilde{F}_{qp} - \delta_{qp} \frac{\partial \varepsilon_p}{\partial y} - \varepsilon_p \widetilde{S}_{qp} + (\varepsilon_q - \varepsilon_p)U_{qp} = 0 \tag{54}$$

Similarly for the orthonormality contraint,

$$U^*_{qp} + \widetilde{S}_{qp} + U_{pq} = 0 \tag{55}$$

The following terms have been defined for the above two equations,

$$\widetilde{S}_{qp} = \sum_{\mu\nu} C^*_{\mu q} \frac{\partial S_{\mu\nu}}{\partial y} C_{\nu p} \tag{56}$$

and $$\widetilde{F}_{qp} = \widetilde{H}_{qp} + \widetilde{G}_{qp} \tag{57}$$

where $$\widetilde{H}_{qp} = \sum_{\mu\nu} C^*_{\mu q} \frac{\partial H_{\mu\nu}}{\partial y} C_{\nu p} \tag{58}$$

and $$\widetilde{G}_{qp} = \sum_i^n \sum_r [U^*_{ri} (qr||pi) + U_{ri} (qi||pr)$$
$$+ \sum_{\mu\nu\lambda\sigma} C^*_{\mu q} C_{\nu p} P_{\lambda\sigma} \frac{\partial (\mu\lambda||\nu\sigma)}{\partial y}] \tag{59}$$

Diagonal elements of eq. (54) give

$$\frac{\partial \varepsilon_p}{\partial y} = \tilde{F}_{pp} - \tilde{S}_{pp}\,\varepsilon_p \tag{60}$$

Off-diagonal elements of eq. (54) lead to

$$U_{qp} = [\tilde{F}_{qp} - \tilde{S}_{qp}\,\varepsilon_p]/[\varepsilon_p - \varepsilon_q] \tag{61}$$

The sum over r in eq. (52) is conveniently separated into an occupied part (r=j=1...n) and virtual part (r=a=n+1,...2N). The occupied part $U_{ij}$ can be simplified using eq. (51)

$$U^*_{ji} + U_{ij} = -\tilde{S}_{ij} \tag{62}$$

$$\text{for } U_{ai} = (Q_{ai} + \sum_j \sum_b [U^*_{bj}(ab||ij) + U_{bj}(aj||ib)/(\varepsilon_i - \varepsilon_a)] \tag{63}$$

where

$$Q_{ai} = \tilde{H}_{ai} - \tilde{S}_{ai}\,\varepsilon_i - \sum_{kl} \tilde{S}_{kl}\,(al||ik)$$

$$+ \sum_{\mu\nu\lambda\sigma} c^*_{\mu a}\, c_{\nu i}\, P_{\lambda\sigma}\, \frac{\partial}{\partial y}(\mu\lambda||\nu\sigma) \tag{64}$$

This set of equations can be used to solve for $U_{ai}$ by some iterative process.

In the second-order derivative expression, eq. (49), we need $\partial P_{\mu\nu}/\partial y$ and $\partial W_{\mu\nu}/\partial y$. Again, separating the contributions from the occupied block and occupied virtual block, we obtain:

$$\frac{\partial P_{\mu\nu}}{\partial y} = \sum_i^n \left(\frac{\partial C^*_{\mu i}}{\partial y} C_{\nu i} + C^*_{\nu i} \frac{\partial C_{\nu i}}{\partial y}\right.$$

$$= - \sum_{ij} C^*_{\mu i}\, C_{\nu j}\, \tilde{S}_{ij} + \sum_{ia} (C^*_{\mu i}\, C_{\nu a}\, U^*_{ai} + C^*_{\mu a}\, C_{\nu i}\, U_{ai}) \tag{65}$$

And similarly

$$\frac{\partial W_{\mu\nu}}{\partial y} = \sum_i^n \left[\frac{\partial \varepsilon_i}{\partial y} C^*_{\mu i}\, C_{\nu i} + \varepsilon_i \frac{\partial C^*_{\mu i}}{\partial y} C_{\nu i} + \varepsilon_i\, C^*_{\mu i} \frac{\partial C_{\nu i}}{\partial y}\right]$$

$$= \sum_{ij} C^*_{\mu i} C_{\nu j} [\widetilde{F}_{ij} - (\varepsilon_i + \varepsilon_j) \widetilde{S}_{ij}] + \sum_{ia} \varepsilon_i (C^*_{\mu i} C_{\nu a} U^*_{ai} + C^*_{\mu a} C_{\nu i} U_{ai}) \tag{66}$$

This completes the evaluation of the terms required for the second derivative expression. The derivative of the second order Møller-Plesset energy can now be obtained. In addition to the derivative of Fock energies, $\varepsilon_p$, we require the derivatives of the transformed integrals $(ij||ab)$

$$\frac{\partial(ij||ab)}{\partial y} = \sum_p [U^*_{pi}(pj||ab) + U^*_{pj}(ip||ab) + U_{pa}(ij||pb) + U_{pb}(ij||ab) + (ij||ab)^{\dagger}_x \tag{67}$$

where

$$(pq||rs)^{\dagger}_x = \sum_{\mu\nu\lambda\sigma}^{N} C^*_{\mu p} C^*_{\nu q} C_{\lambda r} C_{\sigma s} (\partial/\partial x)(\mu\nu||\lambda\sigma) \tag{68}$$

Full algebraic details will not be given. In terms of $a^{ab}_{ij}$ given by eq. (39), the final result for the derivative of the second-order correlation energy $E_{MP2}$ is

$$\partial E_{MP2}/\partial x = \sum_{ij} \sum_{ab} a^{ab}_{ij} X^{ab}_{ij} \tag{69}$$

$$X^{ab}_{ij} = \frac{1}{2}(ij||ab)^{\dagger}_x + \sum_k [(ij||ak)U_{kb} - \frac{1}{2} a^{ab}_{ij} \{\widetilde{F}_{ki} - \varepsilon_i \widetilde{S}_{ki}\}$$

$$- \frac{1}{2}(kj||ab)\widetilde{S}_{ik}] + \sum_c [(cj||ab)U_{ci} + \frac{1}{2} a^{ac}_{ij} \{\widetilde{F}_{cb} - \varepsilon_b \widetilde{S}_{cb}\}$$

$$- \frac{1}{2}[(ij||ac)\widetilde{S}_{bc}]$$

The first derivative of the configuration interaction energy using all double excitations involves considerably more effort. Since the expansion coefficients $a^{ab}_{ij}$ have been optimized variationally we have $\partial E_{CID}/\partial a^{ab}_{ij} = 0$ in eq. (40). Differentiation of eq. (40) with respect to an external parameter x now gives

$$(\partial E_{CID}/\partial x) = \sum_{st} [a_s(\partial \mathcal{H}_{st}/\partial x)a_t]/ \sum_s a_s^2 \qquad (71)$$

Hence the energy derivative can be reduced to the derivative of the Hamiltonian matrix which in turn is given in part by the set of anti-symmetrized two-electron integrals.

After some algebraic manipulation, the final CID energy derivative in terms of spin-orbitals can be given by

$$(\partial E_{CID}/\partial x) = (\partial E_{HF}/\partial x) + \sum_{ijab} a_{ij}^{ab}\ z_{ij}^{ab}/ \sum_{ijab} (a_{ij}^{ab})^2 \qquad (72)$$

where

$$\begin{aligned} z_{ij}^{ab} = {} & \frac{1}{2}(ij||ab)_x^{\dagger} + \frac{1}{8}\sum_{cd} a_{ij}^{cd}(ab||cd)_x^{\dagger} + \frac{1}{8}\sum_{kl} a_{kl}^{ab}(kl||ij)_x^{\dagger} - \sum_{kc} a_{kj}^{cb}(ka||ic)_x^{\dagger} \\ & + \frac{1}{2}\sum_c a_{ij}^{cb}\ [\widetilde{F}_{ca} - \varepsilon_a\ \widetilde{S}_{ca}] - \frac{1}{2}\sum_k a_{kj}^{ab}\ [\widetilde{F}_{ki} - \varepsilon_i\ \widetilde{S}_{ki}] \\ & + \sum_e U_{ei}\ [(ej||ab) + \frac{1}{2}\sum_{kl} a_{kl}^{ab}(ej||kl) - 2\sum_{kc} a_{kj}^{cb}(ka||ec)] \\ & + \sum_m U_{ma}\ [(ij||mb) + \frac{1}{2}\sum_{cd} a_{ij}^{cd}(mb||cd) - 2\sum_{kc} a_{kj}^{cb}(km||ic)] \\ & + \sum_m \widetilde{S}_{mi}\ [-\frac{1}{4} w_{mj}^{ab} + \frac{1}{2}\sum_{cd} (ab||cd)\ a_{mj}^{cd} - \sum_{kc} (kb||jc)\ a_{mk}^{ac}] \\ & + \sum_e \widetilde{S}_{ea}\ [-\frac{1}{4} w_{ij}^{eb} + \frac{1}{2}\sum_{kl} (kl||ij)\ a_{kl}^{eb} - \sum_{kc} (kb||jc)\ a_{ik}^{ec}] \end{aligned} \qquad (73)$$

## Illustrative Examples

### a) Timing

If derivative methods are to represent a significant advance in the technology of quantum chemistry, they must permit more information to be calculated more accurately and/or more quickly than non-derivative methods. To demonstrate that this is indeed true, a series of calculations has been performed on ethylene at the 6-31G* basis set[35] level (this includes d orbitals on carbon). The timing data are collected in Table 2.

To calculate the energy and all of the first derivatives of the Hartree-Fock energy with respect to the positions of the nuclei requires approximately twice as much computer time as is needed to compute the Hartree-Fock energy alone, regardless of the size of the molecule or the number of degrees of freedom. In this highly symmetrical case the timing for the gradient is even more favorable because symmetry was used to reduce the number of integral derivatives calculated. Similar results are obtained with sp basis set. Since 3N-6 derivatives are calculated for an N atom molecule at only double the cost of the energy alone this can represent an order of magnitude increase in efficiency for calculations on larger and less symmetrical molecules. Other gradient programs[36,19] are reported to require 3-4 times as long as the SCF calculations. This difference in speed is directly attributable to the efficiency of the integral derivative algorithms outlined above.

The Hartree-Fock full second derivative calculation takes only 4 to 5 times as much time as the gradients. Early indications were that the direct analytical computation of the second derivatives might be prohibitively difficult[12], and that second derivatives could be calculated more practically by the finite difference method from the analytical gradients.[10,11] The present work shows that even for a small highly symmetrical molecule, the analytical approach is competitive with the finite difference method. For larger, more general, problems the analytical approach should be superior, since the finite difference approach requires at least 3N-5 calculations to determine the full second derivative matrix.

The second order Møller-Plesset energy can be computed very rapidly once the SCF orbitals are available. The derivatives of the MP2 energy take three times a long as the MP2 energy. More integrals must be transformed and the coupled-perturbed Hartree-Fock equations must be solved. Nevertheless, the calculation of gradients at this simple level of electron correlation is practical.

The configuration interaction calculations on ethylene are an order of magnitude more lengthy than the Hartree-Fock calculations. But similar to Hartree-Fock calculations, the timing for the CID energy plus gradients is very favorable, requiring only twice as long as the CID energy alone. Furthermore, this ratio should be relatively independent of the size of the configuration interaction calculation and the number of degrees of freedom in the molecule, since the most difficult sections of the CID gradient calculations scale in the same manner as the CID energy ($N^5$ in the number of functions). The MP2 or CID gradients and the Hartree-Fock second derivatives can be calculated together more efficiently than both separately. The combination results in a very powerful method for geometry optimization that takes electron correlation into account.

b) Geometry Optimization

One of the most important and widespread applications of energy derivatives is geometry optimization. The wealth of information available in a gradient calculation can be used very effectively to locate the minima on a potential energy hypersurface both rapidly and accurately. Furthermore, analytical first and second energy derivatives make it feasible to search for more complicated stationary points on energy surfaces like saddle points and local maxima. Such features may be inaccessible by non-derivative methods.

A typical non-gradient minimization method proceeds by a series of one dimensional searches for each coordinate. Repeated cycles of varying each coordinate may be necessary for acceptable convergence to the minimum. There are more efficient optimization methods available such as the Fletcher-Powell[37] algorithm, but frequently it is still not practical to optimize all of the geometric parameters in a molecule using only the energy.

With the availability of rapidly calculated analytical gradients, more powerful optimization techniques can be used. A class of methods known as conjugate gradient or variable metric algorithms[38] guarantees exact convergence to an extremum in N + 1 steps on an N dimensional quadratic surface. The details of these and other optimization algorithms are discussed in another chapter. In practice, searches for equilibrium geometries seem to require ca. N/2 steps, for well chosen starting geometries and rough estimates of the second derivatives. Furthermore, gradient methods are well suited to optimize systems with strongly coupled coordinates such as polycyclic molecules that might be intractable otherwise.

Gradient optimization algorithms and Hartree-Fock first derivatives for s p and d type gaussian wavefunctions have been incorporated into the publicly available version of GAUSSIAN 80.[30] These gradient methods have made full geometry optimization with any standard basis set practically automatic and routine. In an attempt to organize the increasingly large amount of data on optimized molecular structures and total energies, a computer archiving system has been set up. The Carnegie-Mellon version of GAUSSIAN 80 now records directly into a computer readable file the essential information from all standard-route calculations. The first version of the

Carnegie-Mellon Quantum Chemistry Archive,[39] available in printed form or on magnetic tape, contains about 2000 fully optimized structures using the STO-3G, 321G and 6-31G* basis sets. Figure 1 illustrates a sample entry in the archive.

Gradient methods have also opened the way to locating transition structures. Unless the transition vector can be determined by symmetry, it is normally very difficult, if not impossible, to find the transition structure using only energy based optimization methods. Minimizing the gradient norm has been suggested,[40] but this can lead to false transition structures.[41] A better approach is to use a modified conjugate gradient method. If the starting guess is within the quadratic region of the saddle point, optimization leads directly to the transition structure. However, no overall satisfactory and efficient algorithm seems to exist to solve the general transition structure problem.[41]

Computer programs have recently been developed to compute analytical second derivatives at the Hartree-Fock level.[4] A one step optimization is possible for a quadratic function but quantum mechanically derived energy surfaces are rarely quadratic. In practice, a single second derivative computation followed by one or two gradient calculations are needed to carry out and confirm the optimization. The previous discussion on gradient optimization also pertains to first derivatives of correlated wavefunctions; however, there is a special advantage in combining MP2 or CID gradients with Hartree-Fock second derivatives. Once the MP2 or CID gradients are calculated, it is relatively cheap (see Table 2) to compute the HF second derivatives as well. The second derivatives are not strongly affected by correlation and thus provide an excellent estimate of the curvature of the correlated energy hypersurface. Usually only 2 to 3 additional MP2 or CID gradient calculations are needed to converge to the minimum. Until the advent of MP2 and CID first derivatives and HF second derivatives, geometry optimizations at a level that includes electron correlation, have been infrequent and costly.

Second derivatives are even more vital for transition structures. At a proper saddle point, the second derivative matrix has one and only one negative eigenvalue. Thus to verify that a geometry represents a true transition structure, not only must the gradient be zero, but also the second derivative matrix must be calculated to check that only one negative eigenvalue is present. Similar to the minimization problem, optimization of saddle points can be accomplished readily if the second derivatives are available and if the

```
            HF/6-31G* ATOMS AND FULLY OPTIMIZED STRUCTURES.              225
                          28-MAR-80 16:33:25

-------------------------------------------------------------------------------
MOLECULE  SYMM                      TITLE                       ENERGY     SEQ
-------------------------------------------------------------------------------

B(1-)     KH    B(1-) ... (1S)2 (2S)2 (2PZ)2 ... 6-31G*        -24.39726  4840

                                                                     Key to full entries
C3H7(1+)  C2V   EDGE-PROTONATED CYCLOPROPANE. 6-31G* OPT     -117.35073   265
          CS    C3H7 (+). METHYL-STAGGERED 1-PROPYL CATI     -117.35111  3791
                CORNER-PROTONATED CYCLOPROPANE. 6-31G* O     -117.35916   187
          C2V   2-PROPYL CATION. 6-31G* OPT.                 -117.38076   513

O2(3)     D*H   OXYGEN MOLECULE,6-31G* STRUCTURE; 001; D     -149.61791  2610
                           ----------------------
 FULL ENTRIES.
 -------------

Sequence number          author                calculation type
 265\20161\C3H7(1+)\KRISHNAN\14-OCT-1978\1\\#P OPT 6-31G* NOPOP\\EDGE-PROTO
 NATED CYCLOPROPANE. 6-31G* OPT.\\1,1\X\C,1,R1\C,1,R2,2,90.\C,1,R2,2,90.,3,
 180.,0\H,1,R3,3,90.,2,180.,0\X,2,1.,1,90.,3,0.,0\X,2,1.,6,90.,1,180.,0\H,2     optimized
 ,R4,7,THETA1,6,90.,0\H,2,R4,7,THETA1,6,-90.,0\X,4,1.,1,THETA2,2,180.,0\H,4     geometry
 ,R5,10,THETA3,1,90.,0\H,4,R5,10,THETA3,1,-90.,0\X,3,1.,1,THETA2,2,180.,0\H
 ,3,R5,13,THETA3,1,90.,0\H,3,R5,13,THETA3,1,-90.,0\\R1=1.19717\R3=0.93317\R
 2=0.88307\R4=1.07439\R5=1.07649\THETA1=57.27369\THETA2=150.66634\THETA3=58
 .8063\\HF=-117.3507286\RMSD=0.530D-07\RMSF=0.381D-03\FWG=C02V [C2(H1C1),SI
 GMAV(C2),SIGMAV'(H2),X(H4)]\\
     Total Energy         SCF convergence      Residual forces after geometry optimization
```

FIGURE 1 Sample extract from the Carnegie-Mellon Quantum Chemistry Archive (reference 39)

TABLE 2

Approximate Execution Times[e] (in minutes) for Ethylene Using 6-31G* Basis (38 Basis Functions)

| Program | HF | HF + 1st Deriv. | HF + 1st & 2nd Deriv. | MP2 | MP2 + 1st Deriv. | CID | CID + 1st Deriv. | MP2 1st & HF 2nd Deriv. | CID 1st & HF 2nd Deriv. |
|---|---|---|---|---|---|---|---|---|---|
| Integrals | 10 | 10 | 10 | 10 | 10 | 10 | 10 | 10 | 10 |
| SCF | 3 | 5[a] | 5[a] | 5[a] | 5[a] | 5[a] | 5[a] | 5[a] | 5[a] |
| Transformation | | | 8 | 5 | 15[d] | 15 | 15 | 15 | 15 |
| CID | | | | | | 126 | 126 | | 126 |
| Integral 1st Deriv. | | 7 | 16[bc] | | 16[bc] | | 16[bc] | 16[bc] | 16[bc] |
| Integral 2nd Deriv. | | | 31[c] | | | | | 31[c] | 31[c] |
| CPHF | | | 13 | | 13 | | 13 | 13 | 13 |
| HF 2nd Deriv. Eval. | | | 6 | | | | | 6 | 6 |
| MP2 or CID Deriv. Eval. | | | | | 18 | | 113 | 18 | 113 |
| Total | 13 | 22 | 89 | 20 | 77 | 156 | 298 | 107 | 335 |

[a] SCF convergence on the density matrix tightened to get more significant figures in the M.O. coefficients.

[b] The integral derivatives are written out.

[c] Information about the symmetry of the molecule was used to aid in these parts of the calculation.

[d] More transformed integrals are calculated in this case as compared to a simple MP2 calculation.

[e] All the calculations were performed on a VAX-11/780 computer at Carnegie-Mellon University.

starting structure is in the quadratic region. More general algorithms for transition structure optimization are under study.

c) Force Constants and Vibrational Frequencies

The original applications of first derivatives were to the calculation of vibrational frequencies.[10] Harmonic and anharmonic force constants were calculated by the finite difference method with gradients obtained analytically at the Hartree-Fock level.[10,11,13-18] With the present programs it is possible to compute the Hartree-Fock second derivatives and, the vibrational frequencies in a single calculation. Furthermore, the effect of electron correlation can be assessed by calculating the force constants numerically from the MP2 gradients.

Table 3 lists the harmonic frequencies for ethylene computed at the Hartree-Fock and MP2 level using the 6-31G* basis set. The respective equilibrium geometry was used in each case. Also listed are the experimental frequencies corrected for anharmonicity.[42] The HF theory appears to overestimate the harmonic frequencies by 55-180 $cm^{-1}$. Inclusion of correlation improves the agreement considerably. The MP2 calculation accounts for about 70% of the discrepancy in the HF theory, yielding frequencies that differ from experiment by only 10-90 $cm^{-1}$. Most of the frequencies are still overestimated but this may be due to the considerable uncertainty in the empirical anharmonicity corrections. Corrections to the harmonic vibrational frequencies can be computed from the cubic and quartic force constants. These are difficult to obtain experimentally for molecules larger than tetra-atomic. However, theoretical anharmonic force constants are predicted reliably even at the Hartree-Fock level.

Zero point vibrational energy is frequently ignored when theoretical energy differences are compared with experiment. The analytical second derivative programs permit this quantity to be evaluated in a single calculation, eliminating one more possible source of error in the comparison between theory and experiment.

d) Electrical and Magnetic Properties

With simple atomic orbitals, very large basis sets with extensive polarization are required to obtain second order magnetic and electrical properties to a reasonable accuracy.[33] In contrast, rather modest basis sets can be used, if a field

TABLE 3

Vibrational Frequencies for Ethylene ($cm^{-1}$)

| Symmetry | HF/6-31G*[a] | MP2/6-31G*[b] | Experimental Harmonic Frequencies[c] |
|---|---|---|---|
| $b_{2u}$ | 897.0 | 851.1 | 842.9 |
| $b_{2g}$ | 1099.4 | 942.6 | 958.8 |
| $b_{1u}$ | 1095.0 | 991.8 | 968.7 |
| $a_u$ | 1154.9 | 1085.5 | 1043.9 |
| $b_{1g}$ | 1352.5 | 1265.9 | 1244.9 |
| $a_g$ | 1496.9 | 1415.7 | 1369.6 |
| $b_{3u}$ | 1610.2 | 1520.8 | 1473.0 |
| $a_g$ | 1856.2 | 1721.1 | 1654.9 |
| $b_{3u}$ | 3320.9 | 3213.3 | 3146.9 |
| $a_g$ | 3344.2 | 3230.9 | 3152.5 |
| $b_{1g}$ | 3394.6 | 3300.4 | 3231.9 |
| $b_{2u}$ | 3420.7 | 3323.3 | 3234.3 |

[a] At the HF/6-31G* equilibrium geometry ($r_{CC}$ = 1.317, $r_{CH}$ = 1.076, <HCH = 116.4).

[b] At the MP2/6-31G* equilibrium geometry ($r_{CC}$ = 1.335, $r_{CH}$ = 1.085, <HCH = 116.5).

[c] Anharmonicity corrections are made on the observed experimental frequencies to get these estimated values.[42]

dependence is built into the basis functions.[25-27] Gauge-dependent atomic orbitals have been especially successful in NMR chemical shift calculations.[25,26]

Summary

In the preceding sections, a number of expressions have been derived for the analytical calculation of energy derivatives. Formulae are discussed for the first derivatives or gradients of the Hartree-Fock energy and also for the second order Møller-Plesset and the double excitation configuration interaction energies, which include electron correlation. Timing data for the computer programs written to calculate these derivatives indicate that such computations are not only practical but indeed very efficient. In fact, derivative calculations represent an order of magnitude increase in the information that can be generated about energy hypersurfaces with a given amount of computer time. Geometry optimization and vibrational frequency calculations have been discussed as applications of analytically computed energy derivatives.

Acknowledgement

J. A. Pople, R. Krishnam and J. S. Binkley are gratefully acknowledged for their contributions to energy derivative methods. I wish to thank my wife for typing the draft of this manuscript under adverse conditions.

References

1. Current address: Department of Chemistry, Wayne State University, Detroit, Michigan 48202.
2. H. B. Schlegel, J. Comput. Phys., to be published; Ph.D. thesis, Queen's University, Kingston, Ont. Canada, 1975.
3. H. B. Schlegel and J. S. Binkley, J. Comput. Phys., to be published.
4. J. A. Pople, R. Krishnan, H. B. Schlegel and J. S. B inkley, Int. J. Quantum Chem.: Quantum Chem. Sym. 13, 225 (1979).
5. R. Krishnan, H. B. Schlegel and J. A. Pople, J. Chem. Phys. 72, 4654 (1980).
6. S. Bratoz, Colloq. Intern. Centre Natl. Rech. Sci. (Paris) 82, 287 (1958).
7. D. M. Bishop and M. Randic, J. Chem. Phys. 44, 2480 (1966).
8. R. Moccia, Chem. Phys. Lett. 5, 260 (1970); Int. J. Quantum Chem. 8, 293 (1974).
9. J. Gerratt and I. M. Mills, J. Chem. Phys. 49, 1719 (1968); J. Chem. Phys. 49, 1730 (1968).

10. P. Pulay, Mol. Phys. 17, 197 (1969).
11. P. Pulay, Modern Theoretical Chemistry (Vol. 4), H. F. Schaefer III, Ed., (Plenum Press, New York, 1977).
12. K. Thomsen and P. Swanstrom, Mol. Phys. 26, 735 (1973).
13. P. Pulay and W. Meyer, J. Mol. Spec. 40, 59 (1971).
14. W. Meyer and P. Pulay, J. Chem. Phys. 56, 2109 (1972).
15. P. Pulay and W. Meyer, Mol. Phys. 27, 473 (1974).
16. H. B. Schlegel, S. Wolfe and F. Bernardi, J. Chem. Phys. 63, 3632 (1975).
17. H. B. Schlegel, S. Wolfe and F. Bernardi, J. Chem. Phys. 67, 4181 (1977).
18. H. B. Schlegel, S. Wolfe and F. Bernardi, J. Chem. Phys. 67, 4194 (1977).
19. K. Ishida, K. Morokuma and A. Komornicki, J. Chem. Phys. 66, 2153 (1977).
20. W. Meyer and P. Rosmus, J. Chem. Phys. 63, 2356 (1975).
21. U. Wahlgren, J. Pacansky and P. S. Bagus, J. Chem. Phys. 63, 2874 (1975).
22. B. J. Rosenberg, W. C. Ermler and I. Shavitt, J. Chem. Phys. 65, 4072 (1976).
23. a) P. Pulay, W. Meyer and J. E. Boggs, J. Chem. Phys. 68, 5077 (1978);
    b) P. R. Taylor, G. B. Bacskay, N. S. Hush and A. C. Hurley, J. Chem. Phys. 69, 1971 (1978).
24. a) J. D. Goddard, N. C. Handy and H. F. Schaefer III, J. Chem. Phys. 71, 1525 (1979);
    b) S. Kato and K. Morokuma, Chem. Phys. Lett. 65, 19 (1979);
    c) B. R. Brooks, W. D. Laidig, P. Saxe, J. D. Goddard, Y. Yamaguchi and H. F. Schaefer III, J. Chem. Phys. 72, 4652 (1980).
25. R. Ditchfield, J. Chem. Phys. 56, 5688 (1972); Mol. Phys. 27, 789 (1974).
26. F. R. Prado, C. Giessner-Prettre and B. Pullman, Int. J. Quantum Chem., Quant. Bio. Sym. 6, 491 (1979).
27. A. J. Sadlej, Chem. Phys. Lett. 47, 50 (1977).
28. J. A. Pople and W. J. Hehre, J. Comput. Phys. 27, 161 (1978).
29. H. Taketa, S. Huzinaga and K. O-Ohata, J. Phys. Soc. Japan 21, 2313 (1966).
30. J. S. Binkley, R. A. Whiteside, R. Krishnan, R. Seeger, D. J. DeFrees, H. B. Schlegel, S. Topiol, L. R. Kahn and J. A. Pople, "GAUSSIAN 80, an Ab Initio Molecular Orbital Program", Carnegie-Mellon University, Pittsburgh, PA, USA (1980).
31. a) M. Dupuis, J. Rys and H. F. King, J. Chem. Phys. 65, 111 (1976);
    b) H. F. King and M. Dupuis, J. Comput. Phys. 21, 1 (1976).

32. C. Møller and M. S. Plesset, Phys. Rev. 46, 618 (1934).
33. R. M. Stevens, R. Pitzer and W. N. Lipscomb, J. Chem. Phys. 38, 550 (1963).
34. T. C. Caves and M. Karplus, J. Chem. Phys. 50, 3649 (1969).
35. P. C. Hariharan and J. A. Pople, Theo. Chim. Acta 28, 213 (1973).
36. P. Pulay, Theo. Chim. Acta 50, 299 (1979).
37. a) R. Fletcher and M.J.D. Powell, Compt. J. 6, 163 (1963); b) J. B. Collins, P.v.R. Schleyer, J. S. Binkley and J. A. Pople, J. Chem. Phys. 64, 5142 (1976).
38. K. W. Brodlie, Math Programming 12, 344 (1977) and references cited.
39. R. A. Whiteside, J. S. Binkley, R. Krishnan, D. J. DeFrees, H. B. Schlegel and J. A. Pople, "Carnegie-Mellon Quantum Chemistry Archive", Carnegie-Mellon University, Pittsburgh, PA., USA (1980).
40. J. W. McIver and A. Komornicki, J. Am. Chem. Soc. 94, 2625 (1972).
41. K. Mueller, Angew. Chem. Int. Ed. Engl. 19, 1 (1980).
42. J. L. Duncan, D. C. McKean and P. D. Mallinson, J. Mol. Spec. 45, 221 (1973).

# ANALYTIC ENERGY GRADIENTS FOR OPEN-SHELL RESTRICTED-HARTREE-FOCK, LIMITED MULTICONFIGURATION SCF, AND LARGE SCALE CONFIGURATION INTERACTION WAVEFUNCTIONS.

John D. Goddard*

Institute for Theoretical Chemistry
Department of Chemistry
University of Texas
Austin, Texas 78712 U. S. A.

*Present address: Division of Chemistry
National Research Council of Canada
Ottawa, Ontario, Canada K1A 0R6

## INTRODUCTION

Analytic derivatives of the energy with respect to nuclear coordinates are exceedingly useful in the optimization of equilibrium and transition state geometries and in the characterization of the stationary points on potential energy surfaces via vibrational analyses (1). However, until quite recently the use of such gradients was somewhat restricted as the gradient method had been developed in detail for only closed-shell single determinant SCF (2) or open-shell unrestricted Hartree-Fock wavefunctions (3). The restriction to these methods would not allow even a qualitatively correct treatment of many reactions such as those for which orbital symmetry considerations suggest a "forbiddenness" due to an orbital crossing or of many unusual free radicals of interest in physical organic chemistry. Within the last two years, the analytic energy gradient approach has been extended to the open-shell restricted Hartree-Fock method (4) to avoid difficulties sometimes encountered with the UHF approach in which the wavefunction is not an eigenfunction of $S^2$. Certain limited multiconfiguration self-consistent-field gradient methods (4,5) have also been developed and applied which allow for a qualitatively correct description of Woodward-Hoffmann forbidden processes and of radicals such as trimethylene.

An even more recent and exciting development is the calculation



*I. G. Csizmadia and R. Daudel (eds.), Computational Theoretical Organic Chemistry, 161–174.*

of analytic energy gradients for general large scale configuration interaction wavefunctions (6,7,8,9). Analytic gradients may now be used to calculate equilibrium geometries, transition state structures, reaction pathway and vibrational frequencies for accurate correlated wavefunctions. A very considerable increase in the level of treatment of problems in chemical reactivity is possible due to these methodological advances.

In this review, the gradient method for restricted Hartree-Fock open-shell and limited multiconfiguration SCF wavefunctions is briefly surveyed. In somewhat more detail, an overview of the CI gradient method is presented. Examples related to the geometries and vibrational frequencies of equilibrium and transition state structures are given.

## GRADIENT TECHNIQUES FOR OPEN-SHELL RESTRICTED-HARTREE-FOCK AND LIMITED MCSCF METHODS (4,5).

For a large class of closed-shell, open-shell and limited MCSCF (GVB) systems, the electronic energy may be written as (10):

$$E = 2\sum_{i}^{n} f_i h_i + \sum_{i,j}^{n} (a_{ij}J_{ij} + b_{ij}K_{ij}) \qquad (1)$$

where n is the number of occupied orbitals, $h_i$ the one electron integrals and $J_{ij}$ and $K_{ij}$ the Coulomb and exchange molecular integrals. $f_i$, $a_{ij}$, and $b_{ij}$ are constants (11) for closed- or open-shell wavefunctions but depend on the CI expansion coefficients for the limited MCSCF case.

For these systems the orbitals, $\phi_i$, are eigenfunctions of Fock operators, $F_i$, which are defined by

$$F_i = 2f_i h + 2\sum_{j}(a_{ij}J_j + b_{ij}K_j) \qquad (2)$$

The number of distinct Fock operators depends upon the type of open-shell system considered. For example, in a high spin open-shell case, two Fock operators are required while three are defined in the case of an open-shell singlet (10). The change in the energy due to an orbital change $\phi_k \rightarrow \phi_k + \delta\phi_k$ (with proper accounting of the changes in other orbitals to insure orthogonality) is (12)

$$2\langle\delta\phi_k|F_k|\phi_1\rangle - 2\sum_{j}\langle\delta\phi_k|\phi_j\rangle\langle\phi_k|F_j|\phi_j\rangle \qquad (3)$$

and at convergence

$$<\phi_i|F_i|\phi_j> = <\phi_i|F_j|\phi_j> \tag{4}$$

and a unique symmetric $\varepsilon$ matrix defined by

$$\varepsilon_{ij} \equiv <\phi_i|F_i|\phi_j> \tag{5}$$

results.

Rewriting the energy expression in terms of basis functions through the usual density matrices $D_{irs}$ yields

$$E = 2\sum_{irs} f_i D_{irs}(r|h|s) + \sum_{ij}\sum_{rstu}(a_{ij}D_{irs}D_{jtu} + b_{ij}D_{irt}D_{jsu})(rs|tu) \tag{6}$$

Further defining a density matrix, $D^I_{rs}$, associated with the orbitals which are eigenfunctions of the M Fock operators $F_I$, $F_J$,... and rewriting the summations gives

$$E = \sum_{r\geq s}(2-\delta_{rs})\,(r|h|s)\sum_I^M f_I D^I_{rs}$$
$$+ \sum_{\substack{r\geq s\\ \overline{rs}\geq\overline{tu}}}\sum_{t\geq u}(2-\delta_{rs})(2-\delta_{tu})(2-\delta_{rs,tu})(rs|tu)$$
$$\times \sum_{I,J}^M(a_{IJ}D^I_{rs}D^J_{tu} + \frac{b_{IJ}}{2}(D^I_{rt}D^J_{su} + D^I_{ru}D^J_{st})) \tag{7}$$

The derivative of this expression with respect to a nuclear coordinate, $X^a$, may be found and is greatly simplified by noting that (1)

$$\sum_{rk}\frac{\partial E}{\partial C_{rk}}\frac{\partial C_{rk}}{\partial X^a} = -\,Tr(\underline{\underline{C\varepsilon C}}^+\underline{\underline{S}}^a) \tag{8}$$

where $\underline{\underline{C\varepsilon C}}^+$ is the "energy weighted density matrix" and $S^a$ a matrix of derivatives of the overlap integrals with respect to the nuclear coordinates. The derivatives of the molecular orbital coefficients for an SCF or MCSCF wavefunction, $\partial C_{rk}/\partial X^a$, are not needed as they are effectively removed by equation 8 in the SCF case and by explicit optimization in the general MCSCF case. An expression for the forces which is as general as equation 1 and thus covers a wide range of cases of interest is

$$\frac{\partial E}{\partial X^a} = -Tr(\underline{C}\underline{\varepsilon}\underline{C}^+\underline{S}^a)$$

$$+ \sum_{r \geq s} (2-\delta_{rs})[(r^a|h|s) + (r|h^a|s) + (r|h^a|s)]\sum_I f_I D_{rs}^I$$

$$+ \sum_{\substack{r \geq s \ t \geq u \\ rs \geq tu}} (2-\delta_{rs})(2-\delta_{tu})(2-\delta_{rs,tu})$$

$$x [(r^a s|tu) + (rs^a|tu) + (rs|t^a u) + (rs|tu^a)]$$

$$x \sum_{IJ} (a_{IJ} D_{rs}^I D_{tu}^J + \frac{b_{IJ}}{2}(D_{rt}^I D_{su}^J + D_{ru}^I D_{st}^J)) \tag{9}$$

$(r^a|h|s)$ and $(r^a s|tu)$ are the first derivatives of the basis function integrals (1,13,14) for which efficient computer programs are available (15). For simple but useful two configuration SCF wavefunctions differing in one doubly occupied orbital

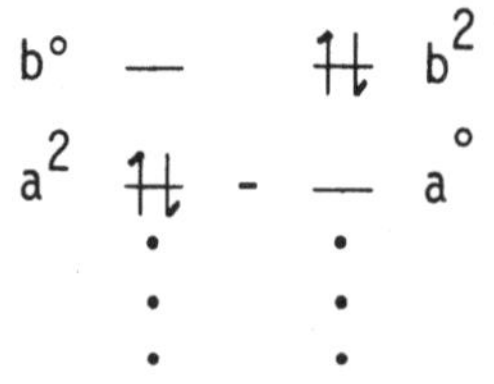

$$
\begin{array}{ll}
b° \ — & ⇅ \ b^2 \\
a^2 \ ⇅ \ - & — \ a° \\
\cdot & \cdot \\
\cdot & \cdot \\
\cdot & \cdot
\end{array}
\tag{10}
$$

the $a_{IJ}$ and $b_{IJ}$ depend upon the CI expansion coefficients, $\sigma_I$:

$$\Phi = [\sigma_1 A(...a^2) - \sigma_2 A(...b^2)] \tag{11}$$

where $A$ is the usual antisymmetrization operator. Such a wavefunction gives a better description of species such as carbenes or silylenes (16) where there is a very low lying unoccupied MO or of barriers which may be considered to arise due to an orbital crossing (4).


ANALYTIC GRADIENTS FROM CI WAVEFUNCTIONS VIA THE TWO-PARTICLE DENSITY MATRIX AND THE UNITARY GROUP APPROACH (6,7,8).

The recent computational implementation of the loop driven graphical unitary group approach (6,7,17) (LDGUGA) to configuration interaction provides a highly efficient scheme for large scale (over 10,000 term) CI calculations. In addition, the efficient generation of the one- and two-particle density matrices via the

LDGUGA algorithm for such CI wavefunctions provides a general and computationally efficient scheme to determine analytic CI energy gradients (18). In this section, the LDGUGA-CI method will be briefly outlined and the features which make it particularly suitable for analytic gradient determination stressed.

The unitary group mathematical formalism has been available for some time (19) but only more recently has work by Paldus (20) and Shavitt (21) clearly demonstrated the potential computational benefits of such an approach. Brooks and Schaefer's first computational implementation of the unitary group CI method (17) proved very efficient relative to state-of-the-art conventional CI techniques. In addition, the LDGUGA approach is particularly amenable to analytic gradient calculation as will be discussed.

The GUGA CI method involves the generation of the distinct row table which determines the configuration space included in the CI and the use of this table as a template to generate all loops which symbolically define the Hamiltonian matrix in a compact form. Each loop defines a set of equivalent matrix element contributions. Each configuration (Gelfand state) may be represented by a three column Paldus array (20) that spans the orbitals used in the CI. The distinct row table is the numerical analog of Shavitt's graphical analysis (21) and in its simplest form is composed of different distinct rows in the Paldus representation. Spatial symmetry for real abelian point groups (21) or the selection of an interacting space (22) may be included by adding complexity to the distinct row table.

Matrix elements between two Gelfand states can be expressed as

$$\begin{aligned} \langle m'|H|m\rangle &= \sum_{i,j} \langle i|\hat{h}|j\rangle \langle m'|E_{ij}|m\rangle \\ &= \frac{1}{2} \sum_{i,j,k,l} [ij;kl] \langle m'|E_{ij}E_{kl} - \delta_{jk}E_{il}|m\rangle \end{aligned} \tag{12}$$

where $E_{ij}$ is a spin independent generator (20). The matrix elements of these generators are determined by a factorization over the levels between and including the indices of the generator or generator product. By combining integral generator products that contribute to the same set of matrix elements, fourteen loop types may be defined.

The loop-driven algorithm is the most important element accounting for the efficiency of the GUGA method. With the loop-driven algorithm the next loop that can be produced with the least effort is generated. Since most loops have large common sections with other loops, savings result in that the algorithm only generates portions of the next loop that differ from the previous loop.

Vast numbers of previously unappreciated relationships between otherwise distinct Hamiltonian matrix elements are revealed by such an algorithm. It also points out relationships between different integrals sharing one or more common indices.

For any CI wavefunction the energy can be written

$$E = \sum_{\substack{IJ \\ ijkl}} C_I C_J (a_{IJ}^{ij} \langle i|h|j\rangle + b_{IJ}^{ijkl} [ij;kl]) \tag{13}$$

and the derivative with respect to a nuclear coordinate

$$E^a = \sum_{\substack{IJ \\ ijkl}} C_I C_J (a_{IJ}^{ij} \langle i|h|j\rangle^a + b_{IJ}^{ijkl} [ij;kl]^a) \tag{14}$$

where $a_{IJ}^{ij}$ and $b_{IJ}^{ijkl}$ are "spin coupling constants" and $\langle i|h|j\rangle^a$ and $[ij;kl]^a$ one-and two-electron derivative integrals in the MO basis

$$\langle i|h|j\rangle^a = \sum_{\lambda\mu} C_{\lambda i} C_{\mu j} \langle\lambda|h|\mu\rangle^a + \sum_{\lambda\mu} (C_{\lambda i}^a C_{\mu j} + C_{\lambda i} C_{\mu j}^a) \langle\lambda|h|\mu\rangle$$

$$= \sum_{\lambda\mu} C_{\lambda i} C_{\mu j} \langle\lambda|h|\mu\rangle + \sum_r (U_{ri}^a \langle r|h|j\rangle + U_{rj}^a \langle i|h|r\rangle) \tag{15}$$

and

$$[ij;kl]^a = \sum_{\lambda\mu\nu\sigma} C_{\lambda i} C_{\mu j} C_{\nu k} C_{\sigma l} [\lambda\mu;\nu\sigma]^a + \sum_r (U_{ri}^a [rj;kl]$$

$$+ U_{rj}^a [ir;kl] + U_{rk}^a [ij;rl] + U_{rl}^a [ij;kr]) \tag{16}$$

where $U_{ri}^a$ denotes the first order changes in the MO (23). $U_{ri}^a$ as well as the derivatives of all atomic integrals are the additional information required to determine the CI analytic derivative beyond that necessary to determine the CI energy alone.

The total energy derivative may be split into two parts

$$E^a = E^{a'} + E^{a''} \tag{17}$$

where

$$E^{a'} = \sum_{\lambda\mu\nu\sigma} \sum_{ijkl} C_I C_J b_{IJ}^{ijkl} C_{\lambda i} C_{\mu j} C_{\nu k} C_{\sigma l} [\lambda\mu;\nu\sigma]^a$$

$$+ \sum_{\lambda\mu} \sum_{ij} \sum_{IJ} C_I C_J a_{IJ}^{ij} C_{\lambda i} C_{\mu j} <\lambda|h|\mu>^a \tag{18}$$

and

$$E^{a''} = \sum_{ijr} \sum_{IJ} C_I C_J a_{IJ}^{ij} (U_{ri}^a <r|h|j> + U_{rj}^a <i|h|r>)$$
$$+ \sum_{ijklr} \sum_{IJ} C_I C_J b_{IJ}^{ijkl} (U_{ri}^a [rj;kl] + U_{rj}^a [ir;kl]$$
$$+ U_{rk}^a [ij;rl] + U_{rl}^a [ij;kr] \tag{19}$$

The optimum method for evaluating these equations involves computing the one- and two-particle density matrices (24):

$$Q_{ij} = \sum_{IJ} C_I C_J a_{IJ}^{ij} \tag{20}$$

$$G_{ijkl} = \sum_{IJ} C_I C_J b_{IJ}^{ijkl} \tag{21}$$

Transforming these density matrices to the AO basis gives

$$Q_{\lambda\mu} = \sum_{ij} C_{\lambda i} C_{\mu j} Q_{ij} \tag{22}$$

and

$$G_{\lambda\mu\nu\sigma} = \sum_{ijkl} C_{\lambda i} C_{\mu j} C_{\nu k} C_{\sigma l} G_{ijkl} \tag{23}$$

Substituting equations 22 and 23 into equation 18, $E^{a'}$ becomes

$$E^{a'} = \sum_{\lambda\mu\nu\sigma} G_{\lambda\mu\nu\sigma} [\lambda\mu;\nu\sigma]^a + \sum_{\lambda\mu} Q_{\lambda\mu} <\lambda|h|\mu>^a \tag{24}$$

$E^{a''}$ in equation 19 is found by first generating

$$X_{ir} = \sum_{jkl} 4G_{ijkl} [rj;kl] + \sum_{j} 2Q_{ij} <r|h|j> \tag{25}$$

When the density matrices are symmetric equation 17 may be reduced to

$$E^{a''} = \sum_{ir} X_{ir} U^a_{ri} \qquad (26)$$

using equation 25.

The use of one- and two-particle density matrices allows the energy contribution to be found for all 3N different derivatives since the density matrices are independent of any particular derivative. The one- and two-particle density matrices defined by equations 20 and 21 require the coupling coefficients $a^{ij}_{IJ}$ and $b^{ijkl}_{IJ}$ which are exactly those used in determining the CI energy and are defined by loops. The density matrices may be generated by generating the loops in a step which computationally involves slightly greater effort than that reauired to generate the diagonalization tape (17). The density matrices must be transformed to an AO basis in a process related to that involved in the four-index integral transformation step.

The integral derivatives are generated and processed to find $E^{a'}$ and $E^{a''}$ given by equations 24 and 26. The integral derivatives are combined with the appropriate CI density matrix element and processed through equation 24. The integral derivatives are also required to solve the coupled-perturbed Hartree-Fock equations (23). For a closed-shell singlet a "Fock matrix" constructed from integral derivatives is set up

$$B^a_{\lambda\mu} = \langle\lambda|h|\mu\rangle^a + \sum_{\nu\sigma} P_{\nu\sigma}(2[\lambda\mu;\nu\sigma]^a - [\lambda\nu;\mu\sigma]^a) \qquad (27)$$

One such matrix is required for each of the 3N nuclear perturbations under consideration. The theoretical formulation of the coupled perturbed Hartree-Fock equations follows the work of Gerratt and Mills (23). The actual solution of the equations follows a method suggested by Pople and co-workers in their work on analytic SCF second derivatives (25). The method which has many similarities to the Davidson diagonalization scheme (26) is outlined elsewhere (7,15,25). At the SCF or MCSCF level it was unnecessary to solve for these first order changes in the molecular orbitals with respect to nuclear perturbations since the orbitals were variationally optimized. Solution of the CPHF equations is a minor step in terms of computational timing (7,8) in the CI gradient calculation.

## EQUILIBRIUM GEOMETRIES, TRANSITION STATE STRUCTURES, AND VIBRATIONAL ANALYSES USING GRADIENT METHODS.

As an example of a limited MCSCF gradient calculation the $C_{2v}$ least-motion insertion of $CH(^2\pi)$ into $H_2(^1\Sigma_g^+)$ to form methyl radical will be considered (4). Under a least motion constraint, methylidyne is constrained to approach hydrogen along the perpendicular bisector of the $H_2$ internuclear axis. Under this constraint of $C_{2v}$ symmetry the reactants are described by

$$^2\pi\ CH + {}^1\Sigma_g^+\ H_2 \quad 1a_1{}^2\ 2a_1{}^2\ 3a_1{}^2\ 4a_1{}^2\ 1b_1 \tag{28}$$

and the product by

$$^2A_2''\ CH_3 \qquad 1a_1{}^2\ 2a_1{}^2\ 3a_1{}^2\ 1b_2{}^2\ 1b_1 \tag{29}$$

Electronic configuration is not conserved in going from reactants to product and thus the least motion insertion is Woodward-Hoffmann forbidden. A two configuration SCF (TCSCF) calculation is required to even qualitatively describe this reaction. Such a calculation was carried out in a double zeta basis set (29) with gradient optimization of reactants, transition state and product. A vibrational analysis was performed on the transition state structure using force constants obtained as numerical differences of the TCSCF analytic gradients (1) to verify that this point was a true transition state (30). At the transition state the CI coefficients of the two configurations were

$$0.5945\ (\ldots 4a_1{}^2 1b_1) - 0.8041\ (\ldots 1b_2{}^2 1b_1)$$

which clearly show the qualitative necessity of a TCSCF treatment. Moreover, the flatness of the energy surface (in the region of the transition state) makes a gradient method nearly obligatory for an accurate location of the transition state geometry.

Table I summarizes the results of geometry optimization and harmonic vibrational frequency analyses on formaldehyde with SCF and CI gradient methods (27). Both double zeta basis sets and such basis sets augmented by polarization functions were employed. For the DZ + P CI calculations the two core MO (corresponding to the C and O 1s orbitals) were held doubly occupied or frozen and the two highest complementary virtual orbitals deleted. The calculations involve 5214 ($C_{2v}$), 10221 ($C_s$) or 18721 ($C_1$) singly and doubly excited configurations and may reasonably be termed large scale CI. The lower symmetry point groups must be used in certain of the nuclear displacement calculations since Cartesian nuclear displacements were used.

As is frequently observed, the DZ SCF results (due to a cancellation of effects produced by the addition of polarization functions and by electron correlation) are in quite reasonable agreement with the experiment. However, it is clear that for more quantitatively accurate results (especially for the vibrational frequencies) the highest level of theory (DZ+P CI) is preferable. The average error in the harmonic frequencies at the DZ SCF level is 8%

Table I. Theoretical predictions for formaldehyde, $H_2CO$. Harmonic vibrational frequencies are given in $cm^{-1}$.

| | DZ SCF | DZ CI | DZ+P SCF | DZ+P CI | EXPERIMENT |
|---|---|---|---|---|---|
| $r_e$(C-H),Å | 1.084 | 1.104 | 1.094 | 1.100 | 1.099±0.009[a] |
| $r_e$(C-O)Å | 1.217 | 1.249 | 1.189 | 1.212 | 1.203±0.003 |
| $\theta_e$(HCH) degrees | 116.8 | 116.8 | 116.2 | 116.3 | 116.5±1.2 |
| E (hartrees) | -113.8307 | -114.0624 | -113.8953 | -114.1954 | --- |
| $\nu_1(a_1)$ | 3223 | 3028 | 3149 | 3074 | 2944[b] |
| $\nu_2(a_1)$ | 1878 | 1703 | 2006 | 1869 | 1764 |
| $\nu_3(a_1)$ | 1651 | 1544 | 1656 | 1596 | 1563 |
| $\nu_4(b_1)$ | 1324 | 1194 | 1335 | 1243 | 1191 |
| $\nu_5(b_2)$ | 3315 | 3112 | 3226 | 3155 | 3009 |
| $\nu_6(b_2)$ | 1349 | 1263 | 1367 | 1306 | 1287 |

[a] K. Yamada, T. Nakagawa, K. Kuchitsu and Y. Morino: 1971, J. Mol. Spectroscopy, 38, pp. 70.

[b] Experimental harmonic frequencies, J. L. Duncan and P. D. Mallinson: 1973, Chem. Phys. Lett., 23, pp. 597

Table II. Selected transition state predictions for $H_2CO \rightarrow H_2 + CO$.

| | DZ SCF | DZ+P CI | DAVIDSON CORRECTED DZ+P CI |
|---|---|---|---|
| Total Energy (hartrees) | -113.64955 | -114.03913 | -114.07230 |
| Barrier Height (kcal/mol) | 113.7 | 98.1 | 94.2 |
| Zero-Point Corrected Barrier (kcal/mol) | 107.4 | 92.8 | 88.9 |
| Harmonic Vibrational Frequencies ($cm^{-1}$) | | | |
| $\nu_1$(a') | 3156 | 3263 | |
| $\nu_2$(a') | 1948 | 1939 | |
| $\nu_3$(a') | 1371 | 1555 | |
| $\nu_4$(a') | 800 | 876 | |
| $\nu_5$(a') | 2320 i | 2124 i | |
| $\nu_6$(a") | 1015 | ~950 | |

while at the DZ+P CI level this quantity is reduced to just less than 4%. Similar results have also been obtained (27) for $CH_4$, HCN, and $H_2O$. Such accurate CI calculations of the structures and harmonic vibrational frequencies allow reliable predictions to be made for yet to be observed species (27) such as gas phase $NH_4^+$.

Theoretical calculations at the CI level allow for a description in unprecedented detail of the transition states for reactions such as the unimolecular dissociation of formaldehyde to molecular hydrogen plus carbon monoxide (28,31). Table II presents DZ SCF and DZ + P S + D CI results for the energies and harmonic vibrational frequencies of the molecular dissociation transition state. The transition state vibrational frequencies are necessary in order to correct classical barrier heights for the effects of zero-point energy. In the present case such a correction amounts to ~5 kcal/mol. The imaginary frequency corresponding to motion along the reaction coordinate (2124 i $cm^{-1}$ in the present case at the DZ + P CI level) also provides an entry into the dynamics of the system. In fact, theory can now provide (particularly with the aid of analytic gradients) accurate values for all the quantities necessary for first order kinetic theories such as RRKM (32). Table II also shows that Davidson's correction (33) for higher excitation effects decreases the barrier height by ~ 4 kcal/mol.

## CONCLUDING REMARKS

The availability of analytic gradients for wavefunctions including correlation effects opens up new areas of application particularly in the area of chemical reactivity. Over a reaction pathway or surface a one configuration SCF treatment will often be inadequate but with CI or MCSCF analytic energy gradients calculable, the quantitatively accurate location and characterization of transition states and reaction paths becomes a reality.

## ACKNOWLEDGEMENTS

The work described in this review was performed by the members of Professor H. F. Schaefer's group in Berkeley and Austin. Their names may be found in the literature cited.

## REFERENCES

1. P. Pulay in Modern Theoretical Chemistry, edited by H. F. Schaefer III (Plenum, New York, 1977), Vol. 4, pp. 153-185.

2. P. Pulay: 1969, Mol. Phys., 17, pp. 197.

3. For example: P. Botschwina: 1974, Chem. Phys. Lett., 29, pp. 98.

4. J. D. Goddard, N. C. Handy and H. F. Schaefer III: 1979, J. Chem. Phys., 71, pp. 1525.

5. S. Kato and K. Morokuma: 1979, Chem. Phys. Lett., 65, pp. 19.

6. B. R. Brooks, W. D. Laidig, P. Saxe, N. C. Handy and H. F. Schaefer III: 1980, Physica Scripta, 21, pp. 312.

7. B. R. Brooks, W. D. Laidig, P. Saxe, J. D. Goddard and H. F. Schaefer III in Proceedings of the Conference on the Unitary Group for the Evaluation of Electronic Energy Matrix Elements. Bielefeld, W. Germany (1979). J. Hinge, editor, Springer-Verlag Lecture Notes Series.

8. B. R. Brooks, W. D. Laidig, P. Saxe, J. D. Goddard, Y. Yamaguchi and H. F. Schaefer III: 1980, J. Chem. Phys., 72, pp. 4652.

9. A more specialized formulation for the gradient of CI wavefunctions which will not be discussed here is also available: R. Krishman, H. B. Schlegel and J. A. Pople: 1980, J. Chem. Phys., 72, pp. 4654.

10. F. W. Bobscowicz and W. A. Goddard III in Modern Theoretical Chemistry edited by H. F. Schaefer III (Plenum, New York, 1977), Vol. 3, pp. 79-129.

11. Reference 10, p. 91.

12. W. A. Goddard III, T. H. Dunning and W. J. Hunt: 1969, Chem. Phys. Lett., 4, pp. 231.

13. For example: P. Pulay, TEXAS. An Ab Initio Gradient (Force) Program, Austin, Texas, 1976.

14. M. Dupuis and H. F. King: 1978, J. Chem. Phys., 68, pp. 3998.

15. See also the article by H. B. Schlegel in this volume.

16. J. H. Meadows and H. F. Schaefer III: 1976, J. Chem. Soc., 98, pp. 4383.

17. B. R. Brooks and H. F. Schaefer III: 1979, J. Chem. Phys., 70, pp. 5092.

18. Although not discussed in this article, the two-particle density matrix also plays a central role in a scheme for large

scale MCSCF computations. See: B. R. Brooks, W. D. Laidig, P. Saxe, and H. F. Schaefer III: 1980, J. Chem. Phys., 72, pp. 3837.

19. For example: M. Moshinsky Group Theory and the Many-Body Problem (Gordon and Breach, New York, 1968).

20. J. Paldus in Theoretical Chemistry: Advances and Perspectives. Vol. 2, H. Eyring and D. J. Henderson, eds. (Academic, New York, 1976), pp. 131.

21. I. Shavitt: 1977, Int. J. Quantum Chem. Symp., 11, pp. 131.

1978, Int. J. Quantum Chem. Symp., 12, pp. 5.

1979, Chem. Phys. Lett., 63, pp. 421.

22. A. Bunge: 1970, J. Chem. Phys., 53, pp. 20.

23. J. Gerratt and I. Mills: 1968, J. Chem. Phys., 49, pp. 1719.

24. R. McWeeny and B. T. Sutcliffe. Methods of Molecular Quantum Mechanics. (Academic Press, London, 1969).

25. J. A. Pople, R. Krishnan, H. B. Schlegel and J. S. Binkley: 1979, Int. J. Quantum Chem. Symp., 13, pp. 225.

26. E. R. Davidson: 1975, J. Comput. Phys., 17, pp. 87.

27. Y. Yamaguchi and H. F. Schaefer III: J. Chem. Phys., in press.

28. J. D. Goddard, Y. Yamaguchi and H. F. Schaefer III: J. Chem. Phys., in press.

29. S. Huzinaga: 1965, J. Chem. Phys.,42, pp. 1293.
T. H. Dunning: 1970, J. Chem. Phys., 53, pp. 2823.

30. J. N. Murrell and K. J. Laidler: 1968, Trans. Faraday Soc., 64, pp. 371.

31. Earlier work is referenced in: J. D. Goddard and H. F. Schaefer III: 1979, J. Chem. Phys., 70, pp. 5117.

32. P. Saxe, Y. Yamaguchi, P. Pulay and H. F. Schaefer III: 1980, J, Am. Chem. Soc., 102, pp. 3718.

33. E. R. Davidson, in The World of Quantum Chemistry, edited by R. Daudel and B. Pullman (Reidel, Dordrecht, Holland, 1974).

# AN INTERNAL INVARIANT REACTION PATHWAY BY THE ACCELERATION METHOD

Michel SANA,

Université Catholique de Louvain,
Laboratoire de Chimie Quantique,
Bâtiment Lavoisier,
Place Louis Pasteur, 1,
B-1348 Louvain-la-Neuve (Belgium)

ABSTRACT.

This work illustrate the fact that some properties of a potential energy surface are not independent of the choice of the coordinate frame. So, the reaction pathway sometime described as a steepest descent way does not correspond to an invariant curve under coordinate transformations. Then we propose to use the concept of instantaneous internal acceleration. Our work is closely related to the one of Fukui (1). It may also be considered as an illustration of the Mezey's lecture (2).

## INTRODUCTION.

A lot of potential energy surface studies use the concept of reaction pathway, but this concept often remains confuse (3). Firstly Laidler and Murell show that any transition point must correspond to a minimax of first kind (4); then McIver and Komornicki request that the reaction pathway must be the minimum energy pathway which is connecting the reactants and the products through the lowest transition point (5). But we have to wait Fukui and coworkers to find a satisfactory definition for the intrinsic reaction pathway (8). Nevertheless many theoretical works seems to ignore the remarks of those papers. In the present work we will try to precise for any set of coordinate frame the concept or reaction pathway and by using some example to clarify such a notion (9).

## GRADIENT AND STEEPEST DESCENT WAY.



*I. G. Csizmadia and R. Daudel (eds.), Computational Theoretical Organic Chemistry, 175–181.*

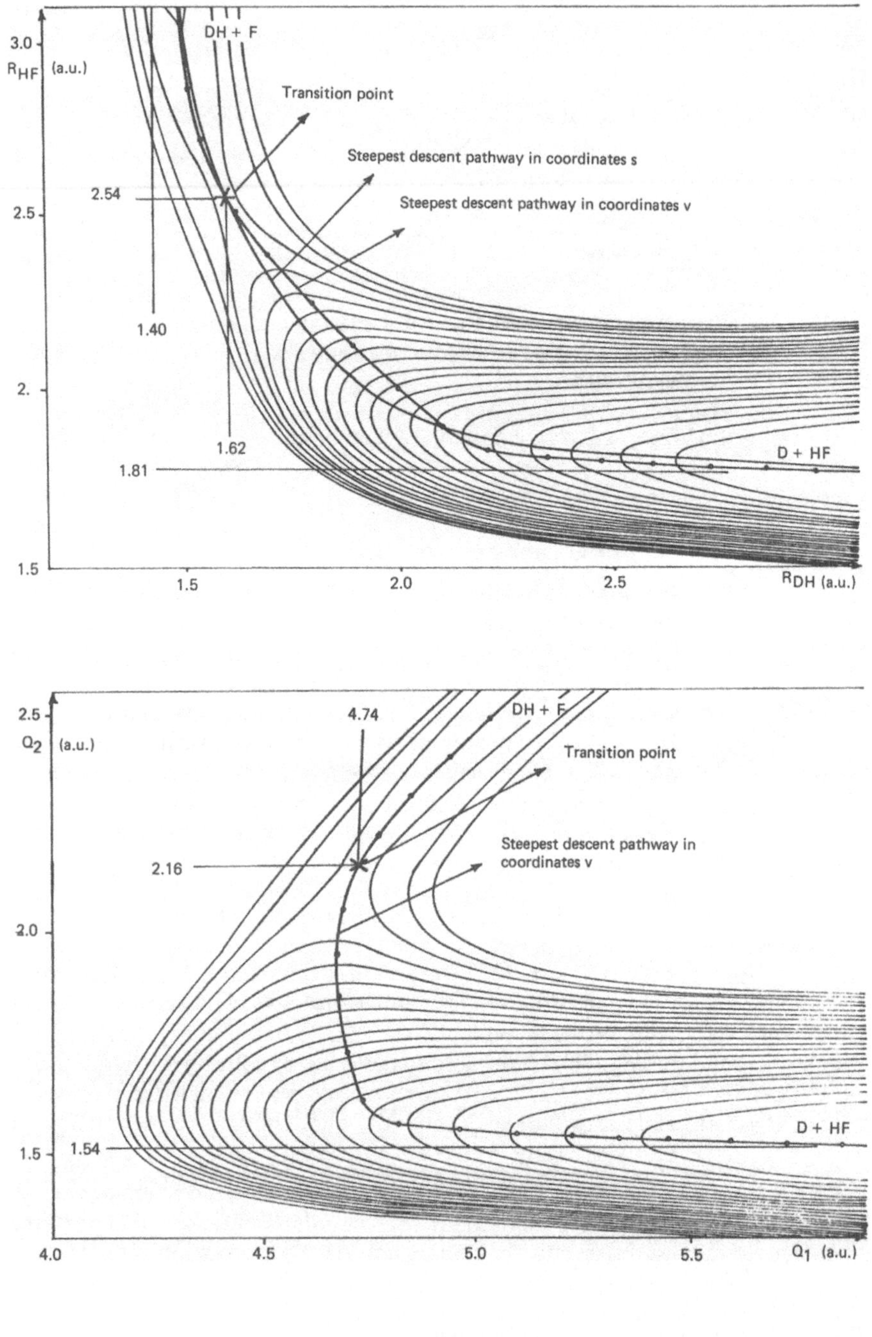


Figure 1. Steepest descent pathway for the colinear D + HF reaction (6).

Let us suppose a potential energy surface of a chemical system of interest which contains N atoms. The potential energy may be expressed as a function of 3N-6 internal coordinates s (continuous in the whole space); thus

$$E = E(s) \qquad [1]$$

Anywhere we can compute the first derivative vector (g) and the second derivative matrix (H) with respect to the internal coordinates :

$$g_s = \nabla_s E_s \qquad [2]$$

$$H_s = (\nabla_s \nabla_s') E_s \qquad [3]$$

The subscript s indicates that the function or
The operator is given in terms of the variables s, and
The superscript ' stands for the transposition operation.

Let us now change of coordinate frame and define a new set v related to the previous one by the relation :

$$v = P^{-1} s \qquad \text{or} \qquad s = Pv \qquad [4]$$

According to the chain rule the derivatives with respect to s are bonded to those with respect to v by the relations :

$$g_v = P' g_s \qquad [5]$$

$$H_v = P' H_s P \qquad [6]$$

On another hand, when we define the reaction pathway as the steepest descent way for connecting the transition point to the reactants and to the products, we have to follow the gradient vector (g). The direction of $g_s$ in v space (noted $d_s^{(v)}$) is given by :

$$d_s^{(v)} = P^{-1} g_s \qquad [7]$$

Comparing the equations [5] and [7] we see that the direction of $g_s$ in v space corresponds to the gradient $g_v$ (everywhere on the potential energy surface) only if :

$$P' = P^{-1} \qquad [8]$$

That means only if the P matrix corresponds to an unitary transformation. When this is not true, $d_s^{(v)}$ equals $g_v$ only at the stationary points (where $g_s$ or $g_v$ is zero). Similarly $H_s$ and $H_v$ may be transformed to each other like a tensor only if we meet the same restriction [8].

So by using any coordinate frame we are enable to find correctly a stationary point location. Nevertheless the gradient direction and the eigendirections of the force constant matrix are not invariant.

As an example let us consider the colinear D + HF reaction. We choose two different internal coordinate sets :

- The first noted s corresponds to the interatomic DH and HF distances :

$$s' = (R_{DH}, R_{HF})$$

- The second noted v is a coordinate frame which diagonalizes the kinetic energy :

$$v' = (Q_1, Q_2)$$

It is related to the previous one by the next P matrix :

$$P = \begin{pmatrix} 1 & -\operatorname{tg}\phi \\ 0 & 1/\beta\cos\phi \end{pmatrix}$$

Where $\sin\phi = \{m_D\, m_F / (m_D + m_H)(m_H + m_F)\}^{1/2}$

and $\beta = \{m_F (m_D + m_H) / m_D (m_H + m_F)\}^{1/2}$

Now starting at the transition point let us follow the gradient vector in order to generate the steepest descent curve which is connecting this point to the products and to the reactants. So we get the curves presented on the figure 1. It clearly appears that the steepest descent way generated in the $E_s$ space does not correspond to the other one generated in the $E_v$ space except at the stationary points.

## THE ACCELERATION METHOD.

The question is now how to find an intrinsic reaction pathway. First of all let us remember that the reaction pathway has no physical significance; this may only be considered as a reference curve with is connecting the transition point to the reactants and the products. Now let us replace the gradient vector by the internal acceleration. The internal acceleration of a system of points corresponds to the evolution capability of the system for a zero instantaneous velocity. Its expression may be found by writing the notion equations :

$$\frac{d}{dt}\frac{\partial T}{\partial \dot{s}} + \frac{\partial E}{\partial s} = 0 \qquad [9]$$

where T and E stand respectively for the kinetic and the potential energy.

Usualy T is given in terms of the cartesian displacements ($\xi$) by :

$$2T = \dot{\xi}' M \dot{\xi} \qquad [10]$$

where M is the $3N \times 3N$ diagonal matrix of the atomic masses.

As $\xi$ may be considered as a function of the internal displacement (S with $S = s - s_o$), by the chain rule we find the formula connecting $\dot{\xi}$ and $\dot{S}$.

$$\dot{\xi} = (\nabla_s \xi')' \dot{S} = A\dot{S} \qquad [11]$$

The A matrix is a first order derivative matrix of $\xi$ with respect to S. So we can introduce such an expression in the kinetic expression :

$$2T = \dot{S}' A' M A \dot{S}' \qquad [12]$$

or

$$2T = \dot{S}' G^{-1} \dot{S} \qquad [13]$$

where G stands for the usual G matrix of Wilson (7).

For the potential energy we use a development in Taylor's series in order to introduce the g vector and the H matrix. We truncate the expansion after the second order assuming that any surface part close to the point of interest has a quadratic form :

$$E_s = E_o + g_s' S + 1/2\, S' H_s S \qquad [14]$$

with $S = s - s_o$

Introducing now T and E in the motion equations [9] and left multiplying by G, we have :

$$\ddot{S} + G_s g_s + G_s H_s S = 0 \qquad [15]$$

such an expression remains invariant with respect to any set of coordinate frame; indeed as by equations [4] and [13] :

$$G_v = (P^{-1})\, G_s (P^{-1})'$$

it follows that G g is transforming like a vector :

$$G_v g_v = (P^{-1})\, G_s g_s$$

and that G H is transforming like a tensor :

$$G_v H_v = (P^{-1})\, G_s H_s (P)$$

So we will adopt the direction of $\ddot{S}$ in order to generate the reaction pathway instead of the gradient direction. This definition corresponds to use a coordinate frame (R) which diagonalizes the kinetic energy and for which the diagonal terms are all one :

if $R = G^{-1/2}\, S$

then $2\,T = \dot{R}'\,E\,\dot{R}$

and $\ddot{R} + g_R + H_R\, R = 0$

The unity condition on the diagonal terms is not necessary to the intrinsic requirement but it is convenient for physical reasons. This reaction pathway definition is close to the one of Fukui and coworkers (1). They use the 3N weighted cartesian coordinates $q(q = M^{1/2}\xi)$ instead of the 3N-6 internal one s; in such a case the $G_q$ matrix is equal to the unity matrix E :

$$2\,T = \dot{q}'\,E\,\dot{q}$$

Practically to find the reaction pathway we start at the transition point and we follow point to point the direction of the acceleration vector :

- As at the transition point ($s_{\neq}$) $g_S$ equals zero, we leave it by following the eigendirection $K^*$ which corresponds to the negative eigenvalue of the GH product. The first displacement is given by :

$$s_{\pm}^{(o)} = s_{\neq} \mp \frac{\varepsilon^2\, K^*}{(K^{*\prime}\, K^*)^{1/2}}$$

  The scalar $\varepsilon^2$ stands for an infinitely small quantity and the term $(K^{*\prime}\, K^*)^{1/2}$ is used to normalize the step size.

- From the new point ($s_{\pm}^{(o)}$) we follow down the direction of the G g product :

$$s_{\pm}^{(i+1)} = s_{\pm}^{(i)} - \frac{\varepsilon^2\,(G\,g)_i}{(g'\,G'\,G\,g)_i^{1/2}}$$

- We stop the search when the current energy is sufficiently close to the energy of the reactants or the products.

CONCLUSION.

Without choosing carefully the coordinate frame we are enable to locate the stationary points and then to compute the activation and the reaction energies. If further we want to characterize a reaction pathway, we have to select a coordinate frame which diagonalizes the kinetic energy. It seems that the choice of the acceleration method here proposed is a suitable choice. If the so generated curve may be considered as an intrinsic curve, it becomes mass depending.

ACKNOWLEDGMENT.

We wish to thank the FNRS for their constant support and for the provision of a research-grant.
We also thank Professors Daudel and Leroy for their interest in this work.

REFERENCES.

(1) Fukui K., Kato S., Fujimoto H.:
(1975) J. Amer. Chem. Soc., 97, p. 1.
(2) Mezey:
This book.
(3) Pechukas P.:
(1976) J. Chem. Phys., 64, p. 1516.
(4) Murrel J.N., Laidler K.J.:
(1968) Trans. Farad. Soc., 64, p. 371.
(5) McIver J.W., Komornicki Jr. A.,
(1972) J. Amer. Chem. Soc., 94, p. 2625.
(6) Reckinger G.:
Ph. D. Thesis, UCL (In preparation).
(7) Wilson E.B., Decius G.C., Cross P.C.:
(1975) "Molecular Vibration", McGraw-Hill, London.
(8) Tachibana A., Fukui K.:
(1978) Theoret. Chim. Acta, 42, p. 311;
(1979) Theoret. Chim. Acta, 51, p. 189 and p. 275.
(9) Sana M., Reckinger G., Leroy G.:
(1980) Theoret. Chim. Acta (Accepted).

# A NUMERICAL APPROACH FOR FINDING STATIONARY POINTS AND FOR COMPUTING FORCE CONSTANT MATRICES: THE EXPERIMENTAL DESIGNS IN LOCAL ANALYTICAL SURFACES. APPLICATION TO THE VIBRATIONAL AND TO THE THERMODYNAMICAL ANALYSIS

Michel SANA

Laboratoire de Chimie Quantique,
Université Catholique de Louvain,
Bâtiment Lavoisier, Place Louis Pasteur, 1,
1348 Louvain-la-Neuve (Belgium)

ABSTRACT.

It exists many reasons to compute the gradient and the force constant matrix of a chemical system. First of all those quantities are interesting in order to locate a stationary point. Further they let us enable to perform the vibrational and the thermodynamical analysis.
As all the available programs and methods are not implemented for performing those computations we propose here a purely numerical approach.

## I. A LOCAL ANALYTICAL SURFACE.

To describe the potential energy hypersurface of a molecular system of interest we are using a set of k internal coordinates noted s. By example such a coordinate frame may contain the bond lengths, the bond angles and the dihedral angles characterizing the molecular geometry. So any nuclear structure is defined by at most 3N-6 components if N is the atom number.

Now let us suppose a region of immediate interest in the k-dimensional internal coordinate space around a central point noted $s_o$. Locally we are enable to develop the potential energy surface in Taylor's serie. If the region of interest is not too large we shall truncate the expansion up to the second order. Calling x a displacement from the expansion center in terms of internal coordinates ($x = s - s_o$) we get :

$$E(x) = E_o + g_o' x + 1/2 \ x' H_o x \qquad [1]$$



*I. G. Csizmadia and R. Daudel (eds.), Computational Theoretical Organic Chemistry, 183–196.*

where $E_o$ stands for the potential energy value at the point $s_o$;
$g_o$ stands for the column vector of the first order derivatives with respect to the internal coordinates at the central point $s_o$;
$H_o$ similarly stands for the second order derivatives matrix.

If we are particularly interested by a stationary point, the gradient vector must be characterized by the next requirement:

$$|g_e| = 0 \qquad [2]$$

As the general expression of the potential gradient in the region of interest is given by :

$$g_x = g_o + H x$$

The stationary point location becomes :

$$x_e = - H_o^{-1} g_o \qquad [3]$$

Due to the second order approximation such an expression remains valid as long $x_e$ is closed enough to zero.

The problem is now to build a local quadratic form which fits as well as possible the potential energy surface around the expected stationary point of interest. So we replace our preceeding expression for the potential energy surface by an estimator of it :

$$\hat{E}(x) = b_o + \sum_i b_i x_i + \sum_{i \leq j} b_{ij} x_i x_j \qquad [4]$$

The unknown expansion coefficients b may be considered respectively as estimator of $E_o$ (for $b_o$), $g_o$ (for $b_i$) and $H_o$ (for $b_{ij}$) :

$$b_o = \hat{E}_o$$

$$b_i = \hat{g}_{o,i}$$

$$b_{ij} = \hat{H}_{o,ij}/(1 + \delta_{ij})$$

In order to find a more compact notation let us introduce the model vector $x^{[2]}$ which contains the functional values at a given point in the k-dimensional space of the internal coordinate frame and the coefficient vector b :

$$x^{[2]'} = \{ 1 , x_1 \ldots x_k , x_1^2 \ldots x_k^2 , x_1 x_2 \ldots x_k x_{k-1} \}$$

$$b' = \{ b_o, b_1 \ldots b_k , b_{11} \ldots b_{kk} , b_{21} \ldots b_{kk-1} \}$$

In matricial form we write now :

$$\hat{E}(x) = x^{[2]'} . b \qquad [5]$$

Extanding such a notation to a set of n different points we have :

$$\hat{E} = X . b \qquad [6]$$

Where X is the global model matrix. Each line of such a matrix corresponds to the vector $x^{[2]'}$ for a particular point of coordinate $x = (x_1 \ldots x_k)$.

To find the value of the b coefficients, the point number must be at least equal to the coefficient number ( $\binom{k+2}{2}$ for a quadratic model), then we have a linear least square fit problem; by using the Gauss criterium we see that the b coefficients are given by the next formula :

$$b = (X' X)^{-1} X'E \qquad [7]$$

II. THE EXPERIMENTAL PLANES.

The quality of the estimator given in [7] may be evaluated by the expression of the b variance. So if β corresponds to the exact value of b we have

$$\mathrm{Var}(b) = \mathcal{E}\,[(b-\beta)(b-\beta)'] = (X'X)^{-1}\sigma^2 \qquad [8]$$

where $\sigma^2$ stands for the error variance.

We find that the variance of the regression coefficients b is closely bonded to the structure of the variance-covariance matrix $(X' X)^{-1}$. If we expect a good estimation of the b values, the points distribution, which determines the model matrix, is not free at all. One call experimental plane the point distribution patern in the coordinate space. In order to improve the quality of the coefficients b we want to expect that the experimental plane has the following properties :

(i) The D-optimality condition :

It should enable to estimate as closely as possible the regression coefficients; that means the determinant of X' X matrix must be maximized.

(ii) The orthogonality condition.

It is convenient to estimate independently all the coefficients. That is done if $(X' X)^{-1}$ is as close as possible of a diagonal matrix.

(iii) The rotatable condition.

We may request that the variance function at every point of the experimental domain must be constant at a constant distance from the origin of the design. Such planes are called rotatable, by example on energy we have the condition :

$$\mathrm{Var}\,[\,\hat{E}(x)\,] = x^{[2]'}\,(X'\,X)^{-1}\,x^{[2]}\,\sigma^2$$

$$= f_o\,(\sum_{i=1}^{k} x_i^2)$$

Similar rotatable conditions may also be found on the successive derivatives.

(iv) The R-efficiency.

The experimental design should not contain an excessively large number of experimental points ; we measure this property using the R-efficiency defined as the ratio between the number of points to be computed and the number of coefficients to be estimated :

$$R_{eff} = 100\,\frac{\text{point number}}{\text{regression coefficient number}}$$

All those requirements cannot be necessarily met at the same time but one must find the best compromize according to the type of information we need.

The classical experimental planes, built to give at the model matrix X a good structure according to our requirements, are known as equiradial planes. One regard such designs as built up from a number of component sets of points, each set having all its points equidistant from the origin. For a second order fitting, one needs two sets of equiradial points at least, only one set provides singularity in the X' X matrix. In the next we show the main known experimental planes. Details concerning such planes may be found elsewhere (1).

First of all we have the composite planes of Box and Hunter (2). In a k-dimensional space they are built by combining the next geometrical figures (figure 1) :

(i) a central point.

(ii) a square ($k = 2$), a cube ($k = 3$) or an hypercube ($k > 3$) with $2^k$ vertices.

(iii) a cross ($k = 2$) or a cross-polytope ($k > 2$) with $2\,k$ vertices. This cross is perpendicular to the faces of the cube.

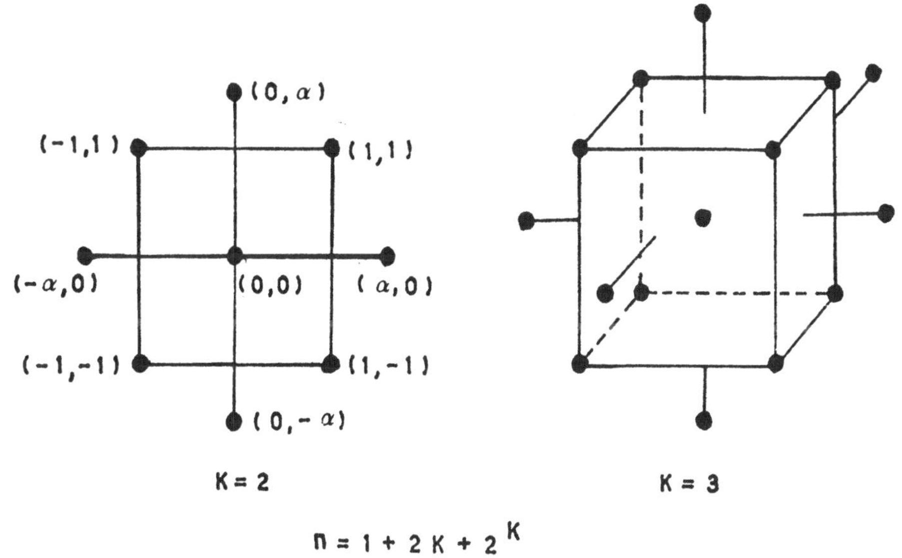


Figure 1. The two- and three- dimensional composite plane.

In those designs the X'X matrix has a quasi diagonal form. We find only two types of off-diagonal terms; the first one corresponds to the covariance terms between the regression coefficients $b_o$ and $b_{ii}$; the second one corresponds to the covariance terms between the coefficients $b_{ii}$ and $b_{jj}$. We dispose of one degree of freedom to improve such a matrix: the length of the arm cross α (figure 1). According to the selected values one can satisfy the orthogonality requirement, the rotatable condition on the energy or on the slope. As shown in reference (1) we have choosen the energy rotatable designs with provide a best value for the determinant of the X' X matrix. Lastly it remains possible to get the orthogonality requirement by replication of the central point $N_o$ times if :

$$N_o = 4(1 + 2^{k/2}) - 2k$$

This fact slightly improves the determinant of the X' X matrix which becomes quasi-orthogonal.
If we use a complete composite design the R efficiency goes down rapidly when k increases. Then we may use a fractional replicate $(1/2)^P$ of the $2^k$ factorial plane in such a way we reduce significatly the number of experimental points (see ref. 1).

In order to explore a larger portion of the potential energy hypersurface we need an experimental plane which can be moved without computing an excessive number of points. The Doehlert's experimental plane meets this requirement (3). The basic figure is a

simplex with k + 1 vertices which generates by rotation a plane of (k + 1)k + 1 vertices. On figure 2a we see the Doehlert's plane in a two dimensional space. Such a design always in more economical that the complete composite plane but not that the fractionnal replicate of such a plane. The Doehlert's plane becomes rotatable only by multiplying the computations at the origin (except for k = 2) and always remains less efficient than the complete composite plane if we speak in terms of D-optimality.
Nevertheless we may translate the plane in the k-dimensional space recomputing only k(k - 1) + 1 points (figure 2b); in this way we generate the uniform shell of Doehlert. So when we have no information concerning the position of the stationary point and when we think we have to explore a certain part of the response surface, such a design could be interesting. Moreover it permits to add new factors after a first investigation without restarting the whole process. To increase the space dimension by one factor we only need to compute 2(k + 1) new points (figure 2c).

Finally it exists a last class of experimental designs called saturated or quasi-saturated planes. Such designs contain the same or approximately the same number of experimental points as the number of regression coefficients (4). They are efficient in order to give information about a stationary point location but sometime are less efficient in order to compute properly the force constant matrix. An example of such planes is given on the figure 3.

Let us now come to some practical considerations. At this time we have studied a few small molecules. The Doehlert's planes are used to the first investigation of the potential energy surface. According to our experience the quadratic approximation often may be considered as efficient in a domain as large as ± 10° on the angles and ± 0.05 Å on the bond lengths.As an example we give in table I the requested work to find the equilibrium structure of the ONNO molecule.
We start with a normal single bond length for NN, with a double bond length for NO, and with an ONN angle slightly less then 120°. Two successive Doehlert's planes are needed to get; a sufficiently stable minimum on the potential energy surface. Finally, we decide to check the accuracy of our result by a composite plane in the full geometrical parameters space. We reduce the search domain approximately by six times in all the space directions. The final result is the same as the preceeding one at $10^{-3}$ Å and 1°. This procedure may also be applied to a transition structure search. When we have to compute a force constant matrix we use the composite planes in order to get a more accurate estimation. In table II we give the experimental plane properties in the case of some small molecules. Always we consider a full geometrical optimization process. Without taking into account the symmetry operator which are present in the normal trial central structure we have

"n" different nuclear structures to be computed for estimating "l" regression coefficients.

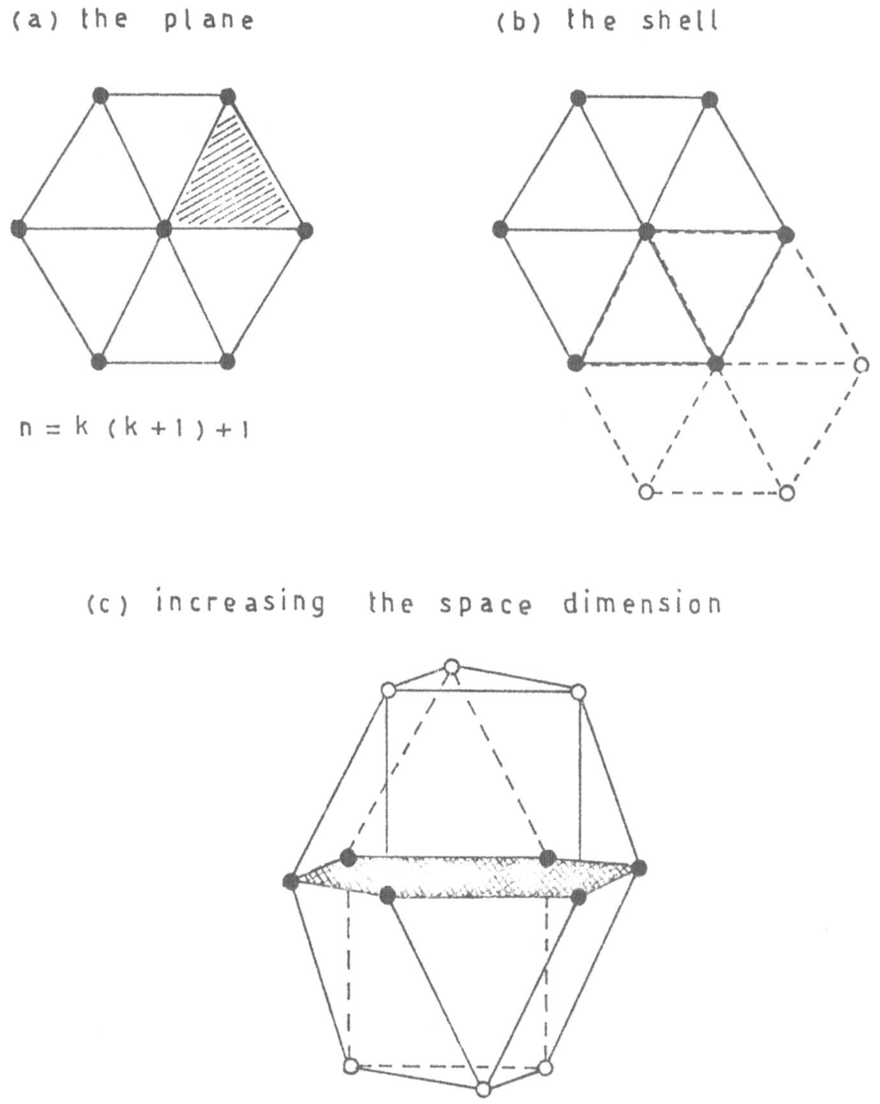


Figure 2. The Doehlert's experimental plane.

(a) A two-dimensional plane.
(b) The evolutive property of the Doehlert's plane.
(c) From a two-dimensional plane to a three-dimensional plane.

Table I. The search of the equilibrium structure for the ONNO molecule (trans configuration) at the 6-31G level.

| Parameter | Trial structure | First Doehlert's plane | Second Doehlert's plane | Final composite plane |
|---|---|---|---|---|
| Space dimension | - | 3 | 3 | 6 |
| $r_{ON}$ (Å) | 1.20 | 1.169 | 1.173 | 1.172 |
| $r_{NN}$ (Å) | 1.50 | 1.566 | 1.577 | 1.577 |
| ONN (°) | 117.00 | 107.79 | 108.86 | 108.74 |
| Step size | - | ± 0.05 Å | | ± 0.006 Å |
| | - | ± 6.0° | | ± 1.0° |

Table II. The composite experimental planes properties for some small molecules (2,5,6).

| Molecule | Space dimension k | Coefficient number 1 | Point number n | Non equal coefficients 1(≠) | Non equal structure n(≠) | Multiple correlation coefficients $\xi$ |
|---|---|---|---|---|---|---|
| OH | 1 | 3 | 5 | 3 | 5 | 0.999 |
| $H_2O$ | 3 | 10 | 15 | 7 | 9 | 0.999 |
| NCCN (linear) | 5 | 21 | 43 | 12 | 16 | 0.998 |
| CNCN (linear) | 5 | 21 | 43 | 21 | 25 | 0.998 |
| ONNO (cis or trans) | 6 | 28 | 45 | 18 | 28 | 0.998 |
| $CH_3$ | 6 | 28 | 45 | 13 | 26 | 0.998 |
| $CH_4$ | 9 | 55 | 83 | 10 | 48 | 0.997 |
| $CH_3$...H...OH | 15 | 136 | 159 | 122 | 151 | 0.998 |

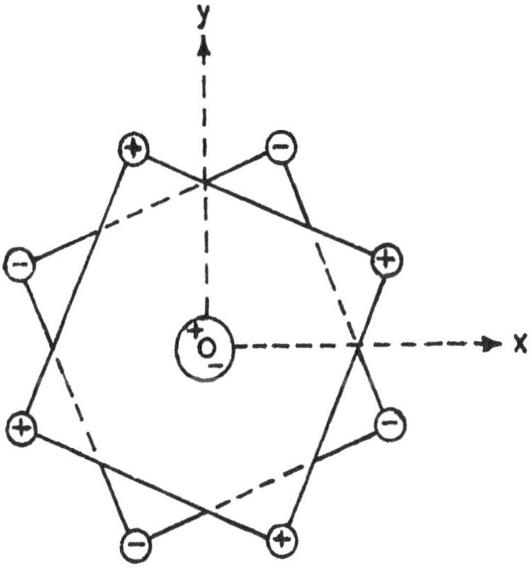


Figure 3. Example of a quasi-saturated plane: the hybrid plane 311 B.

Nevertheless some of the nuclear structures are equivalent and similarly some regression coefficients must be equal that means we must only compute n(≠) different geometries for estimating l(≠) different coefficients. The composite planes given in the table II have an effective R-efficiency of 64 %.

## III. THE VIBRATIONAL ANALYSIS.

Having at our disposal the force constant matrix we can easely find the vibrational properties of the chemical system of interest (7). We will exclude of our consideration the global translation and rotation of the supersystem. So the kinetic energy expression depends only of the 3N-6 internal coordinates "s" or the 3N-6 internal displacement ($S = s - s_e$). As generally T may be expressed in terms of cartesian displacement by the expression :

$$2T = \dot{\xi}' M \dot{\xi}$$

where $\dot{\xi}$ stands for the cartesian speed vector and M for the atomic mass matrix

and as the coordinate frames S and $\xi$ are bonded together by the B matrix of Wilson (7) :

$$S = B\xi$$

then the kinetic energy can be written in terms of the internal speed (S); introducing the usual G matrix notation we get :

$$2\,T = \dot{S}'\,G^{-1}\,\dot{S}$$

where $G^{-1}$ is no other than a metric matrix.

Using the stationary point as the new origin for the Taylor's expansion of the potential energy surface, the equation [1] becomes :

$$V = E(S) - E_e = 1/2\; S'\,H\,S$$

Such an expression is a quadratic approximation for the potential at any stationary point. It is also named harmonic approximation. In the case of polyatomic systems it is experimentaly assumed to be sufficient (8).
Introducing now T and V in the Lagrange's equations and premultiplying by G we find a simple form to the motion equation :

$$\frac{d}{dT}\,\frac{\partial T}{\partial \dot{S}} + \frac{\partial V}{\partial S} = 0$$

gives: $\ddot{S} + G\,H\,S = 0$

We have a system of second order coupled equations. If it exists a matrix L such that this diagonalizes the G H product (9), then we can find a new coordinate frame Q bonded to S by $S = L\,Q$, such that our system of equations becomes uncoupled :

$$\ddot{Q} + \Lambda\,Q = 0$$

where : $\Lambda = L^{-1}\,G\,H\,L$

$\Lambda$ being a square diagonal matrix.

The coordinates Q are the so called normal coordinates; the classical frequency associated to any vibrator is given by the square root of $\Lambda$ divided by two $\pi$. If the classical description leads to a continuous energy spectrum nevertheless it gives a right value for the harmonic frequencies.

Practically we have use such an approach in the preceeding cases. On the figure 4 we show the correlation existing between experimental and theoretical values. A first order regression formula gives a slope of .92 :

$$\tilde{\nu}_{exp} = -\,45.99 + 0.92227\,\tilde{\nu}_{th} \qquad (cm^{-1})$$

A more flexible fitting gives the next regression equation (with a very small independent term) :

$$\tilde{\nu}_{exp} = 1.67 + 851.89\ 10^{-3}\,\tilde{\nu}_{th} + 26.70\ 10^{-6}\,\tilde{\nu}^{2}_{th} - 2.92\ 10^{-9}\,\tilde{\nu}^{3}_{th}\ (cm^{-1})$$

We see that all theoretical frequencies except for CN, are overestimated by a values of about 10 %.

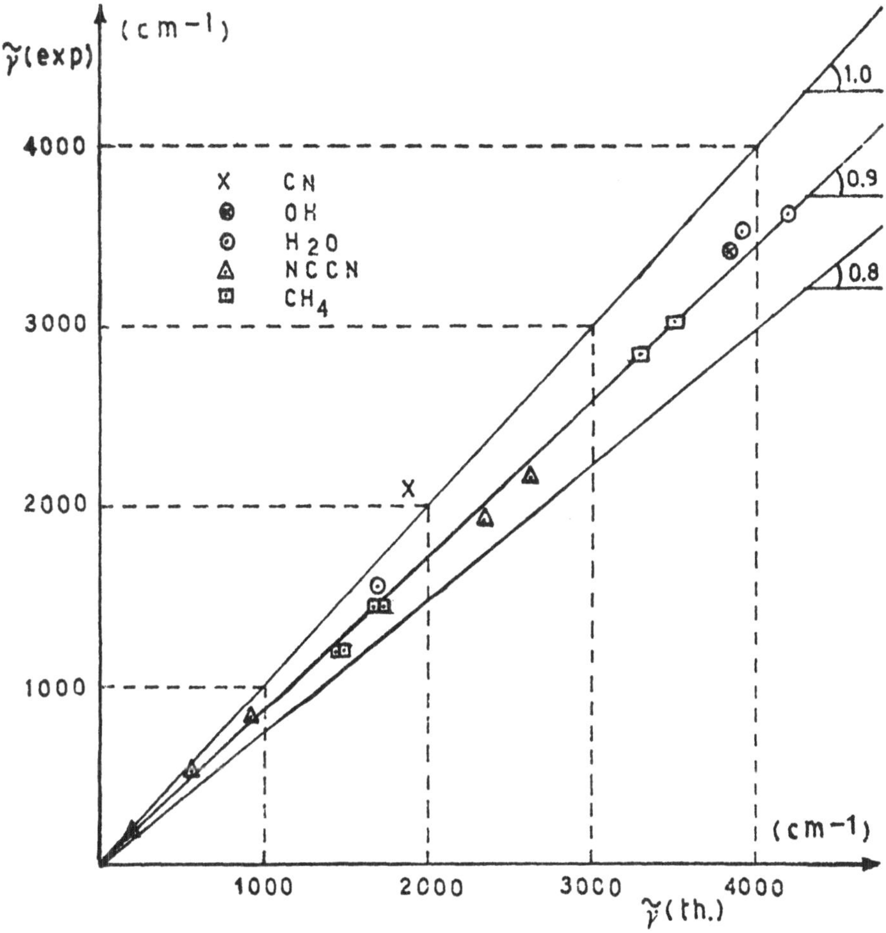


Figure 4. Correlation between the theoretical (at the 6.31G level) and the experimental values for the fundamental frequencies of the nuclear vibrations.

By a similar proceedure as this used to compute the gradient and the force constant matrix, we can also treat the components of the dipole moment. So we are enable to get the first and second order derivatives of the dipolar components in terms of the normal coordinates. Those quantities are bonded to the vibrational probabilities of transition (7). Such an analysis is done elsewhere (6). Its seems to be qualitatively accurate.

## IV. THE THERMODYNAMICAL ANALYSIS.

By using the statistical thermodynamic we can go further if we are disposing of the molecular geometry and of the normal frequencies. In fact if we have a good idea of the energy levels in our compound we are enable to compute the partition function for any temperature range in which our approximations remain valid. That means if the motion separation, the harmonic vibrator approximation and the perfect gas approximation may be considered as accurate. All thermodynamical quantities are bonded to the partition function (see 10 for more details) :

$$H = RT^2 \left( \frac{\partial \ln Q}{\partial T} \right)_V + RT$$

$$S = R\ln Q + U/T$$

$$C_p = \left( \frac{\partial U}{\partial T} \right)_V + R$$

In the table III we give the results we have obtained for some small molecules. The errors are chiefly due to the over-estimation of the fundamental frequencies. Their are mainly important for the zero point energy estimation and for the vibrational-heat capacity and entropy. The geometrical results may also influence the rotational entropy by the way of the inertial tensor; nevertheless our structures are enough accurate. Overall the standard error on the thermal corrections, on the heat capacity and on the entropy are respectively less then 1 kcal/mole and 1 cal/mole °K.

## CONCLUSION.

This work shows that the first and second derivatives with respect to the internal parameters may always be computed with a satisfactory level of accuracy by using a local analytical potential energy hypersurface. For the small molecules our experience is that a purely numerical approach can be done. For larger molecules (up to 9 atoms) we think that an analytical approach may be more efficient (12). So knowledge of the force constant matrix for the quilibrium structure let us enable to perform the classical vibrational analysis and the statistical thermodynamical analysis. This must be of particular interest for molecules experimentaly too difficult to be handled.

Table III. The thermodynamical properties of some small molecules at 298.16°K.

| | Thermal corrections (kcal/mole) | | $C_p^o$ (cal/°K mole) | | S° (cal/°K mole) | |
|---|---|---|---|---|---|---|
| | Th. | Exp. | Th. | Exp. | Th. | Exp. |
| $CH_4$ | 32.94 | 29.49 | - | - | - | 44.48 |
| OH | 7.73 | 7.53 | 8.22 | 7.17 | 44.36 | 43.88 |
| $CH_3$ | 23.52 | 23.18 | 8.81 | 9.25 | 46.05 | 46.38 |
| $H_2O$ | 16.39 | 15.24 | 7.99 | 8.01 | 44.93 | 45.11 |
| CN | 5.03 | 4.82 | 9.00 | 9.00 | 48.40 | 48.40 |
| NCCN | 14.11 | 12.74 | 12.90 | 13.60 | 56.90 | 57.90 |
| CNNC | 13.38 | - | 13.70 | - | 58.10 | - |
| CNCN | 13.88 | - | 12.90 | - | 58.10 | - |
| ONNO (trans) | 11.34 | - | 13.12 | - | 64.35 | - |
| ONNO (cis) | 11.22 | - | 13.30 | - | 64.89 | - |

PROGRAMMING CONSIDERATIONS.

Three specific programs have been written to perform the here presented analysis (11). The first one PLAN generates the nuclear structures according to any type of experimental designs. The second MULFRA performs the regression analysis and locates the stationary point of interest. The last one HVIBR computes the vibrational fundamental frequencies and the thermodynamical properties; It provides too the coriolis coupling matrix and the first and second order distortion on the equilibrium inertial tensor.

ACKNOWLEDGEMENT.

We wish to thank the FNRS for their constant support and for provision of a research-grant.We also thank Professors Daudel and Leroy for their interest in this work.

REFERENCES.

(1) Sana M.:
(1980), Internat. J. Quantum Chem., (in press).
(2) Box G.E.P., Hunter J.S.:
(1957), Ann. Math. Sat., 28 p.195;
(1961), Technometrics, 18 p.311.
(3) Doehlert D.H.:
(1970), Applied Stat., 19 p.231.
(4) Roquemore K.G.:
(1976), Technometrics, 18 p.419.
(5) Villaveces J.-L.:
Ph. D. Thesis in preparation.
(6) Sana M.:
(1980), J. Mol. Structure, (submitted).
(7) Wilson E.B., Decius G.C., Cross P.C.:
(1975), "Molecular Vibration", Mc GrawHill, London.
(8) Cyvin S.J.:
(1972), "Molecular Structures and Vibrations", Elsevier, Amsterdam.
(9) Sana M., Reckinger G., Leroy G.:
(1980), Theoret. Chim. Acta, (in press).
(10) Herzberg G.:
(1950), "Molecular Spectra and Molecular Structure", Van Nostrand Reynold Company;
Hehre W.J., Ditchfield R., Random H., Pople J.A.:
(1970), J. Amer. Chem. Soc., 92 p.4796.
(11) Sana M.:
(1980), Kant Program Package.
(12) See Schlegel's Lectures in this book.

# QUANTITATIVE ORBITAL ANALYSIS OF STRUCTURAL PROBLEMS AT THE AB-INITIO SCF-MO LEVEL

Fernando Bernardi and Andrea Bottoni

Istituto di Chimica Organica, Viale Risorgimento 4, Bologna, Italy.

CONTENTS.



## 1. INTRODUCTION

Qualitative Perturbational Molecular Orbital (PMO) procedures have prooved to be a very useful instrument for analysing the energy effects associated with the orbital interactions occurring between the component fragments and for understanding a variety of chemical problems (1-7). The methodology employed in these studies is founded upon One Electron Molecular Orbital (OEMO) theory and usually involves the following steps: (i) sequential dissection of the molecule under consideration into component fragments; (ii) construction of the group MO's of each fragment; (iii) evaluation of the interaction energy which obtains in the course of the union of the component fragments to yield the composite system in a specified geometry.



*I. G. Csizmadia and R. Daudel (eds.), Computational Theoretical Organic Chemistry, 197–231.*

However, in many cases, a qualitative approach is unsatisfactory and recently there has been a considerable effort in trying to make these orbital analyses more quantitative (8-11). In particular procedures which perform orbital interaction analyses at the ab-initio SCF-MO level seem to be very useful for elucidating a variety of chemical problems and particularly structural problems. In fact, ab-initio SCF-MO theory seems to provide a complete model for molecular structure (12).

It is the purpose of this paper to describe a procedure which allows to evaluate, in the framework of an ab-initio SCF-MO computation, the energy effects associated with the orbital interactions occurring between the component fragments of a closed shell molecule. The whole computational procedure follows as much as possible the line of the qualitative approach and therefore involves a procedure for the computation of the MO's of the component fragments and a procedure for the computation of the energy effects associated with the orbital interactions under examination, based on Perturbational Molecular Orbital (PMO) expressions. Various structural problems are then analyzed with this approach.

## 2. COMPUTATION OF THE FRAGMENTS MO's OF A CLOSED SHELL MOLECULE

The computation of the fragments MO's is the first important computational problem that has to be solved in order to perform a quantitative orbital analysis. The procedure we use for obtaining the fragments MO's of a closed shell molecule is the following. We first apply the procedure recently suggested by Wolfe et al. (10a), here after denoted as the WSW procedure. In this procedure the energies and the eigenvectors of the fragments MO's are obtained from the solution of the following eigenvalue problem:

$$F^\circ C^\circ = S^\circ C^\circ \varepsilon^\circ \qquad [1]$$

where $F^\circ$ and $S^\circ$ are the Fock matrix and the overlap matrix for the composite system with all the non-diagonal matrix elements and overlap integrals between atomic orbitals belonging to the different interacting fragments set equal to zero.

This procedure provides a set of fragments MO's denoted as canonical fragments MO's, which are of limited usefulness in a quantitative analysis: in fact, only the canonical fragments MO's of $\pi$ type are acceptable fragments MO's, while the canonical doubly occupied $\sigma$ MO's do not have correct orbital occupancies. Let we consider, for illustrative purposes, the simple molecule diimide, dissected into two HN- fragments as shown in Scheme I.

The MO's of the HN- fragment obtained with the WSW procedure are shown in Table 1, together with the gross populations of these

Table 1. Energies ($\varepsilon_i$, a.u.) and Gross Populations ($Q_i$) of the Fragment Orbitals of Cis and Trans Diimide Computed with the WSW Procedure and with the Present Procedure.

| | MO's | CIS | | | | TRANS | |
|---|---|---|---|---|---|---|---|
| | | Optimized Model | | Rigid Model | | | |
| | | $\varepsilon_i$ | $Q_i$ | $\varepsilon_i$ | $Q_i$ | $\varepsilon_i$ | $Q_i$ |
| WSW Procedure | | | | | | | |
| | $\sigma_1$ | -15.4187 | 2.0008 | -15.4215 | 2.0008 | -15.4096 | 2.0008 |
| | $\sigma_2$ | -1.1560 | 1.6094 | -1.1669 | 1.6164 | -1.1512 | 1.6063 |
| | $\sigma_3$ | -0.5985 | 1.5758 | -0.5857 | 1.5969 | -0.5869 | 1.6192 |
| | $\sigma_4$ | -0.4285 | 1.7957 | -0.4409 | 1.7739 | -0.4315 | 1.7597 |
| | $\pi$ | -0.1707 | 1.0000 | -0.1734 | 1.0000 | -0.1669 | 1.0000 |
| | $\sigma_5$ | 0.5674 | 0.0184 | 0.5684 | 0.0120 | 0.5847 | 0.0140 |
| Present Procedure | | | | | | | |
| | $1s_N$ | -15.2979 | 2.0004 | -15.3004 | 2.0005 | -15.2889 | 2.0005 |
| | $\sigma_N$ | -0.9505 | 1.0035 | -0.9483 | 1.0108 | -0.9427 | 1.0114 |
| | $\sigma_{NH}$ | -0.7825 | 2.0050 | -0.7912 | 2.0101 | -0.7730 | 2.0044 |
| | $n_\sigma$ | -0.5708 | 1.9727 | -0.5751 | 1.9667 | -0.5745 | 1.9697 |
| | $\pi$ | -0.1707 | 1.0000 | -0.1734 | 1.0000 | -0.1669 | 1.0000 |
| | $\sigma^*_{NH}$ | 0.5674 | 0.0184 | 0.5684 | 0.0120 | 0.5847 | 0.0140 |
| | IS | -34.4236 | | -34.4551 | | -34.3824 | |

orbitals. It can be seen that none of the occupied σ fragment MO's has an electron occupancy close to 1 or 2. This is a serious limitation since the correct application of the PMO expressions requires that the occupied interacting orbitals have integer orbital occupancies.

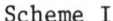

Scheme I

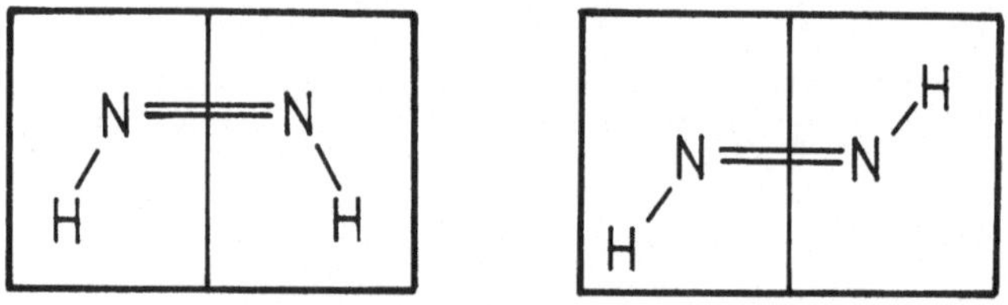


Cis Trans

We have shown (13) that this problem can be satisfactorily solved through the application of a localization procedure to the set of canonical fragments MO's. The localization procedure is applied separately to the set of the occupied and to the set of the vacant fragment MO's. In our applications we have used the Boys' method of localization (14). This new procedure, i.e. the WSW procedure followed by localization, provides for each fragment a new set of MO's that are still orthogonal and have now correct orbital occupancies. We have denoted this new set of MO's as fragment localized MO's.

The localized MO's of the HN- fragment in diimide obtained with this procedure, where we have applied the localization treatment only to the σ MO's, together with the corresponding gross orbital populations are shown in Table 1. It can be seen that now the occupied σ MO's have electron occupancies close to 2, except one whose electron occupancy is close to 1. There is also another positive feature of the fragment localized MO's and this is illustrated in Figure 1, where we have sketched the valence localized MO's of an HN- fragment. It can be seen that these fragment MO's can be easily associated with bond orbitals, lone pairs and antibond orbitals. In fact, the first MO ($\sigma_N$) is the singly occupied orbital localized along the N-N axis and pointing toward the other nitrogen atom, the second MO ($\sigma_{NH}$) is a doubly occupied σ MO associated with the N-H bond, the third MO ($n_\sigma$) is a doubly occupied in plane nitrogen lone pair, the fourth MO (π) is a singly occupied $P_\pi$-type nitrogen atomic orbital and the fifth MO ($\sigma^*_{NH}$) is a vacant orbital associated with the N-H bond.

Therefore with the use of the fragment localized MO's, it becomes possible to discuss at a quantitative level the role of the various factors that determine the structure of organic molecules within a model which is very near to the qualitative descrip-

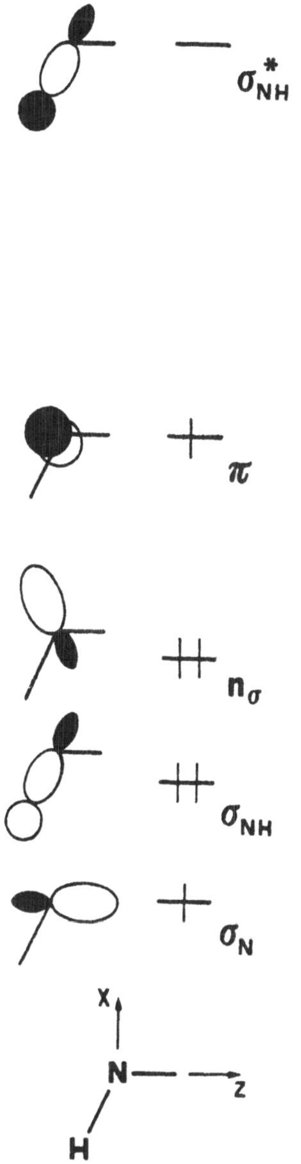


Figure 1. The Valence Localized MO's of a HN- Fragment.

tion used by chemists. In fact the various energy effects can be associated with the repulsions between the bond orbitals and the lone pairs and with the conjugative stabilizations between bond orbitals and antibond orbitals and between lone pairs and antibond orbitals.

## 3. QUANTITATIVE PMO APPROACH

In closed shell species the orbital interactions which play a significant role in determining the molecular structure are either two-electron two-orbital interactions or four-electron two-orbital interactions (7). The associated energy effects are stabilizing and destabilizing, respectively, and are estimated on the basis of the following expressions, valid in the framework of a OEMO model:

$$\Delta E^2_{ij} = 2\,\frac{(H_{ij}-S_{ij}\varepsilon_i)^2}{\varepsilon_i-\varepsilon_j} \qquad [2]$$

$$\Delta E^4_{ij} = 4\,\frac{(\varepsilon_o S^2_{ij}-H_{ij}S_{ij})}{(1-S^2_{ij})} \qquad [3]$$

$$= 4\,(\varepsilon_o S^2_{ij}-H_{ij}S_{ij})\,(1+S^2_{ij}+\ldots) \qquad [3']$$

Here $\Delta E^2_{ij}$ represents the stabilization energy associated with a two-electron two-orbital interaction, i.e. the interaction of a doubly occupied MO, $\phi_i$, with a vacant MO, $\phi_j$, while $\Delta E^4_{ij}$ represents the destabilization energy associated with a four-electron two-orbital interaction, i.e. the interaction between two doubly occupied MO's, $\phi_i$ and $\phi_j$. In these expressions $\varepsilon_i$ and $\varepsilon_j$ denote the energies of the two unperturbed MO's $\phi_i$ and $\phi_j$, $S_{ij}$ their overlap integral, $H_{ij}$ their matrix element, and $\varepsilon_o$ the mean of the energies $\varepsilon_i$ and $\varepsilon_j$. Eq. [2] is a well known result of perturbation theory(15), while eq. [3] is obtained by application of the variational method to the case of a two-orbital interaction problem. The expansion of the term $1/(1-S^2_{ij})$ in power series leads to eq. [3'], whose first term agrees with the expression obtained in terms of a perturbational treatment (16). These energy terms are the only ones appearing in the expression of the interaction energy when the PMO treatment is carried out at the Extended Huckel level of MO theory (1). However, when we operate within the framework of a SCF-MO scheme, other terms appear in the expression of the interaction energy, and among these of particular relevance should be the Coulomb Energy ($E_C$) associated with the interaction of the various fragments (17,18).

Therefore the basic expression used here for the interaction energy $\Delta E$ which obtains in the union of the component fragments to yield the composite system is the following:

$$\Delta E = \Sigma\Delta E^{4}_{ij} + \Sigma\Delta E^{2}_{ij} + E_c \qquad [4]$$

The various terms appearing in eq. [4] are then computed using the results of the ab-initio SCF-MO computations. In particular:

(i) the term $E_c$ is simply computed according to the following expression:

$$E_c = \sum_{r<r'} \frac{q_r q_{r'}}{Q_{rr'}} \qquad [5]$$

which represents the Coulomb interaction between the net charges of the interacting fragments, with the net charges taken from the results of the Mulliken population analysis;

(ii) the matrix elements $H_{ij}$ and the overlap integrals $S_{ij}$ between the interacting MO's are computed according to the following relations:

$$H = (C^{\circ}_{L})^{\dagger}\, F\, C^{\circ}_{L} \qquad [6]$$

$$S = (C^{\circ}_{L})^{\dagger}\, \bar{S}\, C^{\circ}_{L} \qquad [7]$$

where $C^{\circ}_{L}$ denotes the coefficient matrix of the fragment localized MO's, $F$ and $\bar{S}$ the Fock and the overlap matrices for the composite system over the atomic orbital basis;

(iii) the values of the energies of fragment localized MO's, $\varepsilon_i$, are chosen to be the expectation values of the Fock operator over the fragment localized basis, which are the diagonal elements of the H matrix.

## 4. APPLICATIONS.

The quantitative PMO procedure previously described can be applied to the analysis of the orbital interactions occurring between any type of fragment MO's. So far only the effects of $\pi$-type orbital interactions have been investigated quantitatively in some problems of molecular structure (10,11). Here, in order to illustrate the various possibilities of application of the present procedure, we discuss a certain number of structural problems where various different types of orbital interactions play a significant role.

All computations have been carried out at the STO-3G level (19), and the SCF values have been computed with the GAUSSIAN 70 series of programs (20).

4.1. Cis, Trans Isomerism in $N_2H_2$ (13)

The molecule HN=NH (diimide) can exist in the cis and the trans geometries. For this molecule we have performed the computations at three different geometries, i.e. at the trans STO-3G optimized geometry (12), at a cis geometry obtained through a rigid rotation of the trans geometry and at a cis STO-3G optimized geometry. The trans geometry is predicted to be more stable in agreement with the experimental results (21). In particular at the STO-3G level the trans configuration has been found to be more stable by 10.17 kcal/mol than the cis obtained through rigid rotation and more stable by 7.35 kcal/mol than the cis with optimized geometry. The diimide molecule has been dissected as shown in Scheme I: the various localized MO's of a HN- fragment have already been described (see Figure 1). We examine now the energy effects associated with the non-bonded interactions occurring between the two HN- fragments in cis and trans diimide. These interactions are depicted in Figure 2, while the values of the energy effects associated with the various orbital interactions are listed in Table 2.

We compare first the trans configuration with the cis in the geometry obtained with rigid rotation. The following points are of interest:
(i) The overall energy effect associated with the non-bonded interactions ($\Sigma\Delta E$) is destabilizing and more destabilizing in the cis than in the trans geometry. The difference between the two destabilizing effects (12.31 kcal/mol) is of the same order of magnitude of the difference between the corresponding total energies and therefore, at this level, the non-bonded interactions provide a satisfactory rationalization of the preferential stability of trans over cis diimide.
(ii) The results of the quantitative analysis suggest that the dominant factor is the repulsion between the two $\sigma_{NH}$ bond orbitals: this interaction is, in fact, much more destabilizing in the cis than in the trans geometry, and the relative destabilization is much larger than that associated with the destabilizing interaction between the two $\sigma$ lone pairs, which favors the cis geometry. The relative effect of the third destabilizing interaction, $\sigma_{NH}$-$n_\sigma$, is small and favors the trans geometry. Also the energy effect associated with the conjugative stabilizing interactions is small: in particular that associated with the $\sigma_{NH}$-$\sigma^*_{NH}$ interaction is almost negligible, while that associated with the $n_\sigma$-$\sigma^*_{NH}$ interaction is more significant and favors the cis geometry as previously suggested. However its effect is not large enough to change the trend dictated by the $\sigma_{NH}$-$\sigma_{NH}$ interaction.
(iii) We have also attempted to estimate the effect of the electrostatic interaction, by computing the Coulomb energy ($E_c$) between the two HN- fragments in terms of the net atomic charges of the composite molecule obtained with the Mulliken analysis. We have

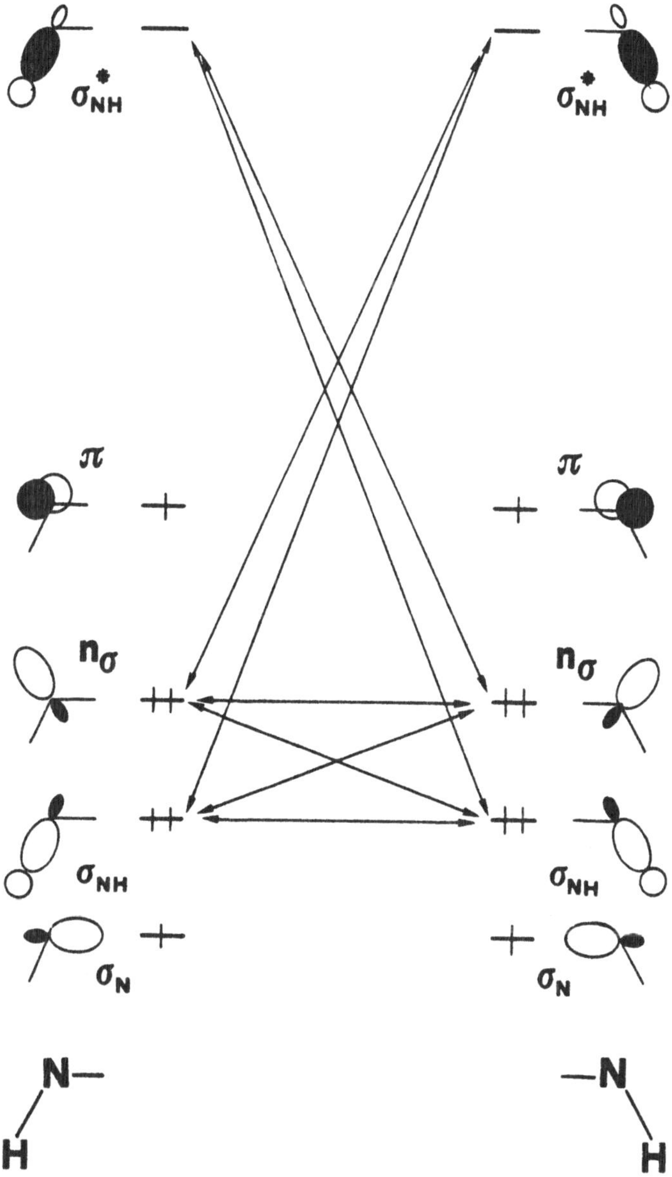


Figure 2. Non-bonded Orbital Interactions Between two HN- Fragments in Diimide.

Table 2. Stabilization ($\Delta E^2$) and Destabilization ($\Delta E^4$) Energies (kcal/mol) Associated with the non-bonded Interactions in Cis and Trans Diimide, together with the Values of the Coulomb Energy ($E_C$, kcal/mol) Associated with the Interaction of the two HN- Fragments.

| | CIS | | TRANS |
|---|---|---|---|
| | Optimized Model | Rigid Model | |
| $\Delta E^4_{\sigma_{NH},\sigma_{NH}}$ | 18.39 | 31.95 | 5.71 |
| $\Delta E^4_{n_\sigma,n_\sigma}$ | 2.78 | 0.44 | 14.75 |
| $\Delta E^4_{\sigma_{NH},n_\sigma}$ | 10.59 | 10.30 | 8.75 |
| $\Delta E^2_{\sigma_{NH},\sigma^*_{NH}}$ | -0.30 | -0.45 | -1.41 |
| $\Delta E^2_{n_\sigma\ \sigma^*_{NH}}$ | -3.75 | -2.71 | -0.39 |
| $\Sigma\Delta E$ | 34.25 | 46.67 | 34.36 |
| $E_C$ | 1.64 | 1.69 | 0.65 |

found that also this term favors the trans geometry, however its relative effect is small (~1 kcal/mol).

We proceed now to compare the results previously discussed with those obtained for the cis optimized geometry. The comparison shows that in the cis configuration the geometry tends to change in order to reduce the large destabilizing effect associated with the $\sigma_{NH}$-$\sigma_{NH}$ repulsion and increase the stabilizing effect associated with the conjugative interaction $n_{\sigma}$-$\sigma^{*}_{NH}$. This result is obtained mainly through an increase of the HNN angle (from 105.3° to 111.5°) and of the N-H bond length (from 1.061 Å to 1.064 Å). At the cis optimized geometry the overall energy effect is much less destabilizing than at the cis geometry obtained with rigid rotation of the trans geometry and becomes of the same order of magnitude of that obtained for the trans geometry. However this reduction of the destabilizing effect has been obtained at the expenses of a destabilization of the various fragment localized MO's which accompanies the geometry relaxation in the cis configuration (see Table 1).

The above comparative discussion shows that the results of the quantitative PMO treatment parallel the total energy behaviour only in the case of the rigid model. This finding can be justified on the basis of the assumptions under which a PMO expression of the type used here is derived. In fact, in the PMO treatment here we have considered explicitly only the energy effects described by the eq. [2] and [3] , i.e. second order effects. Such a treatment is based upon the assumption that there are no changes in the intramolecular Hamiltonian matrix elements (22); this is the restriction which makes the first order effects to become zero. This restriction is expected to be roughly satisfied in a rigid model, but much less in an optimized model where, therefore, also the first order effects become significant.

The inclusion of the Coulomb Energy $E_c$ does not modify significantly the situation. Additional information about the trend of the first order effects can be obtained from the comparison of the following quantity:

$$IS = \sum_{i}^{occ.} n_i \varepsilon_i \qquad [8]$$

where $\varepsilon_i$ denotes the energy of the various fragment localized MO's, $n_i$ the corresponding occupation number (1 or 2), and the sum is taken over all the occupied fragment localized MO's. The values of this quantity for diimide are listed in Table 1. It can be seen that the IS value increases going from the cis rigid to the cis optimized geometry, indicating that the geometry relaxation in the cis configuration is accompanied by a destabilization of the various fragments localized MO's. On the other hand, the comparison of the IS values for the trans and the cis rigid geometries is

not very informative. These results suggest that the IS values provide useful information about the trend of the first order effects only in the comparison of two similar geometries, such as in the present case the two cis geometries. It seems that in order to reproduce the total energy behaviour in the comparison of two non-related geometries such as the cis and the trans optimized geometries, other effects have to be considered. This problem is presently under investigation. However the results here suggest the following two-step procedure which seems to provide a satisfactory understanding of the structural problem under investigation and which will be used also in the other applications described in this paper:
(i) first, a quantitative PMO analysis in the rigid model, choosing as a starting point the less crowded optimized geometry;
(ii) second, a quantitative comparative analysis of the effects of geometry relaxation in the more crowded geometries.

4.2. Cis, Trans Isomerism in $N_2F_2$ (23)

Let us consider now the related species FN=NF (difluorodiazine), where both hydrogen atoms have been replaced by two more electronegative fluorine atoms. Also this molecule can exist in the cis and trans geometries, and in this case the cis configuration is found experimentally to be more stable by 3 kcal/mol (24). For this molecule we have performed the quantitative analysis at the following three different geometries: at the trans STO-3G optimized geometry (25), at a cis geometry obtained through a rigid rotation of the trans, and at the cis STO-3G optimized geometry (25). The cis configuration at the optimized geometry is found to be sligtly more stable than the trans (0.3 kcal/mol) and the latter more stable than the cis configuration at the geometry obtained by rigid rotation (3.7 kcal/mol). Therefore the STO-3G results reproduce the order of relative stability observed experimentally, even if the energy difference between the cis and trans configurations is somewhat underestimated. For each geometry we have considered the two fragmentation modes illustrated in Scheme II for the cis configuration and in all cases we have applied the localization treatment only to the σ fragments MO's.

Scheme II

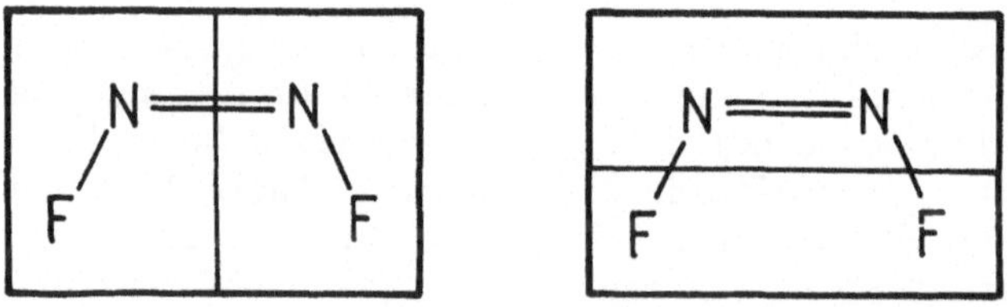

Mode 1 Mode 2

The fragmentation mode 1 is similar to that used for HN=NH and similarly provides information about the $\sigma$ interactions occurring between two FN- fragments. However, in addition to these interactions, in this molecule there are also the $\pi$-type interactions occurring between the $\pi$ MO's associated with the -N=N- fragment and the $\pi$-type fluorine lone pairs. These interactions are better analyzed with the fragmentation mode 2, which provides $\pi$-type fragment MO's with electron occupancies 2 or 0. For similar reasons this fragmentation mode is more suitable also for the analysis of the interactions between the $\sigma$-type fluorine lone pairs and the $\sigma$ and $\sigma^*$ MO's associated with the N-N bond.

We discuss first the results obtained with the fragmentation mode 1. The energies and gross orbital populations of the various localized MO's of a FN- fragment are listed in Table 3. It can be seen that also in this case the occupied $\sigma$ MO's have electron occupancies close to 2, except one whose electron occupancy is close to 1. The various valence localized $\sigma$ MO's of a FN- fragment are shown in Figure 3: the first MO ($\sigma_{NF}$) is the doubly occupied $\sigma$ MO associated with the N-F bond, the second MO ($\sigma_N$) is the singly occupied orbital localized along the N-N axis and pointing toward the other nitrogen atom, the third and fourth MO's ($n_{\sigma_F}$ and $n'_{\sigma_F}$) are the two doubly occupied $\sigma$-type fluorine lone pairs, the fifth MO ($n_\sigma$) is the doubly occupied in plane nitrogen lone pair and the sixth MO ($\sigma^*_{NF}$) is a vacant orbital associated with the N-F bond.

We compare first the trans and cis geometries in the rigid model. In order to understand better the effects due to the replacement of the hydrogen atoms with the fluorine atoms, we consider first the energy effects associated with those orbital interactions which correspond to those occurring in HN=NH. These interactions are shown in Figure 3 (full lines), while the associated energy effects are listed in Table 4. When only these interactions are taken into account, the overall energy effect ($\Sigma\Delta E_I$) in FN=NF is again destabilizing, but roughly of the same order of magnitude in the cis and trans geometries, while in HN=NH this term favors largely the trans geometry. The term responsible for this different behaviour is the repulsion between the MO's associated with the N-F bonds ($\Delta E^4_{\sigma_{NF},\sigma_{NF}}$) which favors the trans geometry at a much smaller extent compared with diimide. This trend is controlled by the matrix elements; it is found, in fact, that in the trans configurations the matrix elements associated with the $\sigma_{NH}$-$\sigma_{NH}$ and the $\sigma_{NF}$-$\sigma_{NF}$ interactions are of similar order of magnitude, but in the cis configurations the matrix element associated with the $\sigma_{NF}$-$\sigma_{NF}$ interaction is about one half the value of that associated with the $\sigma_{NH}$-$\sigma_{NH}$ interaction. It is also found that the matrix elements parallel the trend of the related overlap integrals, so that this different behaviour can be explained in terms of a smaller overlap between the two $\sigma_{NF}$

Table 3. Energies ($\varepsilon_i$, a.u.) and Gross Populations ($Q_i$) of the Localized Orbitals of the FN-Fragment in FN=NF.

| | CIS | | | | TRANS | |
|---|---|---|---|---|---|---|
| | Optimized Model | | Rigid Model | | | |
| | $\varepsilon_i$ | $Q_i$ | $\varepsilon_i$ | $Q_i$ | $\varepsilon_i$ | $Q_i$ |
| $\varepsilon_{1s_F}$ | -25.8393 | 2.0000 | -25.8372 | 2.0000 | -25.8376 | 2.0000 |
| $\varepsilon_{1s_N}$ | -15.4014 | 2.0003 | -15.4074 | 2.0005 | -15.4007 | 2.0005 |
| $\varepsilon_{\sigma_{NF}}$ | -1.0261 | 1.9970 | -1.0344 | 1.9990 | -1.0249 | 1.9978 |
| $\varepsilon_{\sigma_N}$ | -1.0086 | 1.0310 | -1.0553 | 1.0237 | -1.0523 | 1.0221 |
| $\varepsilon_{n_{\sigma_F}}$ | -1.0334 | 1.9983 | -1.0353 | 1.9987 | -1.0261 | 1.9993 |
| $\varepsilon_{n'_{\sigma_F}}$ | -1.0153 | 1.9926 | -1.0160 | 1.9939 | -1.0131 | 1.9929 |
| $\varepsilon_{n_\sigma}$ | -0.6379 | 1.9527 | -0.6455 | 1.9572 | -0.6444 | 1.9633 |
| $\varepsilon_{\pi_1}$ | -0.5985 | 1.9818 | -0.6012 | 1.9787 | -0.5955 | 1.9774 |
| $\varepsilon_{\pi_2}$ | -0.1875 | 1.0182 | -0.1955 | 1.0213 | -0.1922 | 1.0226 |
| $\varepsilon_{\sigma^*_{NF}}$ | 0.4308 | 0.0283 | 0.4318 | 0.0270 | 0.4381 | 0.0241 |
| IS | -92.2999 | | -92.4048 | | -92.3291 | |

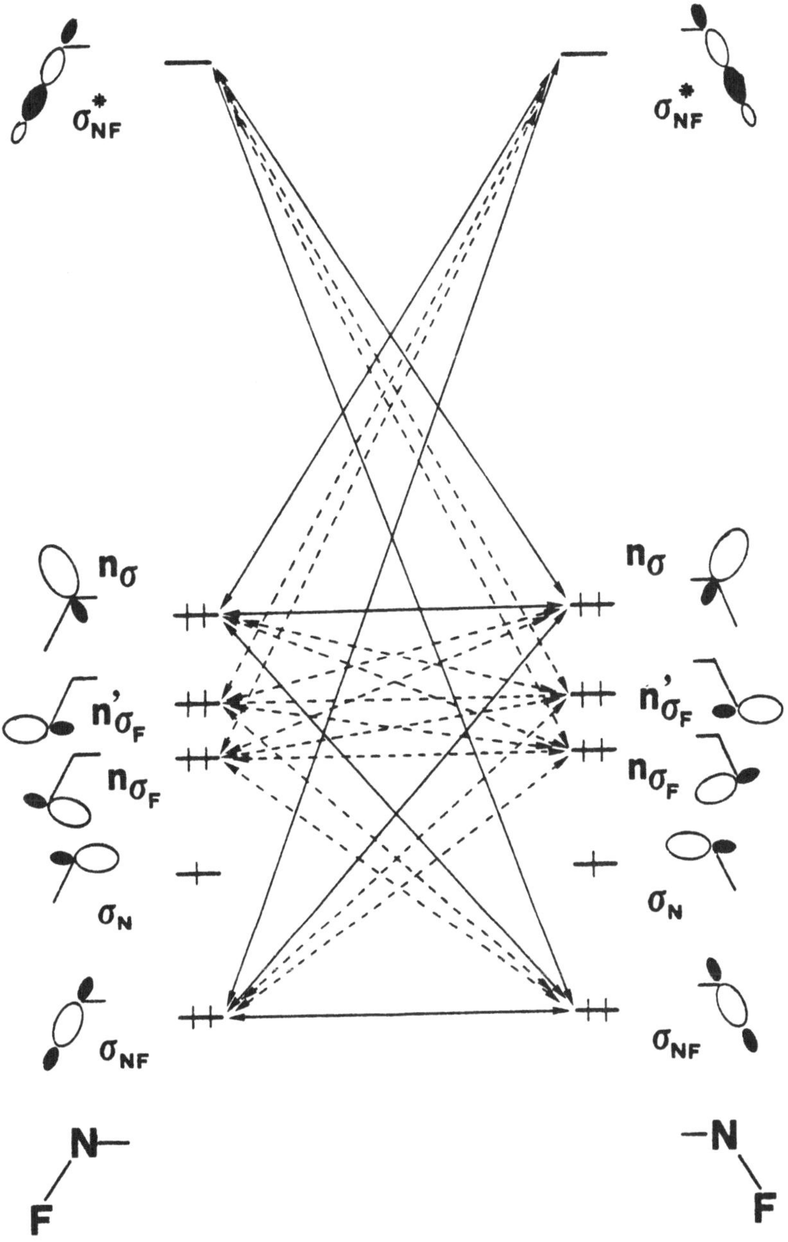


Figure 3. Non-bonded Orbital Interactions Between two FN- Fragments in Difluorodiazine.

Table 4. Stabilization ($\Delta E^2$) and Destabilization ($\Delta E^4$) Energies (kcal/mol) Associated with the Various σ non-bonded Interactions in the Cis and Trans Configurations of FN=NF Obtained with the Fragmentation Mode 1.

| | CIS | | TRANS |
|---|---|---|---|
| | Optimized Model | Rigid Model | |
| $\Delta E^4_{\sigma_{NF},\sigma_{NF}}$ | 4.93 | 9.84 | 5.19 |
| $\Delta E^4_{\sigma_{NF},n_\sigma}$ | 7.28 | 8.32 | 4.01 |
| $\Delta E^4_{n_\sigma,n_\sigma}$ | 1.70 | 0.55 | 11.50 |
| $\Delta E^2_{\sigma_{NF},\sigma^*_{NF}}$ | -0.19 | -0.37 | -0.87 |
| $\Sigma E^2_{n_\sigma,\sigma^*_{NF}}$ | -3.82 | -2.89 | -1.39 |
| $\Sigma\Delta E^4_I$ | 21.19 | 27.03 | 24.71 |
| $\Sigma\Delta E^2_I$ | -8.02 | -6.52 | -4.52 |
| $\Sigma\Delta E_I$ | 13.17 | 20.51 | 20.19 |
| $\Delta E^4_{\sigma_{NF},n_{\sigma_F}}$ | 0.17 | 1.02 | 0.19 |
| $\Delta E^4_{\sigma_{NF},n'_{\sigma_F}}$ | 0.11 | 0.23 | 0.03 |
| $\Delta E^4_{n_{\sigma_F},n_{\sigma_F}}$ | 0.42 | 3.86 | 0.05 |
| $\Delta E^4_{n_{\sigma_F},n'_{\sigma_F}}$ | 0.11 | 0.42 | 0.01 |
| $\Delta E^4_{n'_{\sigma_F},n'_{\sigma_F}}$ | 0.00 | 0.00 | 0.00 |
| $\Delta E^4_{n_{\sigma_F},n_\sigma}$ | 0.13 | 0.36 | 0.27 |
| $\Delta E^4_{n'_{\sigma_F},n_\sigma}$ | 0.00 | 0.00 | 0.18 |
| $\Delta E^2_{n_{\sigma_F},\sigma^*_{NF}}$ | -0.08 | -0.10 | -0.63 |
| $\Delta E^2_{n'_{\sigma_F},\sigma^*_{NF}}$ | -0.02 | -0.04 | -0.03 |
| $\Sigma\Delta E^4_{II}$ | 1.46 | 7.92 | 1.41 |
| $\Sigma\Delta E^2_{II}$ | -0.20 | -0.28 | -1.32 |
| $\Sigma\Delta E_{II}$ | 1.26 | 7.64 | 0.09 |

bond MO's, compared with that between the two $\sigma_{NH}$ bond MO's, in the cis geometry.

We consider now the additional energy effects arising in the course of the union of the two FN-fragments, i.e. those associated with the interactions involving the two σ-type fluorine lone pairs. These interactions are shown in Figure 3 (dashed lines), and the associated energy effects are listed in the second part of Table 4. It is found that all these energy terms are almost negligible in the trans geometry, while some of them are large and destabilizing in the cis geometry. In particular the larger destabilizing effects are due to the repulsion between the two σ fluorine lone pairs pointing toward each other ($n_{\sigma_F}$-$n_{\sigma_F}$) and to the repulsion between this lone pair and the NF bond MO ($\sigma_{NF}$-$n_{\sigma_F}$). Therefore, when all the interactions shown in Figure 3 are considered, the trans geometry is found to be more stable than the cis (compare the $\Sigma\Delta E_I+\Sigma\Delta E_{II}$ values in Table 5. In Table 5 we have also listed the values of the Coulomb Energy ($E_C$) associated with the interaction between the two FN-fragments in difluorodiazine: it is found that in this case this energy contribution is completely negligible.

We proceed now to examine the effects of the interactions involving the MO's associated with the N=N bond. As already pointed out, these interactions are better described with the fragmentation mode 2. The relevant localized MO's of the F...F and -N=N- fragments are shown in Figure 4, together with the various orbital interactions, while the associated energy effects are listed in Table 6. The relevant interactions here are those between the in phase combinations of the fluorine lone pairs ($\chi_1$, $\chi_3$, $n_S$) and the doubly occupied MO's associated with the -N=N- fragment ($\sigma_{NN}$,π) and those between the out of phase combinations of the fluorine lone pairs ($\chi_2$, $\chi_4$, $n_A$) and the vacant MO's associated with the -N=N- fragment ($\sigma^*_{NN}$, $\pi^*$). It is found that all the energy effects associated with these interactions favor the cis geometry.

Therefore the replacement of the two hydrogen atoms in diimide with two fluorine atoms causes the following effects:
(i) a reduction in the cis geometry of the repulsion associated with the interaction between the two N-F bond MO's, compared with that associated with the two N-H bond MO's;
(ii) a preferential stabilization of the cis geometry due to the orbital interactions occurring between the fluorine lone pairs and the MO's associated with the N=N bond.
Both these effects concur to determine the preferential stabilization of cis over trans difluorodiazine.

The comparison of the results obtained in the rigid model with those obtained at the cis optimized geometry shows that also in this case in the cis configuration the geometry tends to change

Table 5. Overall Energy Effects (kcal/mol) Associated with the Various σ non-bonded Interactions in the Cis and Trans Configurations of FN=NF, Obtained with the Fragmentation Mode 1, together with the Values of the Coulomb Energy ($E_c$, kcal/mol) Associated with the Interaction of the Two FN- Fragments.

| | CIS | | TRANS |
|---|---|---|---|
| | Optimized Model | Rigid Model | |
| $\Sigma\Delta E^4_I + \Sigma\Delta E^4_{II}$ | 22.65 | 34.95 | 26.12 |
| $\Sigma\Delta E^2_I + \Sigma\Delta E^2_{II}$ | -8.22 | -6.80 | -5.84 |
| $\Sigma\Delta E_I + \Sigma\Delta E_{II}$ | 14.43 | 28.15 | 20.28 |
| $E_c$ | 0.17 | 0.21 | 0.04 |

Table 6. Additional Stabilization ($\Delta E^2$) and Destabilization ($\Delta E^4$) Energies (kcal/mol) Associated with σ and π non-bonded Interactions in the Cis and Trans Configurations of FN=NF, Obtained with the Fragmentation Mode 2.

| | CIS | | TRANS |
|---|---|---|---|
| | Optimized Model | Rigid Model | |
| $\Delta E^4_{\chi_1,\sigma_{NN}}$ | 2.52 | 6.66 | 8.80 |
| $\Delta E^4_{\chi_3,\sigma_{NN}}$ | 15.20 | 17.24 | 19.35 |
| $\Delta E^2_{\chi_2,\sigma^*_{NN}}$ | -0.01 | 0.00 | -0.01 |
| $\Delta E^2_{\chi_4,\sigma^*_{NN}}$ | -7.81 | -7.23 | -5.50 |
| $\Sigma\Delta E^4$ | 17.72 | 23.90 | 28.15 |
| $\Sigma\Delta E^2$ | -7.82 | -7.23 | -5.51 |
| $\Sigma\Delta E$ | 9.90 | 16.67 | 22.64 |
| $\Delta E^4_{n_{\pi_S},\pi}$ | 28.19 | 29.71 | 29.58 |
| $\Delta E^2_{n_{\pi_A},\pi^*}$ | -39.67 | -39.29 | -36.83 |
| $\Sigma\Delta E(\pi)$ | -11.48 | -9.58 | -7.25 |

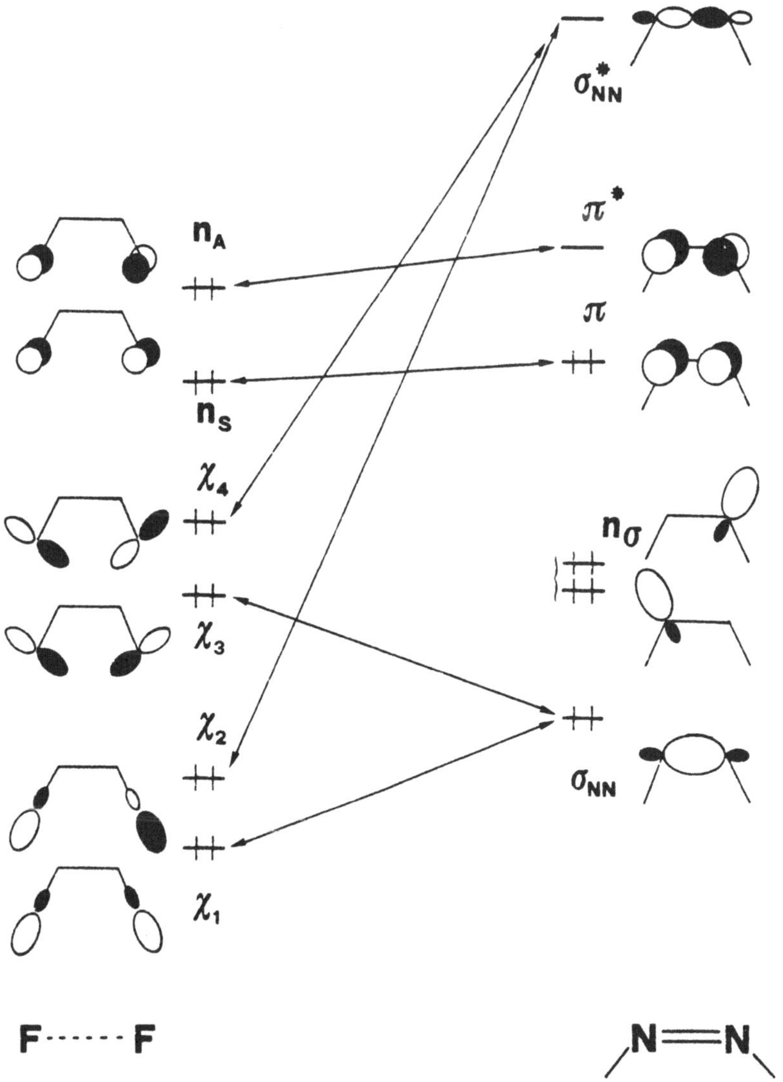


Figure 4. Relevant non-bonded Orbital Interactions Between the -N=N- Fragment and the F.....F Fragment in Difluorodiazine.

in order to reduce the large destabilizing effects and increase the stabilizing effects. This result is obtained mainly through an increase of the FNN angle (from 106.6° to 111.6°) and an increase of the N-F bond length (from 1.373 Å to 1.377 Å). Again, this energy improvement is largely counterbalanced by a destabilization of the various fragment localized MO's, as indicated by the trend of the values of the index IS (see Tables 3), so that the final gain in total energy is again quite small (~3.4 kcal/mol).

4.3. Conformational Isomerism of $O_2H_2$ (23)

The molecule HO-OH (hydrogen peroxide) is one of the simplest systems where conformational preference can be observed. The important conformations are those illustrated in Scheme III

Scheme III

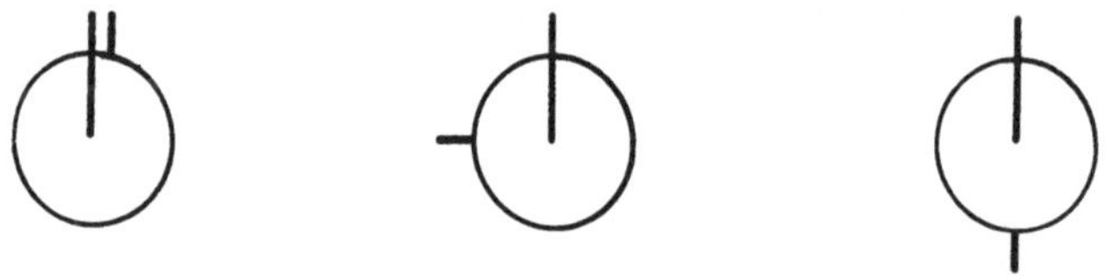

| Cis | Gauche | Trans |
|---|---|---|
| 9.14 | 0.00 | 0.08 |

where we have also listed the STO-3G relative energies (kcal/mol) computed at the corresponding STO-3G optimized geometries (12). At this computational level the gauche conformer is found to be more stable, in agreement with experiment (26). Also the optimum value of the dihedral angle ω between the two HOO planes (112.6°) agrees well with the value experimentally found which is in the range 111°-120°. However while the computed cis - gauche barrier is of the correct order of magnitude (computed value of 9.14 kcal/mol vs. experimental value of 7.9 kcal/mol), the trans - gauche barrier is underestimated (computed value of 0.08 kcal/mol vs. experimental value of 1.1 kcal/mol) and these two conformations are found to have roughly the same energy.

We attempt now to rationalize the various conformational preferences. In all cases the molecule has been dissected as shown in Scheme IV and we have localized only the fragment MO's of σ type. Also in this case the gross orbital populations of the various localized MO's listed in Table 7 have values close to 2, except one whose electron occupancy is close to 1. The localized MO's of the HO- fragment are similar to those of the HN- fragment (see Figure 1): the only significant difference concerns the occupation of the π orbital, which is singly occupied in the HN- fragment and doubly occupied in the HO- fragment. We compare first the

Table 7. Energies ($\varepsilon_i$, a.u.) and Gross Populations ($Q_i$) of the Localized Orbitals of the HO-Fragment in HO-OH.

| | CIS | | | | TRANS | |
|---|---|---|---|---|---|---|
| | Optimized Model | | Rigid Model | | | |
| | $\varepsilon_i$ | $Q_i$ | $\varepsilon_i$ | $Q_i$ | $\varepsilon_i$ | $Q_i$ |
| $\varepsilon_{1s_O}$ | -20.1593 | 1.9996 | -20.1625 | 1.9997 | -20.1509 | 1.9997 |
| $\varepsilon_{\sigma_{OH}}$ | -0.8472 | 1.9962 | -0.8572 | 2.0015 | -0.8411 | 2.0003 |
| $\varepsilon_{\sigma_O}$ | -0.7709 | 1.1591 | -0.7765 | 1.1703 | -0.7704 | 1.1742 |
| $\varepsilon_{n_\sigma}$ | -0.6700 | 1.8435 | -0.6715 | 1.8283 | -0.6653 | 1.8238 |
| $\varepsilon_{n_\pi}$ | -0.4371 | 2.0000 | -0.4406 | 2.0000 | -0.4335 | 2.0000 |
| $\varepsilon_{\sigma^*_{OH}}$ | 0.5754 | 0.0016 | 0.5714 | 0.0002 | 0.5919 | 0.0021 |
| IS | -44.9981 | | -45.0401 | | -44.9520 | |

Scheme IV

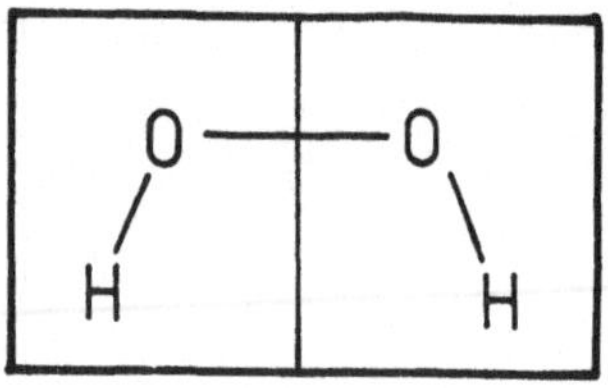


two conformers cis and trans. To this purpose we have performed the quantitative analysis at the following three geometries: at the trans STO-3G optimized geometry (12), at the cis geometry obtained through a rigid rotation of the trans, and at the cis STO-3G optimized geometry (12).

The values of the energy effects associated with the various orbital interactions are listed in Table 8. We compare first the results for the trans and cis geometries in the rigid model: at this level the overall energy effect associated with the non-bonded interactions ($\Sigma\Delta E$) is destabilizing and less destabilizing in the trans conformation. The dominant effect is the repulsion associated with the interaction between the two $\sigma_{OH}$ bond orbitals. The trend of the energy effect associated with the other orbital interactions between $\sigma$ localized MO's is similar to that already observed in diimide. In particular the repulsion between the two in plane $\sigma$ lone-pairs ($n_\sigma$-$n_\sigma$) favors the cis geometry and that between the $\sigma$ lone-pair and the OH bond orbital ($\sigma_{OH}$-$n_\sigma$) favors slightly the trans geometry while the conjugative stabilization associated with the interaction between the $\sigma$ lone pair and the vacant $\sigma^*_{OH}$ antibond orbital favors the cis geometry: however their effect is small compared to that of the $\sigma_{OH}$-$\sigma_{OH}$ repulsion. Another large effect is the destabilization energy associated with the interaction between the two $\pi$ type oxygen lone pairs ($n_\pi$): however, as expected, this repulsive effect is of the same order of magnitude in the cis and trans geometry.

When we compare the results obtained for the cis geometry in the rigid model with those obtained for the cis geometry in the optimized model, we observe a trend similar to that already pointed out in diimide. The cis geometry, in fact, tends to change in order to reduce the large destabilizing effect associated with the $\sigma_{OH}$-$\sigma_{OH}$ repulsion and increase the stabilizing effect associated with the conjugative interaction $n_\sigma$-$\sigma^*_{OH}$. However again this reduction of the destabilizing effect is obtained at the expenses of a destabilization of the various fragment localized MO's as indicated by the trend of the IS values in Table 7.

We can now attempt something more ambitious, specifically we try to rationalize the preferential stability of the gauche over

Table 8. Stabilization ($\Delta E^2$) and Destabilization ($\Delta E^4$) Energies (kcal/mol) Associated with the Various $\sigma$ and $\pi$ non-bonded Interactions in the Cis and Trans Configurations of Hydrogen Peroxide, together with the Values of the Coulomb Energy ($E_c$, kcal/mol) Associated with the Interaction of the Two HO- Fragments.

| | CIS | | TRANS |
|---|---|---|---|
| | Optimized Model | Rigid Model | |
| $\Delta E^4_{\sigma_{OH},\sigma_{OH}}$ | 15.60 | 25.64 | 0.56 |
| $\Delta E^4_{\sigma_{OH},n_\sigma}$ | 2.98 | 3.05 | 2.31 |
| $\Delta E^4_{n_\sigma,n_\sigma}$ | 0.00 | 0.00 | 3.40 |
| $\Delta E^2_{\sigma_{OH},\sigma^*_{OH}}$ | -0.05 | -0.08 | -0.84 |
| $\Delta E^2_{n_\sigma,\sigma^*_{OH}}$ | -1.97 | -1.81 | -0.24 |
| $\Sigma\Delta E^4$ | 21.56 | 31.74 | 8.58 |
| $\Sigma\Delta E^2$ | -4.04 | -3.78 | -2.16 |
| $\Sigma\Delta E$ | 17.52 | 27.96 | 6.42 |
| $\Delta E^4_{n_\pi n_\pi}$ | 21.38 | 21.92 | 21.94 |
| $E_c$ | 2.01 | 2.15 | 0.19 |

the trans conformer. To this purpose we have performed the quantitative PMO analysis at the gauche optimized geometry (12), and at various other geometries obtained through a rigid rotation of the gauche geometry. The values of the various energy effects are shown in Table 9. The following trends are found when the dihedral angle $\omega$ increases from 0° to 180°:

(i) the stabilization energy $\Sigma\Delta E^2$ increases with the increase of $\omega$ and reaches a maximum at 90°, then decreases monotonically and reaches its minimum at 180°. The trend of this energy term is mainly determined by the $n_\pi-\sigma^*_{OH}$ interaction whose energy effect reaches its maximum for $\omega = 90°$ and becomes zero in the cis and trans conformations. The other two stabilizing interactions $n_\sigma-\sigma^*_{OH}$ and $\sigma_{OH}-\sigma^*_{OH}$ vary monotonically with $\omega$, and reach the maximum stabilization for an anti alignement of the two interacting orbitals. This occurs in the cis conformation for the $n_\sigma-\sigma^*_{OH}$ interaction and in the trans conformation for the $\sigma_{OH}-\sigma^*_{OH}$ interaction. The $n_\sigma-\sigma^*_{OH}$ interaction has a larger stabilizing effect than the $\sigma_{OH}-\sigma^*_{OH}$ interaction and therefore the overall stabilizing effect is larger in cis than in trans;

(ii) the destabilizing interaction $\Sigma\Delta E^4$ decreases monotonically with the variation of $\omega$ from 0° to 180°. The trend of this energy term results from the contribution of various interactions. In the planar conformations there are two large destabilizing effects, as already pointed out: that associated with the $\sigma_{OH}-\sigma_{OH}$ interaction decreases rapidly with the increase of $\omega$ and becomes almost negligible for $\omega > 90°$, while that associated with the $n_\pi-n_\pi$ interaction, which is roughly of the same order of magnitude in the cis and trans conformations, becomes zero for $\omega = 90°$. In the non-planar conformations another destabilizing effect becomes significant, that associated with the $n_\pi-\sigma_{OH}$ interaction, which reaches its maximum destabilization for $\omega = 90°$. Therefore in the gauche conformation with $\omega = 90°$ the destabilizing effects associated with the $\sigma_{OH}-\sigma_{OH}$ and $n_\pi-n_\pi$ interactions become negligible, but the destabilizing effect is still large because of the large destabilization associated with the $n_\pi-\sigma_{OH}$ interaction;

(iii) the overall energy effect $\Sigma\Delta E$ decreases monotonically with the variation of $\omega$ from 0° to 180° and its trend is controlled by the destabilizing contribution $\Sigma\Delta E^4$.

Therefore, in this case, the quantitative PMO analysis is not in agreement with the total energy behaviour. However, the present analysis,even if not satisfactory at a quantitative level, nevertheless provides useful suggestions for the rationalization of this problem. In fact the analysis has pointed out that, in addition to the destabilizing contribution $\Sigma\Delta E^4$ which favors the trans conformer, there is also a significant stabilizing contribution which favors a gauche geometry and presumably is responsible for the greater stability of the gauche conformer. Therefore the quantitative analysis supports the qualitative suggestion that the $n_\pi-\sigma^*_{OH}$ interaction is a key factor in determining the preferential

Table 9. Stabilization ($\Delta E^2$) and Destabilization ($\Delta E^4$) Energies (kcal/mol) Associated with the Various non-bonded Interactions in Various Conformations of Hydrogen Peroxide

| | Dihedral angle ω | | | |
|---|---|---|---|---|
| | 0° | 90° | 112.6° | 180° |
| $\Delta E^4_{\sigma_{OH},\sigma_{OH}}$ | 22.41 | 2.31 | 0.00 | 0.96 |
| $\Delta E^4_{\sigma_{OH},n_\sigma}$ | 3.17 | 0.00 | 0.49 | 2.50 |
| $\Delta E^4_{n_\sigma,n_\sigma}$ | 0.00 | 1.09 | 2.06 | 3.72 |
| $\Delta E^4_{n_\pi,n_\pi}$ | 22.86 | 0.04 | 4.39 | 22.89 |
| $\Delta E^4_{n_\pi,\sigma_{OH}}$ | 0.00 | 14.06 | 11.42 | 0.00 |
| $\Delta E^4_{n_\pi,n_\sigma}$ | 0.00 | 4.69 | 3.62 | 0.00 |
| $\Delta E^2_{n_\pi,\sigma^*_{OH}}$ | 0.00 | -1.53 | -1.30 | 0.00 |
| $\Delta E^2_{\sigma_{OH},\sigma^*_{OH}}$ | -0.08 | -0.10 | -0.32 | -0.94 |
| $\Delta E^2_{n_\sigma,\sigma^*_{OH}}$ | -1.93 | -0.83 | -0.51 | -0.18 |
| $\Sigma\Delta E^4$ | 51.61 | 40.94 | 37.51 | 32.57 |
| $\Sigma\Delta E^2$ | -4.02 | -4.92 | -4.26 | -2.24 |
| $\Sigma\Delta E$ | 47.59 | 36.02 | 33.25 | 30.33 |

stabilization of the gauche conformer (7).

## 4.4. Conformational Isomerism of $H_2N$-OH (23)

The molecule $H_2N$-OH (hydroxilamine) has two stable rotational isomers, the cis and trans conformers shown in Scheme V

Scheme V

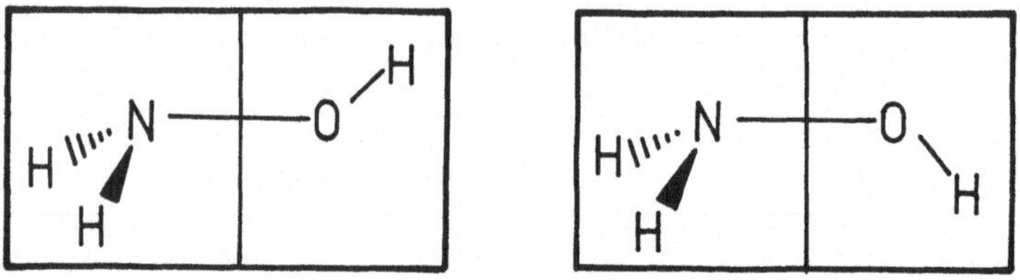


Trans Cis

In this case we have performed the quantitative analysis at the following three geometries: at the trans STO-3G optimized geometry (12), at the cis geometry obtained through a rigid rotation of the trans geometry and at the cis STO-3G optimized geometry (12). The trans conformation has been found to be more stable by 7.00 kcal/mol than the cis obtained through rigid rotation and more stable by 6.04 kcal/mol than the cis with optimized geometry. This molecule has been dissected as shown in Scheme V in the two fragments $H_2N$- and -OH. The gross orbital populations of the various fragments localized MO's are listed in Table 10 and show that this procedure provides in all cases investigated orbital occupancies very close to 2 for the formally doubly occupied MO's, very close to 1 for the formally singly occupied MO's and almost zero for the vacant MO's. The valence localized MO's of the two fragments are shown in Figure 5 together with the relevant non-bonded interactions: the localized MO's of the HO- fragment have already been described, while those of the $H_2N$- fragment, in order of increasing energy, involve:
(i) two degenerate doubly occupied $\sigma$ MO's, localized along the two N-H bonds and bonding between N and H (the bond orbitals $\sigma_{NH}$);
(ii) a singly occupied MO localized along the N-O axis and pointing toward the oxygen atom of the HO- fragment;
(iii) a $\sigma$-type nitrogen lone pair ($n_{\sigma_N}$) and (iv) two degenerate vacant $\sigma$ MO's, localized along the two N-H bonds and antibonding between N and H (the antibond orbitals $\sigma^*_{NH}$).

The energy effects associated with the various non bonded interactions are listed in Table 11. We compare first the results obtained in the rigid model. The overall energy effect associated with the non-bonded interactions ($\Sigma\Delta E$) is destabilizing and less destabilizing in the trans geometry. Therefore, also in this problem, the trend of the overall energy effect parallels the total

Table 10. Energies ($\varepsilon_i$, a.u.) and Gross Orbital Populations ($Q_i$) of the Localized Orbitals of the HO- and $H_2N$- Fragments in $H_2N$-OH.

| | | CIS | | | | TRANS | |
|---|---|---|---|---|---|---|---|
| | | Optimized Model | | Rigid Model | | | |
| Fragment HO- | | $\varepsilon_i$ | $Q_i$ | $\varepsilon_i$ | $Q_i$ | $\varepsilon_i$ | $Q_i$ |
| | $\varepsilon_{1s_O}$ | -20.1454 | 1.9998 | -20.1459 | 1.9998 | -20.1312 | 1.9998 |
| | $\varepsilon_{\sigma_{OH}}$ | -0.8336 | 2.0021 | -0.8431 | 2.0050 | -0.8284 | 2.0043 |
| | $\varepsilon_{\sigma_O}$ | -0.7844 | 1.1602 | -0.7805 | 1.1720 | -0.7713 | 1.1708 |
| | $\varepsilon_{n_{\sigma_O}}$ | -0.6646 | 1.8784 | -0.6643 | 1.8651 | -0.6578 | 1.8632 |
| | $\varepsilon_{n_\pi}$ | -0.4273 | 1.9869 | -0.4296 | 1.9892 | -0.4210 | 1.9888 |
| | $\varepsilon_{\sigma^*_{OH}}$ | 0.5885 | 0.0045 | 0.5915 | 0.0025 | 0.6111 | 0.0018 |
| | IS | -44.9262 | | -44.9463 | | -44.8481 | |
| Fragment $H_2N$- | | | | | | | |
| | $\varepsilon_{1s_N}$ | -15.2529 | 1.9986 | -15.2588 | 1.9988 | -15.2455 | 1.9989 |
| | $\varepsilon_{\sigma_{NH}}$ | -0.7400 | 1.9924 | -0.7448 | 1.9959 | -0.7319 | 1.9963 |
| | $\varepsilon_{\sigma_N}$ | -0.6136 | 0.9801 | -0.6098 | 0.9871 | -0.6031 | 0.9918 |
| | $\varepsilon_{n_{\sigma_N}}$ | -0.5063 | 1.9878 | -0.5058 | 1.9749 | -0.5060 | 1.9739 |
| | $\varepsilon_{\sigma^*_{NH}}$ | 0.6288 | 0.0084 | 0.6240 | 0.0069 | 0.6359 | 0.0072 |
| | IS | -35.0920 | | -35.1182 | | -35.0337 | |

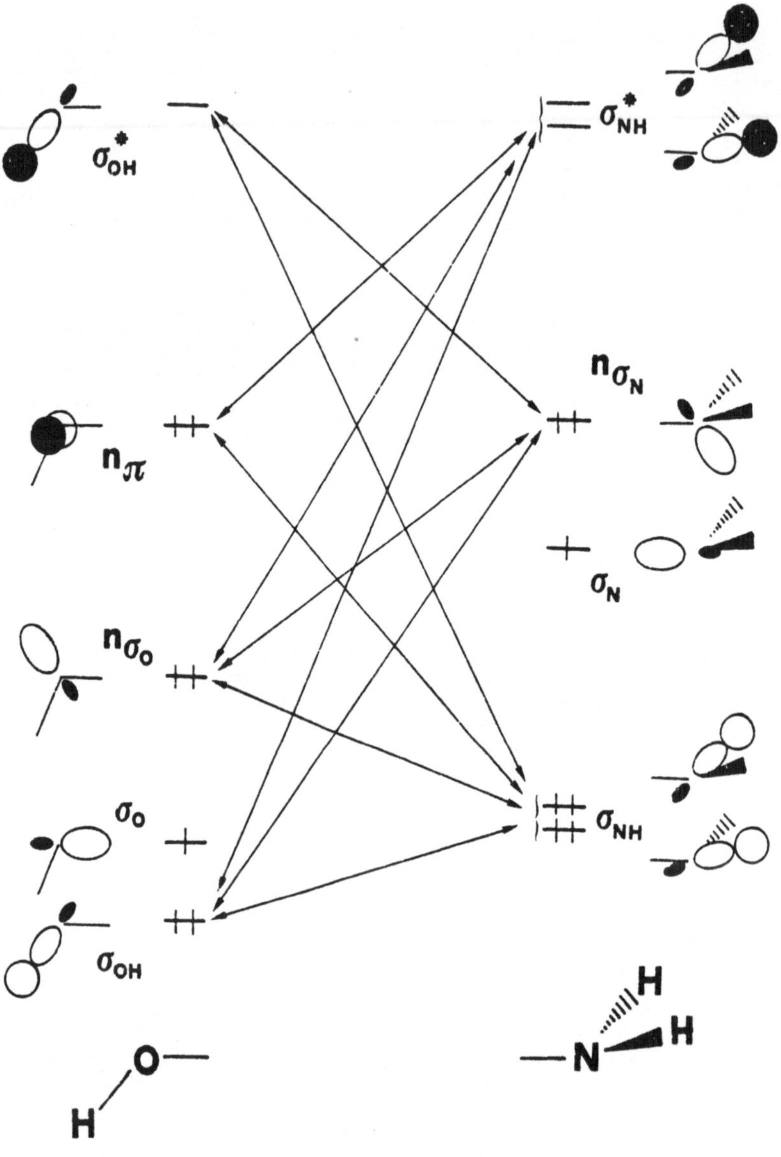


Figure 5. Non-bonded Orbital Interactions Between the HO- Fragment and the $H_2N$- Fragment in Hydroxilamine.

Table 11. Stabilization ($\Delta E^2$) and Destabilization ($\Delta E^4$) Energies (kcal/mol) Associated with the Various non-bonded Interactions in the Cis and Trans Conformations of Hydroxylamine, together with the Values of the Coulomb Energy ($E_c$, kcal/mol) Associated with the Interaction of the $H_2N$- and -OH Fragments.

| | CIS | | TRANS |
|---|---|---|---|
| | Optimized Model | Rigid Model | |
| $\Delta E^4_{\sigma_{NH},\sigma_{OH}}$ | 5.49 | 8.28 | 0.00 |
| $\Delta E^4_{\sigma_{NH},n_{\sigma_O}}$ | 1.26 | 1.25 | 0.57 |
| $\Delta E^4_{\sigma_{NH},n_\pi}$ | 9.13 | 9.30 | 9.32 |
| $\Delta E^4_{n_{\sigma_N},\sigma_{OH}}$ | 5.60 | 6.58 | 10.88 |
| $\Delta E^4_{n_{\sigma_N},n_{\sigma_O}}$ | 1.92 | 0.84 | 6.14 |
| $\Delta E^2_{\sigma_{NH},\sigma^*_{OH}}$ | -0.02 | -0.01 | -0.54 |
| $\Delta E^2_{n_{\sigma_N},\sigma^*_{OH}}$ | -2.69 | -2.28 | -0.02 |
| $\Delta E^2_{\sigma_{OH},\sigma^*_{NH}}$ | -0.02 | -0.04 | -0.21 |
| $\Delta E^2_{n_{\sigma_O},\sigma^*_{NH}}$ | -0.69 | -0.61 | -0.04 |
| $\Delta E^2_{n_\pi,\sigma^*_{NH}}$ | -2.17 | -1.78 | -1.89 |
| $\Sigma\Delta E^4$ | 39.28 | 45.08 | 36.80 |
| $\Sigma\Delta E^2$ | -8.49 | -7.16 | -5.38 |
| $\Sigma\Delta E$ | 30.79 | 37.92 | 31.42 |
| $E_c$ | 2.43 | 2.38 | 1.08 |

energy behaviour. This trend is determined by the overall destabilizing effect ($\Sigma\Delta E^4$), which in turn is controlled by the repulsion between the $\sigma_{NH}$ and $\sigma_{OH}$ bond orbitals. All the other significant effects such as the repulsions between the two σ lone pairs $n_{\sigma_N}$-$n_{\sigma_O}$ and between the σ nitrogen lone pair and the $\sigma_{OH}$ bond orbital and the conjugative stabilization associated with the $n_{\sigma_N}$-$\sigma^*_{OH}$ interaction, favor the cis geometry: however this preference for the cis geometry is not large enough to override the preference for the trans conformation dictated by the $\sigma_{NH}$-$\sigma_{OH}$ interactions.

The comparison of these results with those obtained at the cis optimized geometry shows the same trend already observed in the previous cases. The cis geometry, in fact, tends to change in order to reduce the destabilizing effects, in particular that associated with the $\sigma_{NH}$-$\sigma_{OH}$ interaction, and increase the stabilizing effects of the conjugative interactions. As a consequence the overall destabilizing effect becomes less destabilizing and the overall stabilizing effect more stabilizing, so that the overall energy effect associated with the non-bonded interactions become less destabilizing. However this reduction of the destabilizing effect is accompanied by a destabilization of the various fragments localized MO's, as indicated by the trend of the IS values (see Table 10) and also in this case the overall gain in total energy obtained in the geometry relaxation of the cis conformation is quite small (~1 kcal/mol). Also the trend of the Coulomb Energy ($E_c$) favors the trans geometry, as shown in Table 11.

In this case, the quantitative PMO analysis provides a clear rationalization of the conformational preference. In the rigid model the preference of hydroxilamine for the trans conformation is dictated by the repulsion between the $\sigma_{OH}$ and the $\sigma_{NH}$ bond orbitals and, at a lesser extent, by the Coulomb repulsion between the two interacting fragments. In the optimized model the larger destabilization of the cis conformation has to be associated also to the destabilization of the various fragments localized MO's which accompanies the geometry relaxation.

## 4.5. Stereoelectronic Effects.

Stereoelectronic effects play a major role in stereochemistry and therefore a better understanding of these effects is certainly an important step toward a better understanding of various molecular properties. The PMO analysis previously discussed seems to be one of the first theoretical tools which allows to study the stereoelectronic effects in a quantitative manner. In this section we riexamine in greater detail the effect upon molecular geometry of some types of orbital interactions which display strong directional preferences. We limit our discussion here to the cis and trans geometries.

(i) Interaction between bond MO's.

The interaction between bond MO's is, usually, the largest destabilizing effect and, in all cases investigated, is more destabilizing in a syn than in a anti alignment. The associated destabilization energy, in a given molecule, is proportional to the overlap between the two interacting bond orbitals, while no simple trend appears to exist between overlap and destabilization energy in the comparison of different molecules.

(ii) Interactions between σ-type lone pairs.

The interaction between σ-type lone pairs is again destabilizing and, in all cases investigated, more destabilizing in an anti than in a syn alignment. However, a more detailed investigation reveals that the trend of the destabilization energy associated with this interaction depends on the geometry of the molecule under investigation. In Table 12 we have listed the destabilization energy associated with the $n_\sigma$-$n_\sigma$ interaction computed at the STO-3G level for cis and trans diimide at various values of the HNN angle. It can be seen that, at ∠ HNN = 90° the destabilization energy $\Delta E^4_{nn}$ is negligible in the cis geometry and large in the trans geometry. With the increase of the HNN angle, in the cis geometry $\Delta E^4$ remains negligible till 100° and then increases becoming very large for ∠ HNN > 130°, while in the trans geometry $\Delta E^4_{nn}$ first slightly increases till 100°, then slightly decreases and tends to become small for large values of ∠ HNN. The trend of $\Delta E^4_{nn}$ parallels the trend of the matrix element $H_{nn}$, which in turn roughly parallels the trend of the overlap integral $S_{nn}$. Therefore the understanding of the relative size of the syn and anti overlap integrals between the two lone pairs allows to understand also the behaviour of $\Delta E^4_{nn}$. The overlap integral between the two σ lone pairs in a syn and an anti arrangement can be written, with reference to the coordinate system of Figure 1, as:

$$S_{syn} = |Q - a\langle 2s|2p_z\rangle + b\langle 2p_x|2p_x\rangle| \quad [9]$$

$$S_{anti} = |Q - a\langle 2s|2p_z\rangle - b\langle 2p_x|2p_x\rangle| \quad [10]$$

where Q is a constant positive term for both cases and the overlap integrals are taken to be positive. It is found that for large values of ∠HNN (∠HNN > 130°) the term $Q-a\langle 2s|2p_z\rangle$ is positive because of the small contribution of the $2p_z$ atomic orbital and therefore

$$S_{syn} > S_{anti}$$

However, for smaller values of the HNN angle, the term $Q-a\langle 2s|2p_z\rangle$ becomes negative, because of the larger contribution of the $2p_z$ orbital, and therefore

$$S_{anti} > S_{syn}$$

Table 12. Orbital Energies ($\varepsilon_n$, a.u.) of the $n_\sigma$ Lone Pairs, together with the Overlap Integrals ($S_{nn}$), Matrix Elements ($H_{nn}$, a.u.) and Destabilization Energies ($\Delta E^4_{nn}$, kcal/mol) Associated with the $n_\sigma$-$n_\sigma$ Interaction in Cis and Trans Diimide.

| | CIS | | | | TRANS | | | |
|---|---|---|---|---|---|---|---|---|
| ∠ HNN | $\varepsilon_n$ | $S_{nn}$ | $H_{nn}$ | $\Delta E^4_{nn}$ | $\varepsilon_n$ | $S_{nn}$ | $H_{nn}$ | $\Delta E^4_{nn}$ |
| 90.0° | -0.5851 | 0.0036 | 0.0271 | -0.27 | -0.5809 | 0.0711 | -0.1159 | 13.38 |
| 100.0° | -0.5790 | 0.0230 | -0.0054 | -0.46 | -0.5774 | 0.0807 | -0.1181 | 14.58 |
| 105.3° | -0.5751 | 0.0347 | -0.0250 | 0.44 | -0.5745 | 0.0850 | -0.1175 | 14.75 |
| 120.0° | -0.5641 | 0.0742 | -0.0900 | 9.02 | -0.5656 | 0.0922 | -0.1062 | 12.62 |
| 130.0° | -0.5586 | 0.1089 | -0.1449 | 23.25 | -0.5607 | 0.0899 | -0.0860 | 8.11 |
| 140.0° | -0.5566 | 0.1513 | -0.2100 | 48.89 | -0.5591 | 0.0766 | -0.0486 | 1.11 |

(iii) Interactions involving antibond orbitals.

We consider here two types of such interactions, that between a bond and an antibond orbital ($\sigma$-$\sigma^*$) and that between a $\sigma$-type lone pair and an antibond orbital (n-$\sigma^*$). Both these interactions are stabilizing and in all problems investigated more stabilizing in an anti than in a syn alignment, in agreement with a previous qualitative analysis (7). In all problems investigated the energy effects associated with the $\sigma$-$\sigma^*$ interactions are small, while those associated with the n-$\sigma^*$ interactions are more significant, as expected on the basis of the fact that the $\sigma$ lone pair is usually a better donor than a bond orbital (7). The n-$\sigma^*$ interactions play a significant role in various molecular problems: in particular it is found here that they are responsible for the larger length of the N-H, N-F and O-H bonds observed in the cis geometries of diimide (12), difluorodiazine (25), hydrogen peroxide (12) and hydroxilamine (12).

## 5. CONCLUSIONS.

In this paper we have described a computational procedure which allows to perform a quantitative PMO analysis in the framework of ab-initio SCF-MO computations and we have discussed its application to the analysis of some structural problems. The interesting features of this procedure are the following:

(i) this procedure allows to discuss at a quantitative level the structure of molecules within a model which is very near to the qualitative description used by chemists. Since, in fact, in this procedure we use a basis of fragments localized MO's, which can be easily related to bond orbitals, lone pairs and antibond orbitals, the structure of molecules can be discussed in terms of the repulsions occurring between bond orbitals, lone pairs and bond orbitals and lone pairs and in terms of the stabilizing interactions involving the doubly occupied bond orbitals and lone pairs and the vacant antibond orbitals. The former interactions represent the larger contribution of the "steric repulsions", while the latter interactions correspond to the conjugative interactions;

(ii) this procedure provides a larger number of information, compared with the qualitative PMO approach, and therefore provides a better rationalization of the problem under investigation.

The various applications have also pointed out that this quantitative analysis has still some limitations, as indicated by the fact that it does not provide a quantitative rationalization of the SCF results, but just a semi-quantitative one, or, in some cases, just a qualitative one. However we think that this procedure, also with its limitations, can be a useful instrument for the study of molecular structure.

## REFERENCES AND NOTES

1. Hoffmann, R.: 1971, Acc. Chem. Res. 4, pp. 1-9.
2. Epiotis, N.D.: 1973, J. Am. Chem. Soc. 95, pp. 3087-3096.
3. Lowe, J.P.: 1973, Science 179, pp. 527-532.
4. Hoffmann, R., Levin, C.C., and Moss, R.A.: 1973, J. Am. Chem. Soc. 95, pp. 629-631.
5. Hehre, W.J., and Salem, L.: 1973, Chem. Comm., pp. 754-755.
6. Eisenstein, O., Anh, N.T., Jean, Y., Devaquet, A., Cantacuzene, J., and Salem, L.: 1974, Tetrahedron 30, pp. 1717-1723.
7. Epiotis, N.D., Cherry, W.R., Shaik, S., Yates, R.L., and Bernardi, F.: 1977, Top. Curr. Chem. 70.
8. Baird, N.C.: 1970, Theor. Chim. Acta 16, pp. 239-242.
9. Muller, C., Schweig, A., and Vermeer, H.: 1974, Angew. Chem., Int. Ed. Engl. 13, pp. 273.
10. (a) Whangbo, M.H., Schlegel, H.B., and Wolfe, S.: 1977, J. Am. Chem. Soc. 99, pp. 1296-1304; (b) Whangbo, M.H., and Wolfe, S.: 1977, Can. J. Chem. 55, pp. 2778-2787; Wolfe, S., Mitchell, D.J., and Whangbo, M.H.: 1978, J. Am. Chem. Soc. 100, pp. 3698-3706.
11. Bernardi, F., Bottoni, A., Epiotis, N.D., and Guerra, M.: 1978, J. Am. Chem. Soc. 100, pp. 6018-6022; Bernardi, F., Bottoni, A., and Epiotis, N.D.: 1978, J. Am. Chem. Soc. 100, pp. 7205-7209; Bernardi, F., Bottoni, A., and Tonachini, G.: 1979, Theor. Chim. Acta 52, pp. 37-43.
12. Lathan, W.A., Curtiss, L.A., Hehre, W.J., Lisle, J.B., and Pople, J.A.: 1974, Progr. Phys. Org. Chem. 11, pp. 175.
13. Bernardi, F., and Bottoni, A.: Theor. Chim. Acta, submitted for publication.
14. Boys, S.F.: 1960, Rev. Mod. Phys. 32, pp. 296-299; Boys, S.F., and Foster, J.M.: 1960, ibid. 32, pp. 300-302.
15. The fundaments of perturbation theory may be found in any quantum mechanics text. Applications to quantum chemistry are particularly well represented in: Heilbronner, E., and Bock, H., "Das HMO-Modell und Seine Anwendung", Verlag Chemie, Weinheim/Bergstr., 1968, and Dewar, M.J.S., "The Molecular Orbital Theory of Organic Chemistry", Mc Graw-Hill, New York, N.Y., 1969.
16. Imamura, A.: 1968, Mol. Phys. 15, pp. 225-237.
17. Devaquet, A.: 1970, Mol. Phys. 18, pp. 233-247.
18. Basilevsky, M.V., and Berenfeld, M.M.: 1972, Int. J. Quantum Chem. 6, pp. 555-574.
19. Hehre, W.J., Stewart, R.F., and Pople, J.A.: 1969, J. Chem. Phys. 51, pp. 2657-2664.
20. Hehre, W.J., Lathan, W.A., Ditchfield, R., Newton, M.D., and Pople, J.A., Quantum Chemistry Program Exchange, N. 236, Indiana University, Bloomington, Ind..
21. Trombetti, A.: 1968, Can. J. Phys. 46, pp. 1006-1011.
22. Libit, L., and Hoffmann, R.: 1974, J. Am. Chem. Soc. 96, pp. 1370-1383.

23. Bernardi, F., and Bottoni, A.: to be published.
24. Bohn, R.K., and Bauer, S.H.: 1967, Inorg. Chem. 6, pp. 309-312.
25. Epiotis, N.D., Yates, R.L., Larson, J.R., Kirmaier, C.R., and Bernardi, F.: 1977, J. Am. Chem. Soc. 99, pp. 8379-8388.
26. Redington, R.L., Olson, W.B., and Cross, P.C.: 1962, J. Chem. Phys. 36, pp. 1311-1326; Hunt, R.H., Leacock, R.A., Peters, C.W., and Hecht, K.T.: 1965, J. Chem. Phys. 42, pp. 1931-1946.

# PERTURBATIONAL MOLECULAR ORBITAL ANALYSIS

Myung-Hwan Whangbo

Department of Chemistry
North Carolina State University
Raleigh, North Carolina 27650, U.S.A.

Theoretical and conceptual aspects of PMO analysis appropriate for inter- and intra-system perturbations were discussed within the framework of SCF MO theory.

## 1. INTRODUCTION

A molecular orbital (MO) interpretation of a chemical reaction, A + B → AB, results from the study of the relationship between the MO's of AB and those of A and B. In perturbational molecular orbital (PMO) analysis this relationship is examined from the viewpoint of perturbation theory,[1-4] which leads to the concept of orbital interaction that has provided substantial insight into the specificity of chemical reaction. In describing the electronic structure of a single molecule this orbital interaction analysis has been of great utility as well. That is, when a molecule is regarded as an assembly of molecular fragments, it becomes possible to describe the molecular electronic structure as a result of orbital interactions among the molecular fragments. The perturbation behind those interactions is that separate molecules or molecular fragments were brought together in close proximity, and hence may be termed an inter-system perturbation.

Some chemical phenomena, though not obviously the result of an inter-system perturbation, can also be subject to orbital interaction analysis. For example, when a molecule distorts from one geometry to another, the MO's of this molecule in the two different structures may be related to each other by using perturbation theory.[5] In this case, orbital interactions result from the perturbation that a molecular structure underwent distortion.



*I. G. Csizmadia and R. Daudel (eds.), Computational Theoretical Organic Chemistry, 233–252.*

This perturbation may be called an intra-system perturbation so as to distinguish from the aforementioned inter-system perturbation. In the present work some of the theoretical and conceptual aspects of PMO analysis appropriate for inter- and intra-system perturbations are discussed within the framework of SCF MO theory.

## 2. THEORETICAL DEVELOPMENT

### 2.1. Molecular orbitals in terms of fragment orbitals

Let us consider a composite molecule AB whose fragments A and B are closed-shell molecules, so that the MO's of AB, A and B may be obtained from SCF MO calculations. Typical examples of AB are molecular complexes such as $H_3B$-CO and $(H_2O)_2$. In this section we will be concerned with how the MO's of AB are related to those of A and B.

Given a set of n atomic orbitals $\underline{\underline{\chi}} = (\chi_1\ \chi_2\ \cdots\ \chi_n)$ for AB, its MO's $\underline{\underline{\phi}} = (\phi_1\ \phi_2\ \cdots\ \phi_n)$ may be written as

$$\underline{\underline{\phi}} = \underline{\underline{\chi}}\ \underline{\underline{C}} \quad ; \quad \underline{\underline{\phi}}_i = \underline{\underline{\chi}}\ \underline{\underline{C}}_i \qquad (1)$$

where

$$\underline{\underline{C}} = (\underline{\underline{C}}_1\ \underline{\underline{C}}_2\ \cdots\ \underline{\underline{C}}_n) \qquad (2a)$$

and $\underline{\underline{C}}_i$ is the column vector consisting of the i-th MO coefficients $C_{ji}$ (j = i, 2, ..., n), so that its transpose $\underline{\underline{C}}_i^\dagger$ is given by

$$\underline{\underline{C}}_i^\dagger = (C_{1i}\ C_{2i}\ \cdots\ C_{ni}) \qquad (2b)$$

The MO coefficients $C_{ji}$ are determined from the Fock equation

$$\underline{\underline{F}}\ \underline{\underline{C}} = \underline{\underline{S}}\ \underline{\underline{C}}\ \underline{\underline{e}} \quad ; \quad \underline{\underline{F}}\ \underline{\underline{C}}_i = \underline{\underline{S}}\ \underline{\underline{C}}_i\ e_i \qquad (3)$$

The atomic orbital basis set $\underline{\underline{\chi}} = (\chi_1\ \chi_2\ \cdots\ \chi_m\ \chi_{m+1}\ \cdots\ \chi_n)$ can be arranged such that $\chi_1, \chi_2, \ldots, \chi_m$ belong to A, while $\chi_{m+1}, \chi_{m+2}, \ldots, \chi_n$ belong to B. Then the Fock matrix $\underline{\underline{F}}$ and the overlap matrix $\underline{\underline{S}}$ are partitioned as follows[4]

$$\underline{\underline{F}} = \begin{vmatrix} \underline{\underline{F}}_A & \underline{\underline{F}}_{AB} \\ \underline{\underline{F}}_{AB}^\dagger & \underline{\underline{F}}_B \end{vmatrix} \quad , \quad \underline{\underline{S}} = \begin{vmatrix} \underline{\underline{S}}_A & \underline{\underline{S}}_{AB} \\ \underline{\underline{S}}_{AB}^\dagger & \underline{\underline{S}}_B \end{vmatrix} \qquad (4)$$

When the MO's of A and B are determined by using the basis sets $\underline{\underline{\chi}}_A = (\chi_1\ \chi_2\ \cdots\ \chi_m)$ and $\underline{\underline{\chi}}_B = (\chi_{m+1}\ \chi_{m+2}\ \cdots\ \chi_n)$, respectively, the MO's of A and B can be written as

$$\underline{\underline{\phi}}_A^\circ = \underline{\underline{\chi}}_A\ \underline{\underline{C}}_A^\circ \quad ; \quad \underline{\underline{F}}_A^\circ\ \underline{\underline{C}}_A^\circ = \underline{\underline{S}}_A^\circ\ \underline{\underline{C}}_A^\circ\ \underline{\underline{e}}_A^\circ \qquad (5a)$$

$$\underline{\underline{\phi}}_B^\circ = \underline{\underline{\chi}}_B \underline{\underline{C}}_B^\circ \quad ; \underline{\underline{F}}_B^\circ \underline{\underline{C}}_B^\circ = \underline{\underline{S}}_B^\circ \underline{\underline{C}}_B^\circ \underline{\underline{e}}_B^\circ \tag{5b}$$

where the superscript ° is a reminder that Eq. 5 refers to the isolated fragments A and B. Upon introducing the following matrices

$$\underline{\underline{F}}^\circ = \begin{vmatrix} \underline{\underline{F}}_A^\circ & \underline{\underline{0}} \\ \underline{\underline{0}} & \underline{\underline{F}}_B^\circ \end{vmatrix} , \quad \underline{\underline{S}}^\circ = \begin{vmatrix} \underline{\underline{S}}_A^\circ & \underline{\underline{0}} \\ \underline{\underline{0}} & \underline{\underline{S}}_B^\circ \end{vmatrix}$$
$$\underline{\underline{C}}^\circ = \begin{vmatrix} \underline{\underline{C}}_A^\circ & \underline{\underline{0}} \\ \underline{\underline{0}} & \underline{\underline{C}}_B^\circ \end{vmatrix} , \quad \underline{\underline{e}}^\circ = \begin{vmatrix} \underline{\underline{e}}_A^\circ & \underline{\underline{0}} \\ \underline{\underline{0}} & \underline{\underline{e}}_B^\circ \end{vmatrix} \tag{6}$$

Eqs. 5a and 5b can be combined together to a simpler equation

$$\underline{\underline{\phi}}^\circ = \underline{\underline{\chi}}\, \underline{\underline{C}}^\circ \quad ; \underline{\underline{F}}^\circ \underline{\underline{C}}^\circ = \underline{\underline{S}}^\circ \underline{\underline{C}}^\circ \underline{\underline{e}}^\circ \tag{7a}$$

For an i-th fragment MO $\phi_i^\circ$, Eq. 7a can be rewritten as

$$\phi_i^\circ = \underline{\underline{\chi}}\, \underline{\underline{C}}_i^\circ \quad ; \underline{\underline{F}}^\circ \underline{\underline{C}}_i^\circ = \underline{\underline{S}}^\circ \underline{\underline{C}}_i^\circ e_i^\circ \tag{7b}$$

Because of Eq. 6, the fragment orbital $\phi_i^\circ$ belongs to A if $i \leq m$, but to B if $i > m$. The composite MO's are orthonormal, and so are the fragment MO's. Therefore,

$$\underline{\underline{C}}^\dagger \underline{\underline{S}}\, \underline{\underline{C}} = \underline{\underline{1}} \quad ; \underline{\underline{C}}_i^\dagger \underline{\underline{S}}\, \underline{\underline{C}}_i = 1 \tag{8a}$$

$$\underline{\underline{C}}^{\circ\dagger} \underline{\underline{S}}^\circ \underline{\underline{C}}^\circ = \underline{\underline{1}} \quad ; \underline{\underline{C}}_i^{\circ\dagger} \underline{\underline{S}}^\circ \underline{\underline{C}}_i^\circ = 1 \tag{8b}$$

Examination of the relationship between $\underline{\underline{C}}$ and $\underline{\underline{C}}^\circ$ and that between $\underline{\underline{e}}$ and $\underline{\underline{e}}^\circ$ is in order. The composite MO's $\underline{\underline{C}}$ may be expressed in terms of the fragment MO's $\underline{\underline{C}}^\circ$ as follows

$$\underline{\underline{C}} = \underline{\underline{C}}^\circ \underline{\underline{T}} \tag{9a}$$

so that

$$\underline{\underline{C}}_i = \sum_j \underline{\underline{C}}_j^\circ T_{ji} \tag{9b}$$

From Eqs. 8b and 9a, it can be shown that

$$\underline{\underline{T}} = \underline{\underline{C}}^{\circ\dagger} \underline{\underline{S}}^\circ \underline{\underline{C}} \tag{10}$$

For the examination of the relationship between $\underline{\underline{e}}$ and $\underline{\underline{e}}^\circ$, it is convenient to define $\delta\underline{\underline{F}}$

$$\delta\underline{\underline{F}} = \underline{\underline{F}} - \underline{\underline{F}}^\circ = \begin{vmatrix} \underline{\underline{F}}_A - \underline{\underline{F}}_A^\circ & \underline{\underline{F}}_{AB} \\ \underline{\underline{F}}_{AB}^\dagger & \underline{\underline{F}}_B - \underline{\underline{F}}_B^\circ \end{vmatrix} \tag{11}$$

Premultiplying $\underline{\underline{C}}^{\circ\dagger}$ to both sides of Eq. 3 and using Eqs. 9a and 11, we obtain

$$(\underline{\underline{e}}^{\circ} + \underline{\underline{\Delta}})\,\underline{\underline{T}} = \underline{\underline{\mathcal{S}}}\,\underline{\underline{T}}\,\underline{\underline{e}}\ ;\ (\underline{\underline{e}}^{\circ} + \underline{\underline{\Delta}})\,\underline{\underline{T}}_i = \underline{\underline{\mathcal{S}}}\,\underline{\underline{T}}_i\, e_i \tag{12}$$

where

$$\begin{aligned} \underline{\underline{\Delta}} &= \underline{\underline{C}}^{\circ\dagger}\,\delta\underline{\underline{F}}\,\underline{\underline{C}}^{\circ} \ ;\ \Delta_{ij} = \underline{\underline{C}}_i^{\circ\dagger}\,\delta\underline{\underline{F}}\,\underline{\underline{C}}_j^{\circ} \\ \underline{\underline{\mathcal{S}}} &= \underline{\underline{C}}^{\circ\dagger}\,\underline{\underline{S}}\,\underline{\underline{C}}^{\circ} \ ;\ \mathcal{S}_{ij} = \underline{\underline{C}}_i^{\circ\dagger}\,\underline{\underline{S}}\,\underline{\underline{C}}_j^{\circ} \end{aligned} \tag{13}$$

It is noted that $\Delta_{ij}$ and $\mathcal{S}_{ij}$ are the Fock and the overlap matrix elements between the fragment MO's $\phi_i^{\circ}$ and $\phi_j^{\circ}$, respectively. By using Eq. 9a, Eq. 8a is rewritten as

$$\underline{\underline{T}}^{\dagger}\,\underline{\underline{\mathcal{S}}}\,\underline{\underline{T}} = \underline{\underline{1}}\ ;\ \underline{\underline{T}}_i^{\dagger}\,\underline{\underline{\mathcal{S}}}\,\underline{\underline{T}}_i = 1 \tag{14}$$

## 2.2. Perturbational treatment

An approximate but informative relationship between $\underline{\underline{C}}$ and $\underline{\underline{C}}^{\circ}$ and that between $\underline{\underline{e}}$ and $\underline{\underline{e}}^{\circ}$ may be derived from Eq. 12 by employing perturbation theory.[4,5] By taking $\underline{\underline{C}}^{\circ}$ as the zero-th order approximation to $\underline{\underline{C}}$, the following perturbation may be defined

$$\underline{\underline{h}} = \underline{\underline{h}}^{0} + \underline{\underline{h}}^{1} = \underline{\underline{e}}^{\circ} + \underline{\underline{\Delta}} \tag{15a}$$

$$\underline{\underline{\mathcal{S}}} = \underline{\underline{\mathcal{S}}}^{0} + \underline{\underline{\mathcal{S}}}^{1} = \underline{\underline{1}} + (\underline{\underline{\mathcal{S}}} - \underline{\underline{1}}) \tag{15b}$$

Namely, there exist two sources of perturbation to consider, one from the energy factor, $\underline{\underline{h}}^{1} = \underline{\underline{\Delta}}$, and the other from the overlap factor, $\underline{\underline{\mathcal{S}}}^{1} = (\underline{\underline{\mathcal{S}}} - \underline{\underline{1}})$. The latter arises from the fact that the MO's of A are not necessarily orthogonal to those of B. Let us expand $e_i$ and $\underline{\underline{T}}_i$ as follows

$$e_i = e_i^{0} + e_i^{1} + e_i^{2} + \ldots \tag{16a}$$

$$\underline{\underline{T}}_i = \underline{\underline{T}}_i^{0} + \underline{\underline{T}}_i^{1} + \underline{\underline{T}}_i^{2} + \ldots \tag{16b}$$

Eqs. 15 and 16 may be inserted into Eq. 12 to derive the following series of equations

$$[\,\underline{\underline{h}}^{0} - e_i^{0}\,\underline{\underline{\mathcal{S}}}^{0}\,]\,\underline{\underline{T}}_i^{0} = 0 \tag{17a}$$

$$[\,\underline{\underline{h}}^{0} - e_i^{0}\,\underline{\underline{\mathcal{S}}}^{0}\,]\,\underline{\underline{T}}_i^{1} + [\,\underline{\underline{h}}^{1} - e_i^{0}\,\underline{\underline{\mathcal{S}}}^{1} - e_i^{1}\,\underline{\underline{\mathcal{S}}}^{0}\,]\,\underline{\underline{T}}_i^{0} = 0 \tag{17b}$$

$$\begin{aligned} &[\,\underline{\underline{h}}^{0} - e_i^{0}\,\underline{\underline{\mathcal{S}}}^{0}\,]\,\underline{\underline{T}}_i^{2} + [\,\underline{\underline{h}}^{1} - e_i^{0}\,\underline{\underline{\mathcal{S}}}^{1} - e_i^{1}\,\underline{\underline{\mathcal{S}}}^{0}\,]\,\underline{\underline{T}}_i^{1} \\ &\qquad + [\,- e_i^{1}\,\underline{\underline{\mathcal{S}}}^{1} - e_i^{2}\,\underline{\underline{\mathcal{S}}}^{0}\,]\,\underline{\underline{T}}_i^{0} = 0 \end{aligned} \tag{17c}$$

where the expressions of higher than the second order are not shown for simplicity.

The zero-th order approximation to $\underline{\underline{C}}$ is $\underline{\underline{C}}^{\circ}$, so that $\underline{\underline{C}}^{\circ} = \underline{\underline{C}}^{\circ}\,\underline{\underline{T}}^{0}$. In other words,

$$\underline{\underline{T}}^{0} = \underline{\underline{1}} \quad ; \; T^{0}_{ji} = \delta_{ji} \tag{18a}$$

From Eqs. 15 and 17a,

$$( \underline{\underline{e}}^{\circ} - e^{0}_{i}\,\underline{\underline{1}} )\, \underline{\underline{T}}^{0}_{i} = 0 \tag{19a}$$

which, upon using Eq. 18a, leads to the expected result

$$e^{0}_{i} = e^{\circ}_{i} \tag{18b}$$

By taking the transpose of Eq. 19a, the following equation is obtained

$$\underline{\underline{T}}^{0\dagger}_{i} ( \underline{\underline{e}}^{\circ} - e^{0}_{i}\,\underline{\underline{1}} ) = 0 \tag{19b}$$

Thus, if we premultiply Eq. 17b by $\underline{\underline{T}}^{0\dagger}_{i}$ and use Eqs. 15 and 18, it is found that

$$e^{1}_{i} = \Delta_{ii} - e^{\circ}_{i} ( S_{ii} - 1 ) \tag{20a}$$

Premultiplication of Eq. 17b by $\underline{\underline{T}}^{0\dagger}_{j}$ ($j \neq i$) leads to the expression

$$T^{1}_{ji} = ( \Delta_{ji} - e^{\circ}_{i}\, S_{ji} )/( e^{\circ}_{i} - e^{\circ}_{j} ) \tag{20b}$$

where it was assumed that there is no degeneracy in $\underline{\underline{e}}^{\circ}$.

In order to obtain $T^{1}_{ii}$, Eq. 14 may be rewritten by using Eqs. 15b and 16b. Then the following series of equations are derived.

$$\underline{\underline{T}}^{0\dagger}_{i}\, \underline{\underline{S}}^{0}\, \underline{\underline{T}}^{0}_{i} = 0 \tag{21a}$$

$$\underline{\underline{T}}^{1\dagger}_{i}\, \underline{\underline{S}}^{0}\, \underline{\underline{T}}^{0}_{i} + \underline{\underline{T}}^{0\dagger}_{i}\, \underline{\underline{S}}^{0}\, \underline{\underline{T}}^{1}_{i} + \underline{\underline{T}}^{0\dagger}_{i}\, \underline{\underline{S}}^{1}\, \underline{\underline{T}}^{0}_{i} = 0 \tag{21b}$$

$$\underline{\underline{T}}^{2\dagger}_{i}\, \underline{\underline{S}}^{0}\, \underline{\underline{T}}^{0}_{i} + \underline{\underline{T}}^{0\dagger}_{i}\, \underline{\underline{S}}^{0}\, \underline{\underline{T}}^{2}_{i} + \underline{\underline{T}}^{1\dagger}_{i}\, \underline{\underline{S}}^{1}\, \underline{\underline{T}}^{0}_{i} + \underline{\underline{T}}^{0\dagger}_{i}\, \underline{\underline{S}}^{1}\, \underline{\underline{T}}^{1}_{i} + \underline{\underline{T}}^{1\dagger}_{i}\, \underline{\underline{S}}^{0}\, \underline{\underline{T}}^{1}_{i} = 0 \tag{21c}$$

where the equations of higher than the second order are not shown. Use of Eqs. 15b and 18a in Eq. 21b leads to the result

$$T^{1}_{ii} = - ( S_{ii} - 1 )/2 \tag{20c}$$

Proceeding in a similar manner, we obtain the second order solutions summarized below

$$e_i^2 = - e_i^1 (S_{ii} - 1) + \sum_j{}' (\Delta_{ji} - e_i^\circ S_{ji}) T_{ji}^1 \tag{22a}$$

$$T_{ji}^2 = [ -e_i^1 S_{ji} + \sum_k{}' (\Delta_{kj} - e_i^\circ S_{kj}) T_{ki}^1 ]/(e_i^\circ - e_j^\circ) \tag{22b}$$

$$T_{ii}^2 = - \sum_j{}' S_{ji} T_{ji}^1 - \sum_j{}'(T_{ji}^1)^2/2 - T_{ii}^1 [S_{ii} - 1 + T_{ii}^1/2] \tag{22c}$$

## 2.3. Degenerate perturbation

So far it was assumed that there is no degeneracy in the fragment MO's. Suppose, for example, that the fragment MO's $\phi_m^\circ$ ($m = i_1, i_2$) are degenerate, so that

$$e_{i_1}^\circ = e_{i_2}^\circ = e_i^\circ \tag{23}$$

The fragment MO's $\underline{\underline{C}}_m^\circ$ ($m = i_1, i_2$) may be linearly combined to produce new fragment MO's

$$\underline{\underline{D}}_i^\circ = \sum_m \alpha_m \underline{\underline{C}}_m^\circ \tag{24a}$$

that satisfy the normalization condition

$$\underline{\underline{D}}_i^{\circ\dagger} \underline{\underline{S}}^\circ \underline{\underline{D}}_i^\circ = 1 \tag{24a}$$

Since the MO's $\underline{\underline{D}}_i^\circ$ are also the zero-th order solution to $\underline{\underline{C}}$ like $\underline{\underline{C}}_m^\circ$, the following constraint may be imposed on $\underline{\underline{T}}_i^\circ$ if $i = i_1$ or $i_2$.

$$\begin{aligned} T_{mi}^0 &= \alpha_m \quad , \text{ if } m = i_1 \text{ or } i_2 \\ &= 0 \quad , \text{ if } m \neq i_1 \text{ or } i_2 \end{aligned} \tag{25a}$$

Let us define column matrices $\underline{\underline{U}}_i$ ($i = i_1, i_2$) such that $U_{ji} = \delta_{ji}$ for $j = 1, 2, \ldots, n$. Then, premultiplying Eq. 17b by $\underline{\underline{U}}_i^\dagger$ ($i = i_1, i_2$) gives rise to the following expressions

$$\begin{aligned} (\Delta_{i_1 i_1} - e_i^1)\, \alpha_{i_1} + (\Delta_{i_1 i_2} - e_i^\circ S_{i_1 i_2})\, \alpha_{i_2} &= 0 \\ (\Delta_{i_1 i_2} - e_i^\circ S_{i_1 i_2})\, \alpha_{i_1} + (\Delta_{i_2 i_2} - e_i^1)\, \alpha_{i_2} &= 0 \end{aligned} \tag{25b}$$

and thus

$$\begin{vmatrix} \Delta_{i_1 i_1} - e_i^1 & \Delta_{i_1 i_2} - e_i^\circ S_{i_1 i_2} \\ \Delta_{i_1 i_2} - e_i^\circ S_{i_1 i_2} & \Delta_{i_2 i_2} - e_i^1 \end{vmatrix} = 0 \tag{25c}$$

which yields the first order correction $e_i^1$ for $i = i_1$ and $i_2$. The fragment MO's $\underline{\underline{D}}_i^\circ$ determined from Eqs. 24b and 25 may be employed to evaluate higher order energy corrections.

## 2.4 Intra-system perturbation

Suppose that the geometry of a molecule is distorted by moving some of its atoms from the positions initially adopted, but each atom is represented by the same basis set as before. In MO calculations the information about molecular geometry is contained in the basis set functions since they are given specific positions in the coordinate system. Therefore, the basis sets employed in two different geometries may be distinguished by $\underline{\underline{\chi}}$ and $\underline{\underline{\chi}}^\circ$, and thus the MO's of a molecule in the two structures may be written as

$$\begin{aligned} \underline{\underline{\phi}} &= \underline{\underline{\chi}}\,\underline{\underline{C}} \quad ; \quad \underline{\underline{F}}\,\underline{\underline{C}} = \underline{\underline{S}}\,\underline{\underline{C}}\,\underline{\underline{e}} \\ \underline{\underline{\phi}}^\circ &= \underline{\underline{\chi}}^\circ\,\underline{\underline{C}}^\circ \quad ; \quad \underline{\underline{F}}^\circ\,\underline{\underline{C}}^\circ = \underline{\underline{S}}^\circ\,\underline{\underline{C}}^\circ\,\underline{\underline{e}}^\circ \end{aligned} \tag{26}$$

The relationship between $\underline{\underline{C}}$ and $\underline{\underline{C}}^\circ$ and that between $\underline{\underline{e}}$ and $\underline{\underline{e}}^\circ$ may be analyzed as in Section 2.2. With the definition of $\delta\underline{\underline{F}} = \underline{\underline{F}} - \underline{\underline{F}}^\circ$, it can be easily shown that Eqs. 17-35 remain valid for the intra-system perturbation as well.

In principle, there is no restriction on which of the two structures under consideration should be taken as the unperturbed one, but orbital interaction analysis becomes considerably simpler if the structure of higher symmetry is chosen as the unperturbed one. For example, consider a simplified Walsh diagram 1 for $AH_3$ (A = N, P, As), in which a pyramidal structure of point group $C_{3v}$ is regarded as distorted from the planar structure of point group $D_{3h}$. The HOMO $\phi_i^\circ$ of the planar structure is a pure

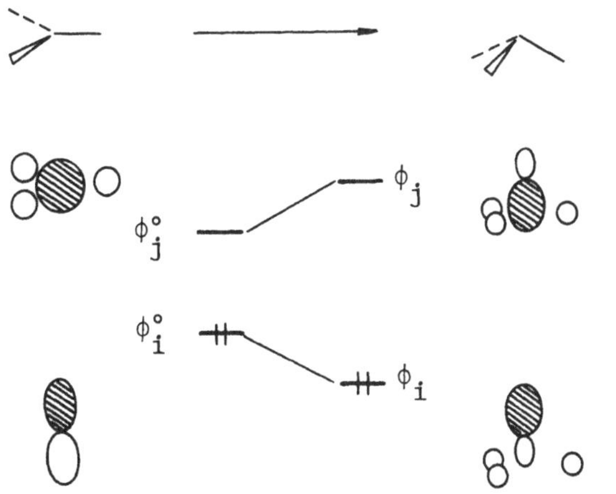


1

p-orbital so that $S_{ii} = \underline{\underline{C}}_i^{\circ\dagger} \underline{\underline{S}}\, \underline{\underline{C}}_j^{\circ} = 1$. Therefore from Eqs. 18b, 20a and 22a, the energy for the HOMO $\phi_i$ of the pyramidal structure can be written as

$$e_i = e_i^{\circ} + \Delta_{ii} + (\Delta_{ij} - e_i^{\circ} S_{ij})^2/(e_i^{\circ} - e_j^{\circ}) \quad (27)$$

Within the framework of one-electron Hamiltonian theory the first order term $\Delta_{ii} = \underline{C}_i^{\circ\dagger}\, \delta\underline{\underline{F}}\, \underline{C}_i^{\circ}$ vanishes, because the HOMO $\phi_i^{\circ}$ is a pure p orbital and further the matrix element of $\delta\underline{\underline{F}}$ for this p orbital is zero. This simplification allows the analysis of the pyramidal inversion barrier in $AH_3$ in terms of the HOMO-LUMO gap $(e_i^{\circ} - e_j^{\circ})$ of the planar structure.[5] Such a simplification would not have been possible had we chosen the pyramidal structure as the unperturbed one. For in the latter case, $S_{ii} \neq 1$ and further $\Delta_{ii} \neq 0$ even in one-electron Hamiltonian theory.

## 3. CONCEPTUAL ASPECTS OF ORBITAL INTERACTION

### 3.1. First order energy correction in correlation diagram

To simplify our discussion, it may be assumed that the coordinates of the isolated fragments A and B are exactly the same as those of the corresponding fragments present in AB. Then,

$$\underline{\underline{S}}_A^{\circ} = \underline{\underline{S}}_A \quad , \quad \underline{\underline{S}}_B^{\circ} = \underline{\underline{S}}_B \quad (28)$$

and thus $S_{im} = \delta_{im}$ if i and m refer to a same fragment. Consequently,

$$e_i^1 = \Delta_{ii} \quad (29a)$$

$$T_{ii}^1 = 0$$

Within the framework of SCF MO theory, the first order term $\Delta_{ii}$ does not necessarily vanish, since each fragment present in AB is under the potential different from that in the isolated state. Namely,

$$\underline{\underline{F}}_A - \underline{\underline{F}}_A^{\circ} \neq \underline{\underline{0}} \quad , \quad \underline{\underline{F}}_B - \underline{\underline{F}}_B^{\circ} \neq \underline{\underline{0}} \quad (30)$$

When the potentials around A and B are significantly different, one might encounter a seemingly unusual correlation diagram such as $\underline{2}$, which does not have the final energy levels $e_i$ and $e_j$ outside the region between the starting energy levels $e_i^{\circ}$ and $e_j^{\circ}$. Such a phenomenon is likely to happen if the first order correction terms $\Delta_{ii}$ and $\Delta_{jj}$ are large. Since $e_i^{\circ} < e_j^{\circ}$ in $\underline{2}$, it is expected that A has a lower potential than does B in the isolated state. When combined together to form AB, therefore, the effective potential around A will become less attractive, but that around B

2

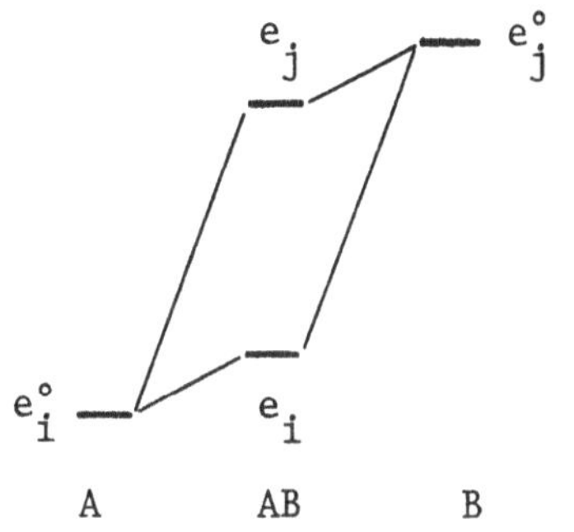


more attractive. Thus it is likely that $\Delta_{ii} > 0$, and $\Delta_{jj} < 0$. Subtraction of these first order terms from the final energy levels may lead to a correlation diagram such as 3a, while addition of these terms to the starting energy levels may provide a correlation diagram such as 3b.

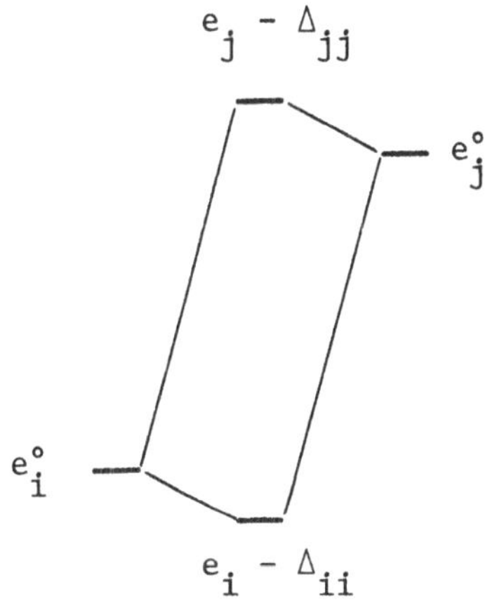


3a

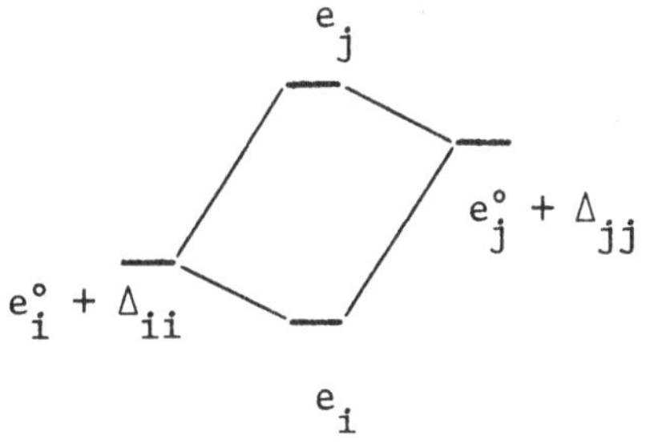


3b

### 3.2. Fock matrix partitioning and functional groups

Within the framework of SCF MO theory, the definition of fragment MO's poses a problem when fragments are merely functional groups,[4,6] since they are not chemical species but conceptual building blocks introduced to simplify the description of molecular electronic structure.[7] The MO's of a molecule result from the eigenvalue problem associated with the Fock matrix, and thus it may be considered that the eigenvalue problem associated with a part of the Fock matrix defines the orbitals pertaining to that part of the molecule.

To gain some insight into the Fock matrix partitioning method for functional groups,[4] let us return to the case of a composite molecule AB whose fragments A and B are closed shells so that separate SCF MO calculations on A and B provide $\underline{\underline{C}}^\circ$ and $\underline{\underline{e}}^\circ$ as in Eq. 7. If there is no degeneracy in $\underline{\underline{e}}^\circ$, the following may be written for $e_i$

$$e_i = e_i^\circ + \Delta_{ii} + \sum_k{}' (\Delta_{ik} - e_i^\circ S_{ik})^2/(e_i^\circ - e_k^\circ) \tag{31a}$$

$$= e_i^\circ + \Delta_{ii} + \sum_m{}' (\Delta_{im})^2/(e_i^\circ - e_m^\circ) + \sum_j{}' (\Delta_{ij} - e_i^\circ S_{ij})^2/(e_i^\circ - e_j^\circ) \tag{31b}$$

where i and m refer to a same fragment, but i and j do not. Eq. 31b follows from Eq. 31a, since $S_{im} = \delta_{im}$ because of Eq. 28.

From the Fock and overlap matrices of AB, we may partition out the following block diagonal matrices and obtain the

$$\underline{\underline{F}}' = \begin{vmatrix} \underline{\underline{F}}_A & \underline{\underline{0}} \\ \underline{\underline{0}} & \underline{\underline{F}}_B \end{vmatrix} , \quad \underline{\underline{S}}' = \begin{vmatrix} \underline{\underline{S}}_A & \underline{\underline{0}} \\ \underline{\underline{0}} & \underline{\underline{S}}_B \end{vmatrix} \tag{32a}$$

corresponding orbitals $\underline{\underline{C}}'$ and orbital energies $\underline{\underline{e}}'$ as the result of the eigenvalue problem

$$\underline{\underline{F}}' \, \underline{\underline{C}}' = \underline{\underline{S}}' \, \underline{\underline{C}}' \, \underline{\underline{e}}' \tag{32b}$$

Because of Eq. 30, it is noted that $\underline{\underline{C}}' \neq \underline{\underline{C}}^\circ$ and $\underline{\underline{e}}' \neq \underline{\underline{e}}^\circ$. As in Section 2.2, the relationship between $\underline{\underline{e}}$ and $\underline{\underline{e}}^\circ$ may be analyzed by defining the following quantities

$$\begin{aligned} \Delta'_{ik} &= \underline{\underline{C}}_i'^{\dagger} \, \delta\underline{\underline{F}}' \, \underline{\underline{C}}_k' \\ S'_{ik} &= \underline{\underline{C}}_i'^{\dagger} \, \underline{\underline{S}} \, \underline{\underline{C}}_k' \end{aligned} \tag{33}$$

where

$$\delta\underline{\underline{F}}' = \underline{\underline{F}} - \underline{\underline{F}}' = \begin{vmatrix} \underline{\underline{0}} & \underline{\underline{F}}_{AB} \\ \underline{\underline{F}}_{AB}^{\dagger} & \underline{\underline{0}} \end{vmatrix} \tag{34}$$

Since the diagonal blocks of $\delta\underline{\underline{F}}'$ have vanishing elements, $\Delta'_{im} = 0$ when i and m refer to a same fragment (i.e., A or B). Further, $S_{im} = \delta_{im}$ because of Eq. 28. Therefore, when there is no degeneracy in $\underline{\underline{e}}'$, the following equation is obtained for $e_i$

$$e_i = e_i' + \sum_j{}' (\Delta'_{ij} - e_i' S'_{ij})^2/(e_i' - e_j') \tag{35}$$

where i and j refer to different fragments.

The meaning of Eq. 35 may now be analyzed by examining the relationship between $\underline{\underline{C}}'$ and $\underline{\underline{C}}^\circ$. Let us define

$$\begin{aligned} \Delta''_{ik} &= \underline{\underline{C}}_i^{\circ\dagger}\ \delta\underline{\underline{F}}''\ \underline{\underline{C}}_k^\circ \\ S''_{ik} &= \underline{\underline{C}}_i^{\circ\dagger}\ \underline{\underline{S}}'\ \underline{\underline{C}}_k^\circ \end{aligned} \tag{36}$$

where

$$\delta\underline{\underline{F}}'' = \underline{\underline{F}}' - \underline{\underline{F}}^\circ = \begin{vmatrix} \underline{\underline{F}}_A - \underline{\underline{F}}_A^\circ & \underline{\underline{0}} \\ \underline{\underline{0}} & \underline{\underline{F}}_B - \underline{\underline{F}}_B^\circ \end{vmatrix} \tag{37}$$

Only the diagonal blocks of $\delta\underline{\underline{F}}''$ have nonzero elements so that $\Delta''_{ij}$ vanishes when i and j refer to different fragments, while $\Delta''_{im} = \Delta_{im}$ if i and m refer to a same fragment. Further $S''_{im} = \delta_{im}$ since $\underline{\underline{S}}' = \underline{\underline{S}}^\circ$ under Eq. 28. Therefore, $e_i'$ is written as

$$e_i' = e_i^\circ + \Delta_{ii} + \sum_m{}' (\Delta_{im})^2/(e_i^\circ - e_m^\circ) \tag{38a}$$

To first order, the following equation can be given for $\underline{\underline{C}}_i'$

$$\underline{\underline{C}}_i' = \underline{\underline{C}}_i^\circ + \sum_m{}' \left[ \Delta_{im}/(e_i^\circ - e_m^\circ) \right] \underline{\underline{C}}_m^\circ \tag{38b}$$

Provided that i and m refer to A, it may be said that the change in potential around A (i.e., $\underline{\underline{F}}_A - \underline{\underline{F}}_A^\circ$), brought about by the neighboring fragment B, leads to intermixing and hence polarization of the MO's of A. This 'static' MO polarization is caused by $\delta\underline{\underline{F}}''$, and is different from the 'dynamic' MO polarization resulting from $\delta\underline{\underline{F}}'$ (i.e., the second term of Eq. 22b which is often called the second order orbital mixing).[8-11] According to Eq. 38a, the level $e_i^\circ$ is adjusted to $e_i'$ so as to accommodate the environmental effect. If the MO polarization in Eq. 38b is not significant, one may derive from Eqs. 33 and 38b that $\Delta_{ij}' \simeq \Delta_{ij}$, and $S_{ij}' \simeq S_{ij}$ when i and j refer to different fragments. Then Eq. 35 can be simplified to

$$e_i = e_i' + \sum_j{}' (\Delta_{ij} - e_i' S_{ij})^2/(e_i' - e_j') \tag{39}$$

In Eq. 39 the interaction between two fragment orbitals are evaluated in terms of the 'environment-adjusted' levels, $e_i'$ and $e_j'$.

In the case of a composite molecule AB whose fragments A and B are functional groups, $\underline{\underline{C}}'$ and $\underline{e}'$ defined by Eq. 32b may be adopted as their orbitals and orbital energies, respectively.[4,7] These fragment orbitals contain environmental effects as described above, but they are found to be nearly transferable from conformation to conformation and from molecule to molecule since the Fock matrix elements of a functional group are nearly transferable.[12]

3.3. Charge redistribution and energy lowering

In orbital interaction analysis the sign of the energy term $(\Delta_{ij} - e_i^\circ S_{ij})$ becomes important.[13-15] Consider a typical example of inter-system interaction between nondegenerate levels shown in 4. For simplicity, Eq. 28 is assumed to be satisfied,

4

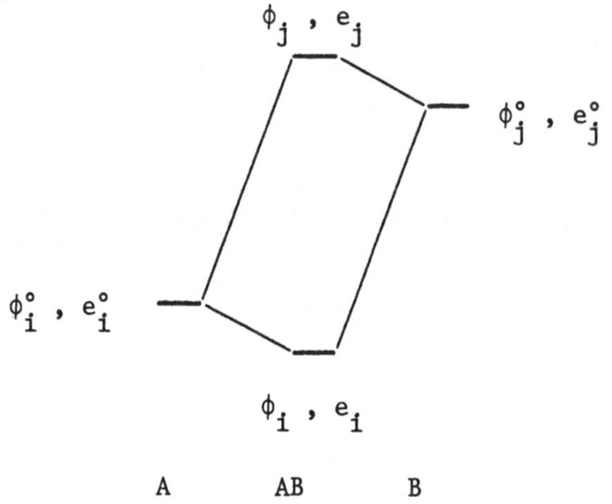


so that $S_{ii} = 1$. Then the following expressions can be given for $e_i$ and $\underline{\underline{C}}_i$.

$$e_i = e_i^\circ + \Delta_{ii} + (\Delta_{ij} - e_i^\circ S_{ij})^2/(e_i^\circ - e_j^\circ) \quad (40a)$$

$$\underline{\underline{C}}_i = (1 - t\, S_{ij} - t^2/2)\, \underline{\underline{C}}_i^\circ + t\, \underline{\underline{C}}_j^\circ \quad (40b)$$

where

$$t = (\Delta_{ij} - e_i^\circ S_{ij})/(e_i^\circ - e_j^\circ) \quad (41)$$

Without loss of generality the following conditions may be adopted for our discussion:[13]

$$S_{ij} > 0, \text{ and thus } \Delta_{ij} < 0 \quad (42a)$$

$$e_i^\circ < 0 \quad (42b)$$

Since $e_i^\circ - e_j^\circ < 0$ in 4, Eqs. 41 and 42 lead to the conclusions

$$t > 0, \text{ if } (\Delta_{ij} - e_i^\circ S_{ij}) < 0 \quad (43a)$$

$$t < 0, \text{ if } (\Delta_{ij} - e_i^\circ S_{ij}) > 0 \quad (43b)$$

Eq. 40a reveals that the second order energy term is negative, regardless of whether $\phi_i^\circ$ and $\phi_j^\circ$ combine in-phase $(t > 0)$ or out-of-phase ($t < 0$). Eqs. 43a and 43b refer to normal and counterintuitive orbital mixings, respectively.

In order to examine the meaning of Eq. 43, we take note of the following equations

$$\begin{aligned} e_i &= \underline{\underline{C}}_i^\dagger \, \underline{\underline{F}} \, \underline{\underline{C}}_i \\ &= \Delta_{ii} + (1 - 2t\, S_{ij} - t^2)\, e_i^\circ + 2t\, \Delta_{ij} + t^2\, e_j^\circ \end{aligned} \tag{44a}$$

$$\begin{aligned} 1 &= \underline{\underline{C}}_i^\dagger \, \underline{\underline{S}} \, \underline{\underline{C}}_i \\ &= (1 - 2t\, S_{ij} - t^2) + 2t\, S_{ij} + t^2 \end{aligned} \tag{44b}$$

Eq. 44b shows the charge density redistribution induced by the orbital interaction (normalized to unity), and Eq. 44a represents the corresponding energy change. Normal orbital mixing ($t > 0$) causes charge buildup in the region between A and B at the expense of charge loss from the region of A ($2t\, S_{ij} > 0$), as depicted in 5a. This charge redistribution leads to the destabilization of $-2t\, S_{ij}\, e_i^\circ$ ( $> 0$) from the region of A, and to the

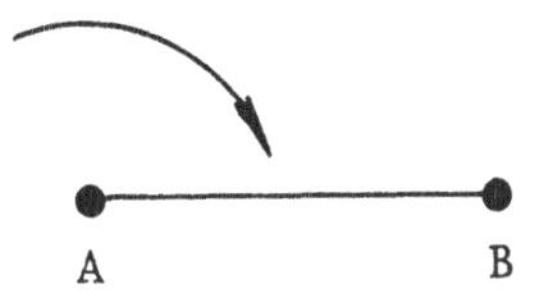


5a

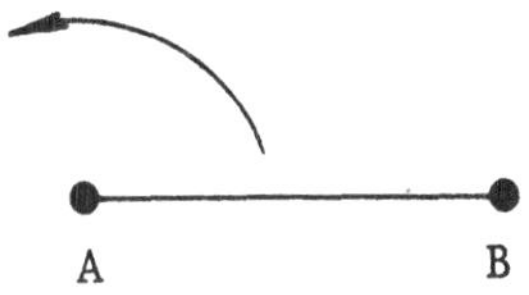


5b

stabilization of $2t\, \Delta_{ij}$ ( $< 0$) from the region between A and B. According to Eq. 43a the net result of these energy contributions is energy-lowering, namely,

$$-\,2t\, S_{ij}\, e_i^\circ + 2t\, \Delta_{ij} \; < \; 0 \tag{45}$$

As shown in 5b, counterintuitive orbital mixing ($t < 0$) causes charge buildup in the region of A at the expense of charge loss in the region between A and B ($2t\, S_{ij} < 0$). This charge shift causes the stabilization of $-2t\, S_{ij}\, e_i^\circ$ ( $< 0$) from the region of A and the destabilization of $2t\, \Delta_{ij}$ ( $> 0$) from the region between A and B. According to Eq. 43b the net result of these energy contributions is again energy-lowering, and Eq. 45 remains valid for the case of counterintuitive orbital mixing.

Counterintuitive orbital mixing occurs when two interacting orbitals differ greatly in energy but overlap significantly, and it provides a mechanism for a diffuse orbital to work as a polarization function for a contracted orbital. Experimentally observable consequences of counterintuitive orbital mixing may be found from the unusual bonding lengthening that occurs when a bond between electropositive atoms is surrounded by highly electronegative ligands (e.g., P-C bonds in $P(CF_3)_3$ and the Si-B bond in $F_3Si$-$BF_2$):[15] Charge shift into the region of the highly electronegative ligands, which might occur at the expense of some charge loss from the bond region between the electropositive atoms, may become energetically favorable. This would lead to a concomitant weakening and hence lengthening of the bond.

The experimental Si-B bond length in $F_3Si$-$BF_2$ is 2.04A,[16] which is quite longer than the expected value of 1.92-1.96A[15,16] for a single Si-B bond. To examine the effect of fluorine atoms on the Si-B bond length, we carried out ab initio SCF MO calculations on $F_3Si$-$BF_2$ and $H_3Si$-$BH_2$ using the STO-3G basis set[17] of GAUSSIAN 70.[18] In this study all the geometrical parameters except for r(Si-B) were frozen, i.e., r(Si-F) = 1.57, r(B-F) = 1.30, r(S-H) = 1.486, and r(B-H) = 1.194A. The structures around Si and B were also taken to be tetrahedral and trigonal planar, respectively. The optimized Si-B bond lengths r(Si-B) and their overlap populations p(Si-B) shown below are compatible with the interpretation that, in going from $H_3Si$-$BH_2$ to $F_3Si$-$BF_2$, the Si-B bond loses some electron density into the region of the surrounding fluorine atoms and hence becomes elongated.

| molecule | r(Si-B) | p(Si-B) |
|---|---|---|
| $H_3Si$-$BH_2$ | 1.965 | 0.359 |
| $F_3Si$-$BF_2$ | 2.013 | 0.299 |

### 3.4. Orbital interaction energy in molecular complexes

Let us examine orbital interaction energies in molecular complexes AB, whose fragments A and B are closed shells, under the condition that Eq. 28 is satisfied. Thus for the interactions between nondegenerate orbitals such as those shown in 6a and 6b, the following can be written

$$e_i = e_i^\circ + \Delta_{ii} + (\Delta_{ij} - e_i^\circ S_{ij})^2/(e_i^\circ - e_j^\circ) \tag{46a}$$

$$e_j = e_j^\circ + \Delta_{jj} + (\Delta_{ij} - e_j^\circ S_{ij})^2/(e_j^\circ - e_i^\circ) \tag{46b}$$

Therefore, in 6a and 6b, the net changes in the occupied orbital energies resulting from interactions may be given by

$$\delta E(\underline{6a}) = 2(\Delta_{ij} - e_i^\circ S_{ij})^2/(e_i^\circ - e_j^\circ) \tag{47a}$$

$$\delta E(\underline{6b}) = -4S_{ij}(\Delta_{ij} - \bar{e}^\circ S_{ij}) \tag{47b}$$

where $\bar{e}^\circ = (e_i^\circ + e_j^\circ)/2$.

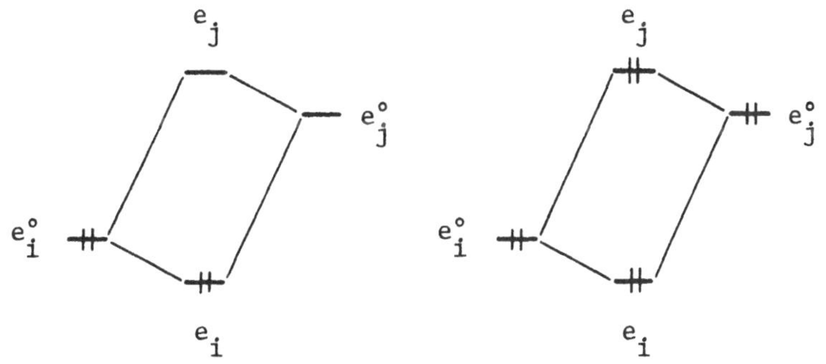


6a 6b

The first order terms $\Delta_{ii}$ and $\Delta_{jj}$ in Eq. 46 originate from the change in potential around A and B that occurs as a result of complex formation. According to Eq. 39, the environment-adjusted levels $e_i^\circ + \Delta_{ii}$ and $e_i^\circ + \Delta_{jj}$ may also be employed to approximate $e_i$ and $e_j$.

$$e_i = e_i' + (\Delta_{ij} - e_i' S_{ij})^2/(e_i' - e_j') \tag{48a}$$

$$e_j = e_j' + (\Delta_{ij} - e_j' S_{ij})^2/(e_j' - e_i') \tag{48b}$$

where $e_i' = e_i^\circ + \Delta_{ii}$, and $e_j' = e_j^\circ + \Delta_{jj}$. Eqs. 48a and 48b are approximate solutions of the secular equation[2]

$$\begin{vmatrix} e_i' - e & \Delta_{ij} - e\,S_{ij} \\ \Delta_{ij} - e\,S_{ij} & e_j' - e \end{vmatrix} = 0 \tag{49}$$

With respect to the levels $e_i'$ and $e_j'$, the orbital interaction energies for 6a and 6b may be written as

$$\delta E'(\underline{6a}) = 2(\Delta_{ij} - e_i' S_{ij})^2/(e_i' - e_j') \tag{50a}$$

$$\delta E'(\underline{6b}) = -4S_{ij}(\Delta_{ij} - \bar{e}' S_{ij}) \tag{50b}$$

where $\bar{e}' = (e_i' + e_j')/2$. For strong interactions in which perturbation treatment is not satisfactory, Eq. 49 may be employed to calculate their interaction energies. In one-electron Hamiltonian theory, $\Delta_{ii} = \Delta_{jj} = 0$, so that Eqs. 47 and 50 become identical.

## 3.5. Effect of charge on orbital interaction

In this section we will examine the effect of charge on orbital interaction[19] by analyzing the relative importance of the σ-donation and π-back donation in the isoelectronic series $H_3B$-XY (XY = $NO^+$, CO, $CN^-$). We have carried out ab initio SCF MO calculations on $H_3B$-XY using the STO-3G basis set, and optimized the B-X bond lengths and HBX bond angles. The X-Y bond lengths obtained from separate SCF MO calculations (i.e., 1.127, 1.147 and 1.165A for $NO^+$, CO and $CN^-$, respectively) and the B-H bond length of 1.194A were used without further change. The results of this study summarized below indicates that, in going from $H_3B$-$NO^+$ to $H_3B$-$CN^-$, the HBX bond angle increases gradually, and the energy of complexation, $\Delta E = E(H_3B\text{-}XY) - E(BH_3) - E(XY)$, increases sharply.

| XY | r(B-X) | ∠HBX | ΔE |
|---|---|---|---|
| | A | degree | kcal/mol |
| $NO^+$ | 1.452 | 100.6 | -7.1 |
| CO | 1.618 | 103.4 | -35.7 |
| $CN^-$ | 1.613 | 108.2 | -131 |

Consider now the two stabilizing orbital interactions of importance in $H_3B$-XY, i.e., the σ-donation (σ - p) and the π-back donation (π - π*) shown in 7a and 7b, respectively. 7a is just one of the two equivalent (π - π*) interactions present in $H_3B$-XY. The initial levels $e_i^\circ$ of the fragment orbitals σ, p, π and π* and their first order correction terms $\Delta_{ii}$, which were calculated by

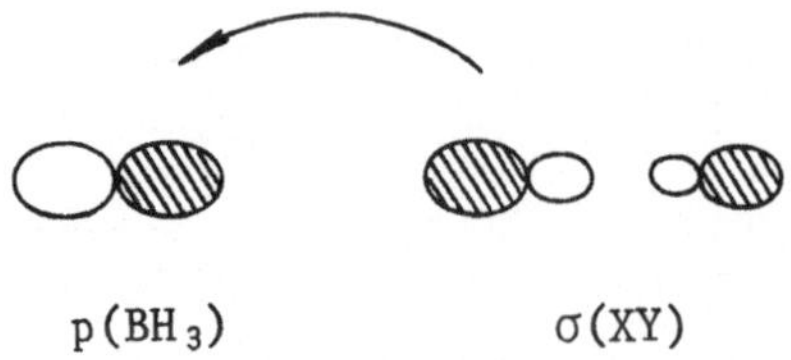


7a

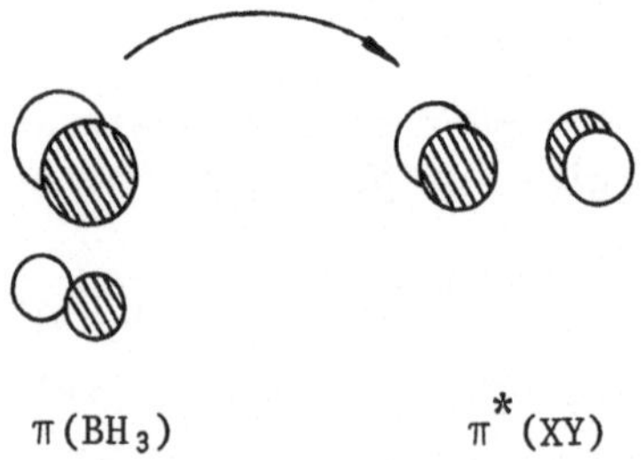


7b

employing the fragment geometries found in $H_3B$-XY, are listed below in atomic units. Note that the first order correction terms are quite significant. The environment-adjusted levels $e'_i$ (= $e^\circ_i$ + $\Delta_{ii}$) of the aforementioned fragment orbitals are shown in 8.

| orbital | $H_3B$-$NO^+$ | | $H_3B$-CO | | $H_3B$-$CN^-$ | |
|---|---|---|---|---|---|---|
| | $e^\circ_i$ | $\Delta_{ii}$ | $e^\circ_i$ | $\Delta_{ii}$ | $e^\circ_i$ | $\Delta_{ii}$ |
| σ | -1.0994 | -0.0881 | -0.4450 | -0.2106 | 0.0449 | -0.2359 |
| p | 0.2227 | -0.8086 | 0.2142 | -0.3637 | 0.1974 | -0.1166 |
| π | -0.4545 | -0.2381 | -0.4504 | 0.0449 | -0.4412 | 0.2878 |
| π* | -0.3142 | -0.0393 | 0.3054 | -0.1362 | 0.8164 | -0.1905 |

In $H_3B$-CO the orbital energy gaps for the (σ - p) and (π - π*) interactions are roughly the same, so that the σ-donation and

8

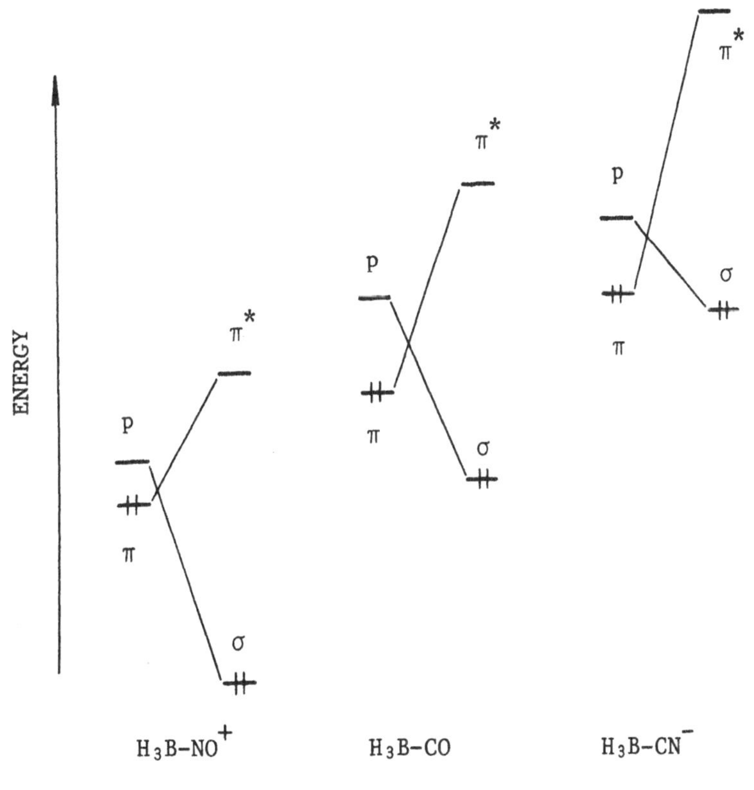

π-back donation are expected to be equally important. In $H_3B\text{-}NO^+$ the σ and π* orbitals of $NO^+$ are shifted downward relative to the p and π orbitals of $BH_3$, due to the presence of positive charge on $NO^+$.[19] The level shift in the opposite direction happens in $H_3B\text{-}CN^-$ because of the negative charge on $CN^-$. Based upon the orbital energy gaps alone, it is expected that the π-back donation would dominate in $H_3B\text{-}NO^+$, while the σ-donation would dominate in $H_3B\text{-}CN^-$. The values of $\Delta_{ij}$ (in a.u.) and $S_{ij}$ for the (σ - p) and (π - π*) interactions are found to be large, so that their interaction energies $\delta E'_{ij}$ (in kcal/mol) were calculated by using Eq. 49. Our expectations based upon the orbital energy gaps are confirmed by these interaction energies.

| interaction | | $H_3B\text{-}NO^+$ | $H_3B\text{-}CO$ | $H_3B\text{-}CN^-$ |
|---|---|---|---|---|
| | $\Delta_{ij}$ | -0.6745 | -0.4825 | -0.2764 |
| (σ - p) | $S_{ij}$ | 0.4575 | 0.5069 | 0.4591 |
| | $\delta E'_{ij}$ | -29.2 | -41.5 | -89.2 |
| | $\Delta_{ij}$ | -0.3180 | -0.1761 | -0.0883 |
| (π - π*) | $S_{ij}$ | 0.1718 | 0.1713 | 0.1514 |
| | $\delta E'_{ij}$ | -102 | -22.7 | -6.61 |

## 4. DISCUSSION AND CONCLUDING REMARKS

Perturbations causing orbital interactions may be divided into two classes, i.e., inter- and intra-system perturbations, depending upon whether or not they involve molecular fragmentation. In either case, the concept of orbital interaction originates naturally from the study of the relationship between two sets of MO's, so that the Fock and overlap matrices play a central role in PMO analysis. It should be emphasized that the perturbation in PMO analysis is not defined by the molecular Hamiltonian operator, and it consists of two factors, namely, the energy factor $\underline{\underline{\Delta}}$ and the overlap factor $(\underline{\underline{S}} - \underline{\underline{1}})$.

Complex formation between two molecules brings about a change in potential around each molecule, which is largely reflected in the first order energy terms $\Delta_{ii}$. Incorporation of these terms into the initial levels $e_i^\circ$ leads to the environment-adjusted levels $e_i'$, which may be used to calculate orbital interaction energies. Since $\delta\underline{\underline{F}} = \delta\underline{\underline{F}}' + \delta\underline{\underline{F}}''$, it is convenient to consider the perturbation of molecular complexation in two stages: The perturbation of $\delta\underline{\underline{F}}''$, which provides environment-adjusted fragments, and that of $\delta\underline{\underline{F}}'$, which leads to the interaction between these environment-adjusted fragments. Consequently, the fragment orbitals defined from the Fock matrix partitioning method, which are frequently employed when molecular fragments refer to functional groups,[4,6,7,11] are the environment-adjusted orbitals.

It is evident from the series of $H_3B$-XY (XY = $NO^+$, CO, $CN^-$) that the nature of orbital interaction can be modified quite significantly (e.g., from the dominance of π-back donation in $H_3B$-$NO^+$ to that of σ-donation in $H_3B$-$CN^-$). Further when a bond is surrounded by highly electronegative ligands, counterintuitive orbital mixing may become significant enough to cause an experimentally observable consequence such as the lengthening of that bond.

It is hoped that the theoretical and conceptual aspects of PMO analysis discussed in the present work will become a useful guideline for analyzing SCF MO wave functions of inorganic and organometallic compounds, and also it will become a convenient starting point for further development of PMO theory necessary for handling such problems as orbital interactions in open-shell molecules within the framework of SCF MO theory.

ACKNOWLEDGMENT

Some parts of the present work were initiated while the author was a member of Professor R. Hoffmann's research group at Cornell University. The author is thankful to Professor R. Hoffmann for his kind hospitality extended to the author during that period. This work was partially supported by Research Corporation through a Cottrell Research Program, which is gratefully acknowledged.

REFERENCES

1. K. Fukui and H. Fujimoto, Bull. Chem. Soc. Japan, 41, 1989(1968)
2. L. Salem, J. Am. Chem. Soc., 90, 543(1968)
3. A. Imamura, Mol. Phys., 15, 225(1968)
4. M. -H. Whangbo, H. B. Schlegel and S. Wolfe, J. Am. Chem. Soc., 99, 1296(1977)
5. C. C. Levin, J. Am. Chem. Soc., 97, 5649(1975)
6. F. Bernardi, A. Bottoni, N. D. Epiotis and M. Guerra, J. Am. Chem. Soc., 100, 6018(1978)
7. M. -H. Whangbo and S. Wolfe, Israel J. Chem., 20, 36(1980), and references cited therein
8. A. Imamura and T. Hirano, J. Am. Chem. Soc., 97, 4192(1975)
9. S. Inagaki, H. Fujimoto and K. Fukui,J. Am. Chem. Soc., 98, 4054(1976)

10. L. Libit and R. Hoffmann, J. Am. Chem. Soc., 96, 1370(1974)
11. M. -H. Whangbo, D. J. Mitchell and S. Wolfe, J. Am. Chem. Soc., 100, 3698(1978)
12. B. O'Leary, B. J. Duke and J. E. Eilers, Adv. Quant. Chem., 9, 1(1975)
13. M. -H. Whangbo and R. Hoffmann, J. Chem. Phys., 68, 5498(1978)
14. J. H. Ammeter, H. -B. Bürgi, J. C. Thibeault and R. Hoffmann, J. Am. Chem. Soc., 100,3686(1978)
15. C. J. Marsden and L. S. Bartell, Inorg. Chem., 15, 2713(1976)
16. T. Ogata, A. P. Cox, D. L. Smith and P. L. Timms, Chem. Phys. Lett., 26, 186(1974)
17. W. J. Hehre, R. F. Stewart and J. A. Pople, J. Chem. Phys., 50, 2078(1969)
18. W. J. Hehre, W. A. Lathan, R. Ditchfield, M. D. Newton and J. A. Pople, GAUSSIAN 70, Quantum Chemistry Program Exchange, Indiana University, Bloomington, Indiana, No. 236
19. M. -H. Whangbo and S. Wolfe, Can. J. Chem., 54, 949(1976)

# STRUCTURE AND PROPERTIES OF FREE-RADICALS. A THEORETICAL CONTRIBUTION.

Georges LEROY

Laboratoire de Chimie Quantique,
Université Catholique de Louvain,
Place Louis Pasteur, 1,
1348 Louvain-la-Neuve (Belgium)



## INTRODUCTION.

Today it is well recognized that a large number of chemical reactions, organic and inorganic, and also many biological processes involve free-radical species as reactive intermediates. The chemistry of free-radicals is now an important branch of organic chemistry which has many useful applications in everyday life. It owes much to the discovery of powerful techniques as electron spin



*I. G. Csizmadia and R. Daudel (eds.), Computational Theoretical Organic Chemistry, 253–334.*

resonance (ESR) which not only provided the proof of the existence of free-radicals but also has been used for determining their structure. As many excellent text-books and reference papers have been devoted to free-radicals, (1) we shall limit ourselves in the first part to a brief review of experimental data concerning the structure and some static and dynamic properties of these species. In part II, we shall compare these results to the theoretical ones. In this way, we shall try to explain the main properties of free-radicals by considering their electronic structure. We shall particularly be concerned with carbon-centered radicals.

## I. EXPERIMENTAL DATA.

### I.1. The basic principles of ESR spectroscopy.

As many informations on free-radicals have been obtained by using ESR spectroscopy, we shall first recall the basic principles of this technique (2). It is well known that the interaction energy between the spin magnetic moment of an electron and an external homogeneous magnetic field can be written:

$$E = g\,\beta\, m_S\, H \qquad (I.1)$$

where g is the spectroscopic splitting factor, $\beta$ is the Bohr magneton, H is the intensity of the homogeneous magnetic field and $m_S$ is the magnetic quantum number whose values are $\pm 1/2$. This interaction is responsible for the anomalous Zeeman effect.
At the position of a magnetic nucleus, the local magnetic field will be slightly different from H:

$$H_{eff} = H + a m_I \qquad (I.2)$$

where a is the hyperfine splitting constant and $m_I$ is the nuclear spin ( $\pm 1/2$ for the proton).
The isotropic interaction energy between a single electron and a magnetic nucleus can be written:

$$E_{iso} = -\frac{8\pi}{3}\,\mu_Z^e\,\mu_Z^N\,|\Psi(0)|^2 \qquad (I.3)$$

where $\mu_Z^e$ et $\mu_N^N$ are respectively the components of the electronic and nuclear spin magnetic moments along the direction of the external magnetic field which is taken in the Z direction and $\Psi(0)$ represents the wave function of the electron evaluated at the nucleus. By substituting $\mu_Z^e$ and $\mu_Z^N$ in I.3 by their explicit expressions, one obtains :

$$E_{iso} = \frac{8\pi}{3}\, g\,\beta\,\gamma_N\hbar\, |\Psi(0)|^2\, m_S m_I \qquad (I.4)$$

or $$E_{iso} = a_N m_S m_I \tag{I.5}$$

with : $$a_N = \frac{8\pi}{3} g \beta \gamma_N \hbar |\Psi(0)|^2 \tag{I.6}$$

where $\gamma_N$ is the nuclear gyromagnetic ratio of the nucleus. $a_N$ is called the isotropic hyperfine coupling constant measured in energy units.
For hydrogen atom, the electronic energy levels are the following ones :

$$-\frac{1}{2} g \beta H - \frac{1}{4} a_H$$
$$-\frac{1}{2} g \beta H + \frac{1}{4} a_H$$
$$\frac{1}{2} g \beta H - \frac{1}{4} a_H$$
$$\frac{1}{2} g \beta H + \frac{1}{4} a_H$$

They are represented in figure I.1.

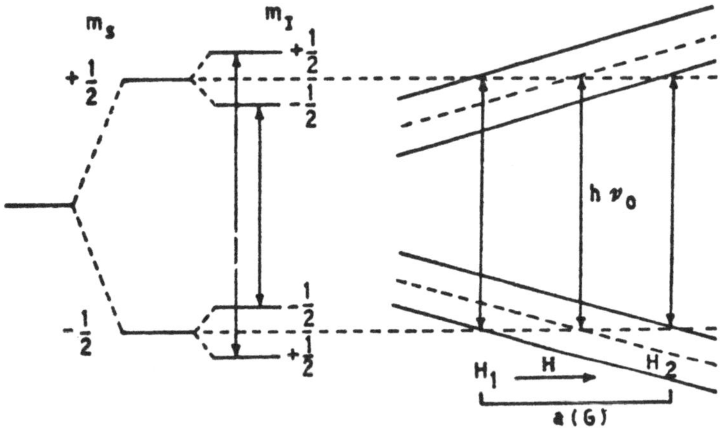


Figure I.1. Electronic energy levels for hydrogen atom.

The allowed electronic transitions correspond to the selection rules $\Delta m_S = \pm 1$; $\Delta m_I = 0$. They can be observed in ESR spectroscopy. The ESR spectrum obtained at constant frequency of the electromagnetic wave which induces the transitions consists of two equally intense lines at external fields satisfying the resonance conditions :

$$h\nu_o = g \beta H_1 + \frac{1}{2} a_H \tag{I.7}$$

and $$h\nu_o = g \beta H_2 - \frac{1}{2} a_H \tag{I.8}$$

These two fields :

$$H_1 = \frac{h\nu_o}{g\beta} - \frac{1}{2}\frac{a_H}{g\beta} \quad (I.9)$$

and

$$H_2 = \frac{h\nu_o}{g\beta} + \frac{1}{2}\frac{a_H}{g\beta} \quad (I.10)$$

are separated by :

$$a(H) = \frac{a_H}{g\beta} \quad (I.11)$$

which is the so-called isotropic hyperfine splitting constant and is measured in magnetic field units, directly in the ESR spectrum. In general, for a free-radical, equations (I.6) and (I.11) become respectively :

$$a_N = \frac{8\pi}{3} g\beta\gamma_N \hbar \rho_S(N) \quad (I.12)$$

and

$$a(N) = \frac{8\pi}{3} \gamma_N \hbar \rho_S(N) \quad (I.13)$$

where $\rho_S(N)$ is the spin density at the position of nucleus N. If the electronic g factor is known, the isotropic hyperfine coupling constants $a_N$ can be calculated from equation (I.11). The spin density at the nuclei are obtained from the splitting constants by using equation (I.13).

## I.2. The structure of alkyl radicals.

ESR studies can provide informations about the configuration of free-radicals (3). Indeed, the carbon 13 hyperfine splitting, $a(^{13}C)$, of the α carbon of an alkyl radical is strongly dependent on the hybridization of this atom which is related to the degree of non planarity of the radical. The splitting constant $a(^{13}C_\alpha)$ has its minimal value for the planar configuration and increases with increasing s character of the unpaired electron orbital.
If the potential energy curve, as a function of the angle θ which measures the degree of non polarity of the radical, has only one minimum for θ = 0 (planar radical, $a(^{13}C_\alpha)$ will increase with increasing temperature. But if the curve has two minima (non planar radical), $a(^{13}C_\alpha)$ will decrease with increasing temperature. When the energy barrier for the inversion of the radical is not too large, the splitting constant can reach a well defined minimum and then increase at higher temperatures.
A careful study of the temperature dependence of the splitting constant $a(^{13}C_\alpha)$ provides not only the configuration of alkyl radicals but also the out of plane vibrational frequency and the corresponding force constant (4-5). In the case of non planar radicals the inversion barrier can also be estimated.

Table I.1 summarizes the results obtained by Griller and co-workers (5) for the first terms of the alkyl radicals series. Methyl, ethyl and isopropyl radicals appear to be planar. Their results suggest that there is a small energy barrier for the inversion of the tert-butyl radical ( ~ 0.5 kcal/mole) (5c,7). The out of plane vibrational frequency is small and decreases from methyl to tert-butyl.
The work of Griller and co-workers needs to be generalized for determining accurately the most stable configuration of carbon-centered radicals by means of ESR experiments. We shall show in part II that quantum mechanical calculations can also provide very detailed informations on the structure of free-radicals.

The main conclusions which can be deduced from ESR studies of alkyl radicals are the following :

- Atoms or substituents which possess electron pairs induce bending at the radical center; for example $CH_2F$ and $CH_2OH$ are non planar.
- The bending at the radical center increases with the number of this type of substituents ($\pi$ donor).
- Methyl group behaves as weak $\pi$-donor from the structural point of view.

I.3. Electronic structure of free-radicals.

No direct informations on the electronic structure of free-radicals can be obtained experimentally. However ESR spectra enable us to calculate some spin properties of these species.
As shown before, spin densities at the nuclei can be deduced from the isotropic splitting constants by using equation (I.13). However, in order to obtain significant results, the splitting constants might be corrected for vibrational effects. This correction has been done for the first alkyl radicals (see table I.1).
Equation (I.13) can also be written :

$$a(N) = \frac{8\pi}{3} \gamma_N \hbar \rho_S(N) = k_N \rho_S(N) \qquad (I.14)$$

The proportionality constants $k_N$ are given in table I.2 for some magnetic nuclei. Using the results of table I.1, one obtains for alkyl radicals :

$$\rho_S(C_\alpha) = 0.083 \text{ e.au}^{-3} \quad \text{in } CH_3$$

$$\rho_S(C_\alpha) = 0.087 \text{ e.au}^{-3} \quad \text{in } CH_3CH_2$$

$$\rho_S(C_\alpha) = 0.093 \text{ e.au}^{-3} \quad \text{in } (CH_3)_2CH$$

Another spin property which is often deduced from ESR spectra is the unpaired spin density at a carbon atom also called $\pi$ (orbital) spin density.

Table. I.1. Vibrational analysis of alkyl radicals.

| Radical | $a(^{13}C_\alpha)$,G | $a(^{13}C_\alpha)$,G (6) | $\nu(cm^{-1})$ | $F_\theta$(mdyn Å $rad^{-2}$) |
|---|---|---|---|---|
| $CH_3$ | 31.9[a] | 38.34[b] ( 96°K) | 612 | - |
| $CH_3CH_2$ | 34.8[a] | 39.07[b] ( 96°K) | 500 | - |
| $(CH_3)_2CH$ | 37.1[a] | 41.3 [b] (203°K) | 380 | 0.0656 |
| $(CH_3)_3C$ | - | 45.2 [b] (203°K) | ~ 150 | - |

[a] Calculated for planar configuration ;
[b] Measured at a given temperature.

Table I.2. Proportionality constants $k_N$ (Equation I.14).

| Nucleus | I | $g_N$ | $k_N = \frac{8\pi}{3} \gamma_N \hbar a^{-3}$ |
|---|---|---|---|
| $^{1}H$ | 1/2 | 5.58536 | 1594.493275 |
| $^{13}C$ | 1/2 | 1.40440 | 400.924265 |
| $^{14}N$ | 1 | 0.40358 | 115.212913 |
| $^{19}F$ | 1/2 | 5.25460 | 1500.068815 |

In 1956, McConnell suggested that the hyperfine splitting of a proton of an aromatic radical is proportional to the unpaired spin density at the adjacent carbon atom. He proposed the so-called McConnell relation (8) :

$$a(H_i) = Q\,\rho^{\pi}(C_i) \qquad (I.15)$$

where Q is a proportionality constant.
This relation has also been used for alkyl radicals. Its interpretation is based on the model of spin polarization according to which there is a preferred electronic configuration in a >C-H fragment as shown in figure I.2. One often refers to Hund's rule to justify this model which predicts the existence of a negative spin density at the proton of the fragment. This spin density and the corresponding splitting constant obviously depend on the unpaired spin density in the π orbital of the carbon atom as assumed by the McConnell relation.

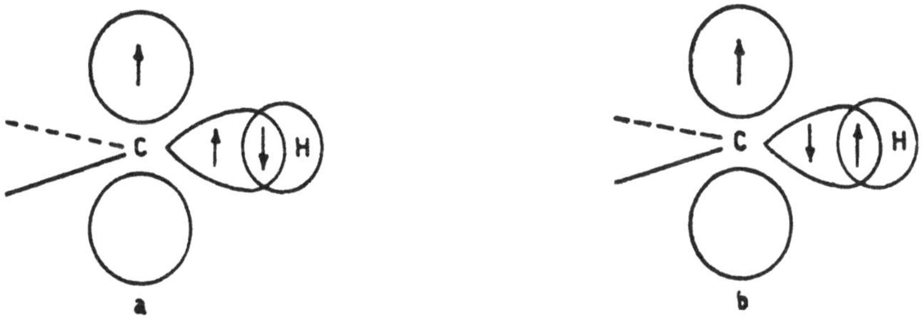


a : Preferred configuration

Figure I.2. Spin polarization of a σ CH bond.

It is easy to show that the proportionality constant Q is not a true constant. For a conjugated radical, the sum of the π spin densities is equal to 1. Then we can write :

$$\sum_i a(H_i) = Q \sum_i \rho^{\pi}(C_i) = Q \qquad (I.16)$$

As shown in table I.3, the Q values lie between 22.5 and 30 G in a series of monocyclic radicals.
Nevertheless, on the base of theoretical calculation of π spin densities (see section II), it is reasonable to assume that Q remains approximately constant in a series of related radicals. For example, in the series of alkyl radicals :

$$Q_{H_\alpha} \simeq 23.04 \text{ G} \qquad (I.17)$$

Table I.3. Q values for monocyclic radicals (9).

| Radical | a(H)(G) | $\sum_i a(H_i) = Q$ |
|---|---|---|
| $C_5H_5$ | 6.00 | 30.0 |
| $C_6H_6^-$ | 3.75 | 22.5 |
| $C_6H_6^+$ | 4.28 | 25.7 |
| $C_7H_7$ | 3.95 | 27.7 |
| $C_8H_8^-$ | 3.21 | 25.7 |

β-Proton splitting of the fragment $CH_3$-C< is often used for determining the unpaired spin density at the α carbon atom of an alkyl radical. These β-proton splittings are also appropriate for studying conformational problems but this point will not be discussed here. The non-zero spin density at a β-proton of an alkyl radical is explained by the hyperconjugation model. According to this model the overlap between the π orbital of the α carbon and the pseudo-π orbitals of the methyl group makes the delocalization of the unpaired electron on the $CH_3$ protons possible leading to a positive spin density at these nuclei (fig. I.3).

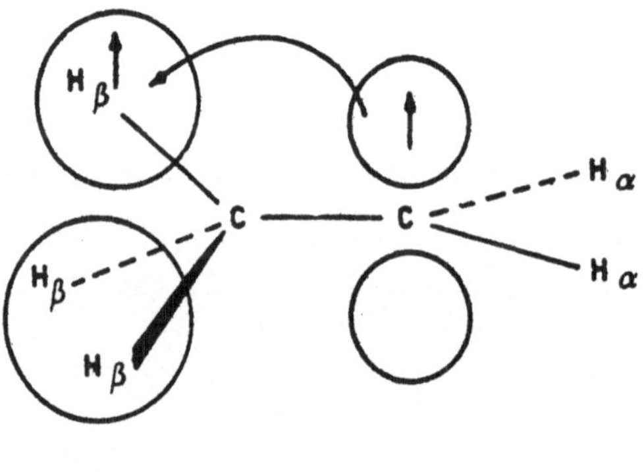


Figure I.3. Hyperconjugation model.

There is evidence that the β-proton splitting is proportional to the unpaired spin density at the α carbon atom :

$$a(H_\beta) = Q_\beta \, \rho^\pi(C_\alpha) \qquad (I.18)$$

The assignement of $Q_\beta$ must be preceded by the evaluation of $\rho^\pi(C_\alpha)$.
The following expression has been proposed by different authors (10) :

$$\rho^\pi(C_\alpha) = (1 + \Delta X)^m \qquad (I.19)$$

where $\Delta X$ is a parameter which measures the influence of a methyl group on $\rho^\pi(C_\alpha)$ and m is the number of methyl groups attached to the $\alpha$ carbon atom.
$Q_\beta$ and $\Delta X$ can be determined by using the following equations :

$$\begin{aligned} 26.87 &= Q_\beta(1 + \Delta X) \quad \text{for} \quad CH_3CH_2 \\ 24.68 &= Q_\beta(1 + \Delta X)^2 \quad \text{for} \quad (CH_3)_2CH \\ 22.72 &= Q_\beta(1 + \Delta X)^3 \quad \text{for} \quad (CH_3)_3C \end{aligned} \qquad (I.20)$$

which lead to :

$$Q_\beta = 29.3 \text{ G} \qquad (I.21)$$

$$\Delta X = -0.081 \quad (X \equiv CH_3) \qquad (I.22)$$

It is important to note that the results obtained by the relations (I.15) and (I.18) are non consilient as shown in table I.4.

Table I.4. Spin properties of alkyl radicals.

| Radical | $a(H_\alpha)$,G | $Q_\alpha$,G | $\rho^\pi(C_\alpha)$ | $a(H_\beta)$,G | $Q_\beta$,G | $\rho^\pi(C_\alpha)$ |
|---|---|---|---|---|---|---|
| $CH_3$ | 23.04 | 23.04 | 1.000 | - | - | - |
| $CH_3CH_2$ | 22.38 | 23.04 | 0.971 | 26.87 | 29.3 | 0.917 |
| $(CH_3)_2CH$ | 22.11 | 23.04 | 0.960 | 24.68 | 29.3 | 0.842 |
| $(CH_3)_3C$ | - | - | - | 22.72 | 29.3 | 0.775 |

The equation (I.19) has been generalized by Fischer (11-12) for a wide range of substituents:

$$\rho^\pi(C_\alpha) = (1 + \Delta X_1)\ (1 + \Delta X_2)\ (1 + \Delta X_3) \qquad (I.23)$$

where $\Delta X_1$, $\Delta X_2$, $\Delta X_3$ are the characteristic increments of the substituents $X_1$, $X_2$, $X_3$ attached to the $\alpha$ carbon atom. Obviously at least one of these substituents is always a methyl group.
The parameters $\Delta X$ can be determined by the following procedure.
Let us consider the radical $CH_3$-CH-X; the relation :

$$a(H_\beta) = 29.3\ \rho^\pi(C_\alpha) = 29.3(1-0.081)\ (1+\Delta X) \qquad (I.24)$$

will give immediately the value of ΔX.
Example : $CH_3$-CH-CN.

$$a(H_\beta) = 22.99 = 29.3(1-0.081)\ (1+\Delta CN) \qquad (I.25)$$

Then ΔCN = -0.146.

Table I.5 shows some typical values of ΔX obtained by the simple procedure just described. Strictly speaking, the relation (I.23) does not apply to non planar radicals. The parameters of substituents which possess electron pairs must then be used with caution.

Table I.5. Typical Fischer parameters.

| Substituents | ΔX |
|---|---|
| H | 0.000 |
| $CH_2OH$ | -0.060 |
| $COOCH_3$ | -0.075 |
| COOH | -0.079 |
| $CH_3$ | -0.081 |
| CN | -0.146 |
| COH | -0.202 |
| OH | -0.168 |
| $OC_6H_5$ | -0.169 |
| $OC_2H_5$ | -0.173 |
| $OCH_3$ | -0.183 |
| $NH_2$ | -0.250 |
| SEt | -0.265 |
| $C_6H_5$ | -0.335 |

It is generally assumed that ΔX value gives an indication of the spin-withdrawing influence of a substituent X attached to the α carbon atom. One can then conclude from table I.5 that any substituent is able to delocalize the unpaired electron of an alkyl radical.

These results suggest to write resonance formulae such as :

$$CH_3 - \overset{\cdot}{C}H - \overset{\cdot\cdot}{O} - H \quad\longleftrightarrow\quad CH_3 - \overset{\cdot\cdot}{\underset{(-)}{C}}H - \overset{(+)}{O} - H$$

and

$$CH_3 - \overset{\cdot}{C}H - C\overset{\diagup O}{\diagdown_H} \quad\longleftrightarrow\quad CH_3 - CH = C\overset{\diagup \overset{\cdot}{O}}{\diagdown_H}$$

The mutual interaction of two substituents can be deduced from the difference between the observed and calculated β-proton splittings:

$$\Delta = a(H_\beta)(obs) - a(H_\beta)(calc) \qquad (I.26)$$

If $\Delta$ is equal to zero, one can conclude that the substituents act independently. An example of this type is given by the radical $(CH_3)_2 CNH_2$ for which $a(H_\beta)(exp) = 18.27$ G and $a(H_\beta)(calc) = 18.6$ In contrast, $\Delta$ is negative for capto-dative substituted radicals (13-14) as shown in table I.6 where we compare the observed and calculated spin densities on the $\alpha$ carbon atom obtained by using equations (I.18) and (I.23) with the parameters of table I.5 and $Q_\beta = 29.3$ G.

Table I.6. $\pi$ orbital spin densities at $C_\alpha$.

| Radical | $\rho^\pi (C_\alpha)$ (obs) | $\rho^\pi (C_\alpha)$ (calc) |
|---|---|---|
| $CH_3 C(OH)CN$ | 0.610 (14) | 0.656 |
| $CH_3 C(OH)COOH$ | 0.584 (15) | 0.704 |
| $CH_3 C(SEt)CN$ | 0.512 (16) | 0.577 |

We must conclude that ESR studies provide very few but interesting informations about the electronic structure of free-radicals.


I.4. The stability of free-radicals.

Before to describe the methods which have been proposed for estimating the stability of free-radicals it is necessary to precise the meaning of this concept.
From a thermodynamical point of view, the stability of a chemical system can be measured by its total energy (reference state: separated electrons and nuclei) or by its heat of atomization (reference state: separated atoms). Unfortunately, by using these definitions, the stabilities of two systems can only be compared

if they contain the same number of "particles" (electrons and nuclei or atoms). It is precisely the case when the systems are the reactants and the products of a chemical reaction: the heat of reaction measures the relative stability of the reactants with respect to the products.

In order to define the stability or more exactly the stabilization energy of a given species, it is interesting to consider reactions where the number and the nature of all chemical bonds are conserved. They are called by Pople isodesmic reactions (15). If all the bonds would remain exactly the same during the transformation, the reaction would be thermoneutral. In fact, in most cases, the reaction is endo- or exothermic.
By an appropriate choice of the isodesmic reaction, the corresponding $\Delta H_o$ can be identified with the (de)stabilization energy of a choosen species. The best choice corresponds to the bond separation reaction where all bonds between heavy atoms are separated from one another (15). If we assume that reference molecules containing only one bond between heavy atoms are not stabilized, the thermal effect of the bond separation reaction is to be considered as the intrinsic stabilization energy (or stability) of the choosen species.

As an illustration of this procedure, let us calculate the stabilization energy of two carbon-centered radicals: $CH_2CN$ and $CF_3$. We have to use the two following bond separation reactions :

$$CH_2CN + 2\,CH_4 \longrightarrow CH_3 + CH_3CH_3 + HCN \qquad [1]$$

$$CF_3 + 3\,CH_4 \longrightarrow CH_3 + 3\,CH_3F \qquad [2]$$

The corresponding standard heats of reaction are :

$$\Delta H^o_o = 23.66 \text{ kcal/mole for } [1]^*$$

$$\Delta H^o_o = 30.20 \text{ kcal/mole for } [2].$$

Assuming that $CH_3$, $CH_3CH_3$, HCN and $CH_3F$ are not stabilized, we obtain the intrinsic (or absolute) stabilization energy of the two radicals :

$$AS(CH_2CN) = 23.66 \text{ kcal/mole}$$

$$AS(CF_3) = 30.20 \text{ kcal/mole} .$$

It is to be noted that $\sigma$ and $\pi$ stabilization effects are included in these results.

---

* The thermochemical data used in this paper are collected in the tables of part II.

One can also define a stabilization energy which resembles the classical resonance energy measuring the stabilizing effect of the delocalization of $\pi$ electrons. For this purpose isodesmic reactions without bond separation will be used. In the case of carbon-centered radicals, these isodesmic reactions can be written in the general form :

$$R^{\cdot} + CH_4 \longrightarrow RH + CH_3 \qquad [3]$$

Examples:

$$CH_2CN + CH_4 \longrightarrow CH_3 + CH_3CN \qquad [4]$$

$$CF_3 + CH_4 \longrightarrow CH_3 + CHF_3 \qquad [5]$$

The corresponding heats of reaction :

$$\Delta H^o_o = 11.3 \text{ kcal/mole for } [4]$$

and

$$\Delta H^o_o = -2.2 \text{ kcal/mole for } [5]$$

measure the stabilization energy of these radicals (with respect to $CH_3$) due to the delocalization of the unpaired electron. Here, the stabilization of the reference molecule ($CH_3CN$ or $CHF_3$) is not taken into account. It is easy to show that the thermal effect of reaction [3] is equal to the difference between the dissociation energy of C-H bonds respectively in $CH_4$ and in RH :

$$\Delta H^o_o\,[3] = BDE(CH_4) - BDE(RH) \qquad (I.27)$$

It corresponds to the resonance energy (RE°) introduced by Szwarc in 1948 (16). In this perspective, BDE(RH) can be used alternatively for measuring the delocalization energy in free-radicals.
In the case of delocalized radicals on can also adopt the Benson's definition of stabilization energy (17) :

$$SE°(R_\pi CH_2) = BDE(R_S CH_3) - BDE(R_\pi CH_3) \qquad (I.28)$$

where $R_S CH_3$ is the fully hydrogenated compound corresponding to the radical.
$SE°(R_\pi CH_2)$ is the difference between the resonance energies of $R_\pi CH_2$ and $R_\pi CH_3$ estimated by means of their heat of hydrogenation :

$$R_\pi CH_3 \xrightarrow{H_2} R_S CH_3 \qquad [6]$$

$$R_\pi CH_2 \xrightarrow{H_2} R_S CH_2 \qquad [7]$$

$$R_S CH_3 \longrightarrow R_S CH_2 + H \qquad [8]$$

$$R_\pi CH_3 \longrightarrow R_\pi CH_2 + H \qquad [9]$$

$$SE°(R_{\pi}CH_2) = \Delta H^o_o\ [8] - \Delta H^o_o\ [9]$$
$$= \Delta H^o_o\ [7] - \Delta H^o_o\ [6] \qquad (I.29)$$

As BDE ($R_SCH_3$) is not significantly affected by changes in the alkyl moiety $R_S$, Rodgers and al. suggested in 1972 to use systematically ethane as reference alkane (18). Then SE° becomes :

$$SE°(R_{\pi}CH_2) = BDE(CH_3CH_3) - BDE(R_{\pi}CH_3) \qquad (I.30)$$

It can be shown that only the stabilization energy defined by (I.28) or (I.30) is invariant with respect to the nature of the bond being broken :

$$SE°(R_{\pi}CH_2) = BDE(R_SCH_2X) - BDE(R_{\pi}CH_3) \qquad (I.31)$$

where X is any substituent including H.
The resonance energy of delocalized radicals can also be estimated from rotational barriers obtained by ESR (19). As a matter of fact the rotational barrier measures the difference between the stabilization (or resonance) energies of the planar and perpendicular forms of the delocalized radical.

As we have seen, many definitions of the thermodynamic stability of chemical systems can be used. There is no reason to prefer one of them because they are all more or less arbitrary. However, if we want to measure the stabilization of a free-radical due to the delocalization of the unpaired electron, RE° and SE° are more convenient or rotational barriers when they are available. The results obtained by using equations (I.27), (I.28) and (I.30) are obviously correlated but they do not seem to vary like rotational barriers. After all, the C-H bond dissociation energies remain a good tool for estimating the relatives stabilities of carbon-centered radicals. Unfortunately many of them are not yet determined.

From a kinetic point of view the stability of a chemical species can be measured by its lifetime under given experimental conditions. In the case of free-radicals, we shall adopt the definitions proposed by Griller and Ingold in 1976 (20).
A radical is persistent if its lifetime is significantly greater than methyl under the same conditions. It is transient if its lifetime is of the same order of magnitude as methyl, always under the same conditions. The persistence of a radical can be quantitatively described by the rate constant of a well-defined reaction, for example the process by which the radical decays. Finally, a pure radical is stable when it can be handled and stored whitout special conditions.
It is to be noted that the kinetic stability is not necessarily related to the thermodynamic stability (21). We shall return to

this point in part II, in the case of free-radicals. Intrinsic and relative stabilization energies of some carbon centered radicals are collected in table I.7 with the corresponding BDE(C-H).

Table I.7. Stabilization energies and BDE (kcal/mole).

| Radical | AS | RE°(I.27) | SE°(I.30) | BDE(C-H) |
|---|---|---|---|---|
| $CH_3$ | - | - | - | 104.30 |
| $CH_3CH_2$ | 6.36 | 6.36 | - | 97.94 |
| $(CH_3)_2CH$ | 11.42 | 9.17 | 2.81 | 95.13 |
| $(CH_3)_3C$ | 19.28 | 11.78 | 5.42 | 92.52 |
| $CH_2F$ | 3.30 | 3.30 | -3.06 | 101.00 |
| $CHF_2$ | 14.78 | 2.35 | -4.01 | 101.95 |
| $CF_3$ | 29.60 | -2.10 | -8.40 | 106.40 |
| $CH_2CN$ | 23.66 | 11.30 | 4.94 | 93.00 |
| $CH_2NH_2$ | 9.70 | 9.70 | 3.34 | 94.60 |
| $CH_2OH$ | 10.33 | 10.33 | 3.97 | 93.97 |
| $CH_2CHCH_2$ | 22.91 | 17.70 | 11.34 | 86.60 |
| $C_6H_5CH_2$ | 86.09 | 16.40 | 10.04 | 87.90 |

## I.5. Abstraction of atoms by free-radicals.

The abstraction of atoms by free-radicals has been the subject of a tremendous amount of experimental work (22) and several models have been proposed for rationalizing the activation energies of these radical transfer reactions. The principal factors which have been considered are : the heats of reaction, the polar effects, the molecular polarizabilities, the steric effects and the solvent effects. We shall analyze briefly some of these factors.

1° Heats of reaction.

Let us consider the following atom abstraction involving the attack of the radical X on the molecule R-Y :

$$R - Y + X \longrightarrow R...Y...X \longrightarrow R + X - Y \qquad [10]$$

The heat of reaction of this concerted process is the difference between the dissociation energies of the bonds R-Y and X-Y :

$$\Delta H_o^o = BDE(R\text{-}X) - BDE(X\text{-}Y) \qquad (I.32)$$

Three types of reactions can be considered :

a) exothermic if BDE(X-Y) > BDE(R-Y), e.g. the reaction of radical OH with methane :

$$CH_4 + OH \longrightarrow CH_3 + H_2O \; ; \; \Delta H_o^o = -15 \text{ kcal/mole} \quad [11]$$

b) thermoneutral if BDE(X-Y) = BDE(R-Y), e.g., $CH_3$ radical attacking methane :

$$CH_4 + CH_3 \longrightarrow CH_3 + CH_4 \; ; \; \Delta H_o^o = 0 \quad [12]$$

c) endothermic if BDE(X-Y) < BDE(R-Y), e.g. the bromination of methane :

$$CH_4 + Br \longrightarrow CH_3 + HBr \; ; \; \Delta H_o^o = 17.3 \text{ kcal/mole} \quad [13]$$

Qualitative reaction paths for these hydrogen abstractions are given in figures I.4, I.5 and I.6.
The activation energies of these reactions considered in the exothermic direction are clearly related to the corresponding enthalpy changes as shown below.

| | $\Delta H_o^o$(kcal/mole) | $E_a$(kcal/mole) | |
|---|---|---|---|
| $CH_4 + OH \longrightarrow CH_3 + H_2O$ ; | -15.0 | 2.4-7.0 | [14] |
| $CH_4 + CH_3 \longrightarrow CH_3 + CH_4$ ; | 0.0 | 14.1-14.9 | [15] |
| $HBr + CH_3 \longrightarrow Br + CH_4$ ; | -17.3 | ~1 | [16] |

This relation bas been recognized for a long time and formalized by means of an Evans-Polanyi equation such as (23) :

$$E = \alpha \Delta H + \beta \qquad (I.33)$$

Where E and ΔH are respectively the (intrinsic) activation energy ans exothermicity of the reaction and α and β, empirical constants. Evans-Polanyi relations are restricted to very narrow series of reactions as exemplified by the following examples fow which $\Delta H_o^o \simeq 0$ :

| | $E_a$(kcal/mole) | |
|---|---|---|
| $CH_4 + CH_3 \longrightarrow CH_3 + CH_4$ ; | 14.1-14.9 | [17] |
| $CH_4 + CH_3O \longrightarrow CH_3 + CH_3OH$ ; | 8.9 | [18] |
| $H_2 + H \longrightarrow H + H_2$ | 7.5 | [19] |

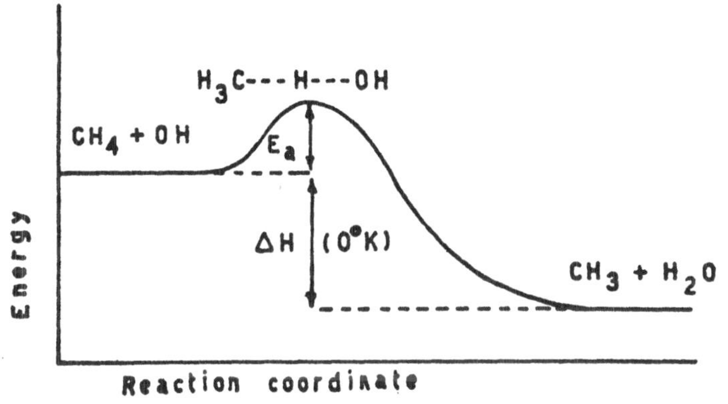


Figure I.4. Reaction path for $CH_4$ + OH reaction.

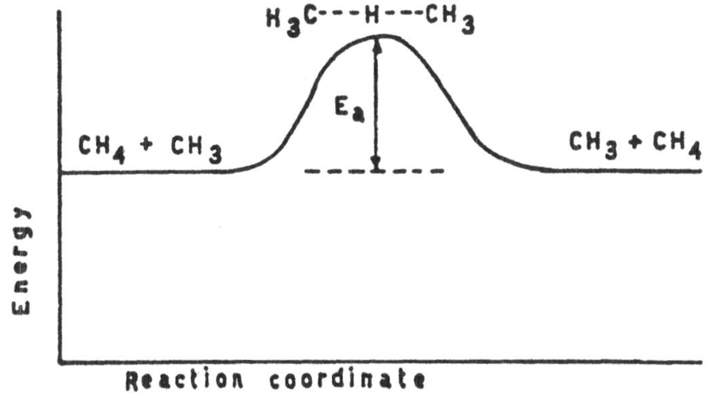


Figure I.5. Reaction path for $CH_4$ + $CH_3$ reaction.

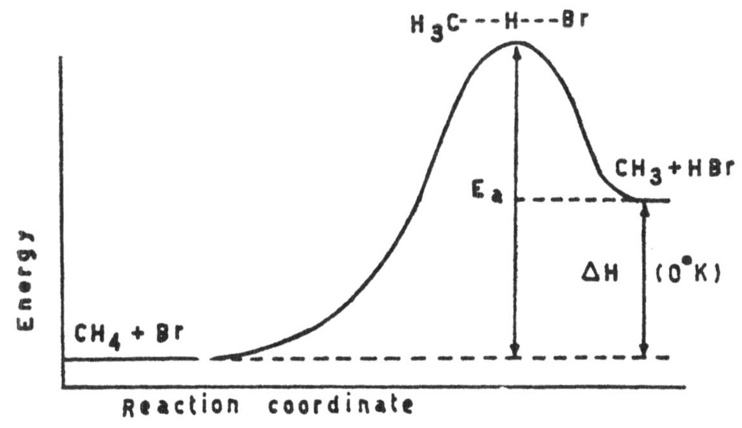


Figure I.6. Reaction path for the bromination of $CH_4$.

Then, other effects have to be considered for predicting the activation energies of radical transfer reactions.

    2° Polar effects.

Strictly speaking, the polar effects are related to the dipole-dipole interactions between the attacking radical and the substrate. They are generally discussed at the transition state by considering the permanent polarities of the isolated reacting species. Dipole-dipole repulsions are considered to account for the relatively low reactivity of $CF_3$ radicals in attack on polar bonds :

|  | $\Delta H_o^o$ (kcal/mole) | $E_a$ (kcal/mole) |  |
|---|---|---|---|
| $CH_3 + HCl \longrightarrow CH_4 + Cl$ ; | -1.2 | 2.4 | [20] |
| $CF_3 + HCl \longrightarrow CF_3H + Cl$ ; | -3.3 | 5.3 | [21] |

Although the second reaction is exothermic, the dipole-dipole repulsions at the transition state destabilize the activated complex and raise the activation energy relative to that for the attack by methyl radicals :

$$
\begin{array}{c}
F \diagdown \\
F \!\!-\!\! C \ldots H \ldots Cl \\
F \diagup \\
\longleftarrow \!\!-\!\! \quad \!\!-\!\! \longrightarrow
\end{array}
$$

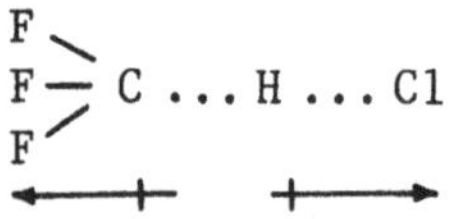

Conversely, dipole-dipole interactions are considered to be responsible for the relative ease of iodine abstraction from $CF_3$-I versus $CH_3$-I, by methyl radicals :

$$
\begin{array}{cc}
F \diagdown \qquad\qquad H & \qquad H \diagdown \qquad\qquad H \\
F \!\!-\!\! C \ldots I \ldots C \!\!-\!\! H & \qquad H \!\!-\!\! C \ldots I \ldots C \!\!-\!\! H \\
F \diagup \qquad\qquad H & \qquad H \diagup \qquad\qquad H \\
\longleftarrow \!\!-\!\! \quad \longleftarrow \!\!-\!\! & \qquad \!\!-\!\! \longrightarrow \quad \longleftarrow \!\!-\!\!
\end{array}
$$

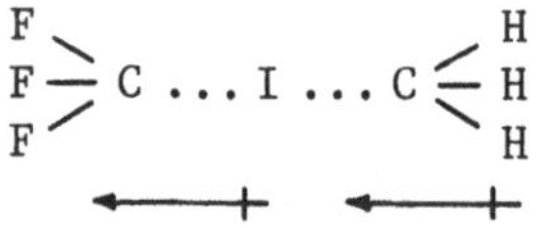

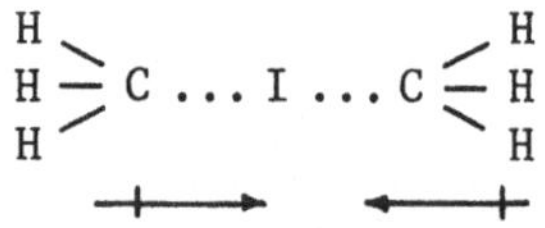

The polarity of a carbon-centered radical can be related to its nucleophilicity or electrophilicity. Methyl radical whose carbon atom is negatively charged is nucleophilic but trifluoromethyl radical, the carbon of which is positive, is electrophilic compared with $CH_3$. An alternative definition of these characters is commonly used (24). It is based on the representation of the transition state of an hydrogen abstraction reaction by resonance structures such as :

$$
\underset{I}{R \overset{\cdot}{.} \overset{\cdot}{H} . X} \;\longleftrightarrow\; \underset{II}{\overset{-}{R} : \overset{\cdot \,+}{H} X} \;\longleftrightarrow\; \underset{III}{\overset{+}{R} \overset{\cdot}{H} : \overset{-}{X}}
$$

Radicals (X) for which structure III is more important are electrophilic, while those for which structure II is more important are nucleophilic. Accordingly the reactivity of free-radicals in hydrogen abstraction reactions would be dependent on physico-chemical properties of the endgroups R and X, such as their ionization potential and their electron affinity. One could also consider their electronegativity defined, according to Mulliken, as the sum of the ionization potential and the electron affinity devided by an arbitrary constant :

$$x = \frac{I + A}{K} \qquad (I.34)$$

Alfassi and Benson (25) have shown that the activation energies of some radical transfer reactions are correlated to the sum of the electron affinities of the endgroups R and X. The same type of correlation could presumably be observed by using the electronegativities instead of the electron affinities. Unfortunately all these quantities are not very accurately determined. Some typical values of I, A and $(I+A)/K$ are collected in table I.8 (26). The constant K has been choosen in order to obtain for the fluorine atom an electronegativity equal to $4 (K = 5.21)$. The electron affinity is taken positive when the attachment of one electron to the species is an exothermic process. Electronegativities of atoms and radicals could be used for estimating their relative electrophilicity (or nucleophilicity).

A general conclusion comes from the model based on the polarity concept. The activation energies of nearly thermoneutral radical transfer reactions tend to be lower if the attacking and leaving radicals are respectively nucleophilic and electrophilic or vice versa. This can be interpreted in terms of dipole-dipole interactions or in terms of a stabilization of the transition state by an important contribution of resonance structures II and III. The results collected in table I.9 show clearly the dependence of activation energy with respect to polar effects and exothermicity of the reaction. These two factors act always simultaneously except of course in the case of thermoneutral reactions. The activation energy of a hydrogen abstraction can be also correlated to the electron density on the hydrogen atom to be abstracted. This can be easely understood in terms of electronegativity of the leaving group or atom.

3° Molecular polarizability of the substrate.

The polarizability of the substrate seems to play a role in radical transfer reactions. Krech and McFadden have shown (27) that the activation energies are inversely correlated with the molecular polarizability of the substrate for a homologous series of atom abstraction reactions.

In this section, we have summarized some of the empirical

methods which have been developed for predicting the activation energies for radical-molecule (or atom - molecule) reactions. We have recognized that these activation energies depend on several factors, the most important of which being, at least in the gas phase, the exothermicity of the reaction, the polar effects and presumably, the polarizability of the substrate. Furthermore the tunnel effect can play an important role especially in hydrogen abstraction reactions. In general these factors are never considered all together in empirical methods.

Table I.8. Electronegativities of atoms and radicals.

| Atom or radical | I(eV) | A(eV) | x |
|---|---|---|---|
| F | 17.42 | 3.40 | 4.00 |
| CN | 14.20 | 3.82 | 3.46 |
| Cl | 12.97 | 3.62 | 3.18 |
| Br | 11.81 | 3.36 | 2.91 |
| H-C≡C | 12.40 | 2.00-2.51 | 2.76-2.86 |
| OH | 12.94 | 1.83 | 2.83 |
| H | 13.60 | 0.75 | 2.75 |
| I | 10.45 | 3.06 | 2.59 |
| SH | 10.50 | 2.32 | 2.46 |
| $CH_2CHO$ | (10.85) | 1.81 | (2.43) |
| $NH_2$ | 11.40-11.70 | 0.744 | 2.33-2.39 |
| $CH_2CN$ | 10.87 | 1.51 | 2.38 |
| $C_6H_5$ | 9.20 | 2.20 | 2.19 |
| $CF_3$ | 9.17 | 2.01 | 2.15 |
| $CCl_3$ | 8.78 | 1.44-2.10 | 1.96-2.09 |
| $CH_3$ | 9.86 | 0.08 | 1.91 |
| $CH_3S$ | 8.06 | 1.86 | 1.90 |
| $CH_3O$ | 8.30 | 1.59 | 1.89 |
| NO | 9.27 | 0.10 | 1.80 |
| $C_2H_5$ | 8.39 | $\leqslant$ 0.34 | $\lessapprox$ 1.68 |
| $C_6H_5CH_2$ | 7.20 | 0.88 | 1.55 |
| $CH_2CHCH_2$ | 8.13 | 0.55 | 1.67 |

Table I.9. Dependence of $E_a$ with respect to exothermicity and polar effects.

| Reaction $X + N - R \rightarrow HX + R$ | Character of X | Character of R | Electronegativities X | Electronegativities R | Resonance structure | Dipole-dipole interactions | $\Delta H_o^o$ (kcal/mole) | $E_a$ (kcal/mole) |
|---|---|---|---|---|---|---|---|---|
| $HO + H - CH_3 \rightarrow HOH + CH_3$ | E | N | 2.83 | 1.91 | $H\bar{O}:\ \dot{H}\ \overset{+}{C}H_3$ | $HO...H...CH_3$ (←+ ←+) | −15.0 | $\gtrsim 2.4$ |
| $F_3C + H - CH_3 \rightarrow CF_3H + CH_3$ | E | N | 2.15 | 1.91 | $F_3\bar{C}:\ \dot{H}\ \overset{+}{C}H_3$ | $F_3C...H...CH_3$ (←+ ←+) | − 2.1 | 10.2 |
| $H_3C + H - CH_3 \rightarrow CH_4 + CH_3$ | N | N | 1.91 | 1.91 | $H_3C.\ \dot{H}\ .CH_3$ | $H_3C...H...CH_3$ (+→ ←+) | 0.0 | 14.7 |
| $H_3C + H - Br \rightarrow CH_4 + Br$ | N | E | 1.91 | 2.91 | $H_3\overset{+}{C}\ \dot{H}:\bar{B}r$ | $H_3C...H...Br$ (+→ +→) | −17.3 | $\lesssim 1.4$ |
| $H_3C + H - Cl \rightarrow CH_4 + Cl$ | N | E | 1.91 | 3.18 | $H_3\overset{+}{C}\ \dot{H}:\bar{C}l$ | $H_3C...H...Cl$ (+→ +→) | − 1.2 | $\gtrsim 1.7$ |
| $F_3C + H - Br \rightarrow CF_3H + Br$ | E | E | 2.15 | 2.91 | $F_3\overset{+}{C}\ \dot{H}:\bar{B}r \leftrightarrow F_3\bar{C}:\ \dot{H}\ \overset{+}{B}r$ | $F_3C...H...Br$ (←+ +→) | −19.4 | 2.9 |
| $F_3C + H - Cl \rightarrow CF_3H + Cl$ | E | E | 2.15 | 3.18 | $F_3\overset{+}{C}\ \dot{H}:\bar{C}l \leftrightarrow F_3\bar{C}:\ \dot{H}\ \overset{+}{C}l$ | $F_3C...H...Cl$ (← →) | −3.3 | 5.3 |

Only elaborate quantum mechanical calculations allow us to predict activation energies with a high accuracy. All the factors are implicitly taken into account in these calculations which unfortunately are yet impossible to perform for many electrons systems.
Besides the empirical methods using very simple correlation equations we have also to mention here the rather elaborate but still empirical BEDO method proposed by Johnston and Parr (28) and refined by several authors (29). This method is based on the model of the conservation of the sum of the bond orders (X...H...R) along the reaction path.

None of the empirical methods is completely satisfactory but the evolution of activation energies in homologous series of reactions is generally well predicted. Furthermore, substituent effects and relative selectivities of atoms and radicals for hydrogen abstraction are qualitatively well understood in terms of exothermicity, polarity and steric hindrance.
It is also to be noted that activation entropies of radical transfer reactions are much easier to predict than activation energies, by using the formalism of statistical thermodynamics.
Finally, it is to be mentioned that a major difficulty for predicting kinetic parameters by empirical methods comes from the imprecision (or the lack) of the experimental data needed for the calculations.

I.6. Dimerization of radicals.

The dimerization reactions of radicals have been also extensively studied. Their Arrhenius parameters have been calculated thanks to thermochemistry and pyrolysis kinetics of the parent molecules. The rate constants of some typical recombination reactions are collected in table I.10. It is often assumed that there is no activation energy for the dimerization of radicals. Then one could write :

$$\log k_c = \log A_c \qquad (I.35)$$

and the activation entropy could be directly deduced from the rate constant by means of the relation :

$$A_c = \frac{ekT}{h} \exp \frac{\Delta S^{\neq}}{R} R'Te \ (1/\text{mole-s}) \qquad (I.36)$$

where k is the Boltzmann constant and R' is the ideal gas constant expressed in units of liter-atm/mole -°K.
Using equation (I.36), the activation parameters for the dimerization of methyl radicals are :

$$\Delta S^{\neq} = -20.7 \text{ e.u.} \qquad \text{and} \qquad E_a = 0$$

for a standard state of 1 atm at 300°K.

Table I.10. Rate constants of dimerization reactions[a] (30).

| Radical | $k_c$ (1/mole-s) | log $k_c$ |
|---|---|---|
| $CH_3$ | 24.0-32.0 $10^9$ | 10.38-10.51 |
| $C_2H_5$ | 7.8 $10^9$ | 9.89 |
| $CH(CH_3)_2$ | 5.1 $10^9$ | 9.71 |
| $C(CH_3)_3$ | 2.4 $10^9$ | 9.38 |
| $C(=O)H$ | 14.0 $10^9$ | 10.15 |
| $C(=O)CH_3$ | 45.0 $10^9$ | 10.65 |
| $CF_3$ | 5.0-10.0 $10^9$ | 9.70-10.00 |
| $CCl_3$ | 4.57-7.24 $10^9$ | 9.66- 9.86 |
| $CH_2$-$CH$=$CH_2$ | 6.6- 8.5 $10^9$ | 9.81- 9.93 |

[a] In the gas phase at 25°C

All the rate constants of table I.10 are of the same order of magnitude : $k_c \simeq 10^{10}$ 1/mole-s. However if one assumes their reliability, the following order of reactivity is found for alkyl radicals :

$$Me > Et > Pr^i > Bu^t$$

Besides, a σ carbon-centered radical such as $CH_3CO$ combines more rapidly than $CH_3$. This result suggests the existence of an activation energy for radical-radical recombination. Some experimental data are in favour of this hypothesis. For example, according to Benson, the dimerization of t-butyl radical would have an activation energy of 5.6 kcal/mole. However, the lower reactivity of this radical is most often attributed to steric factors.
We need certainly more precise experimental data and elaborate quantum mechanical calculations to rationalize the Arrhenius parameters of radical dimerizations and to elucidate their mechanism. In part II we shall try to understand the available kinetic results in terms of structural and energetic properties of free-radicals.

## I.7. Conclusion.

In this first part, we have briefly described the experimental determination of static and dynamic properties of free-radicals. ESR spectroscopy provides informations not only on the configuration and conformation of radicals but also on the energetic barriers associated with structural modifications. Frequencies and corresponding force constants of peculiar normal modes of vibration can also be determined. Very few careful vibrational analysis by ESR spectroscopy have been performed at the present time. A lot a work is yet to be done and quantum mechanical calculations have certainly an important role to play in this field in connection with experimental studies. Let us remind an important rule deduced from the interpretation of ESR spectra: alkyl radicals are generally planar except when they are substituted by atoms or groups which possess electron pairs.

ESR spectroscopy can also yield a few informations on the electronic structure of free-radicals. McConnell type relations based on the models of spin polarization and hyperconjugation enable us to calculate the $\pi$ spin population of the $\alpha$ carbon in alkyl (substituted) radicals. If we assume that Fischer parameters express the spin-withdrawing influence of the substituents attached to the $\alpha$ carbon, the data of table I.5 indicate that any substituent, whatever its type may be, delocalizes the unpaired electron of an alkyl radical. The mechanism of this effect must certainly depend on the nature of the group involved. The following resonance formulae summarize the informations provided by ESR spectroscopy on the electronic structure of carbon-centered radicals :

$$CH_3 - \dot{C}H - \ddot{X} \longleftrightarrow CH_3 - \overset{(\ddot{-})}{C}H - \overset{(\dot{+})}{X}$$

$$CH_3 - \dot{C}H - \underset{|}{C} = X \longleftrightarrow CH_3 - CH = \underset{|}{C} - \dot{X}$$

Here again, theoretical calculations will give complementary results and a more detailed description of the electronic properties of free-radicals.
The stability of these species can be defined either from a thermodynamical point of view or from a kinetic point of view. We have recognized that the resonance energy introduced by Szwarc is nothing else than the thermal effect of an isodesmic reaction. The systematic use of these reactions in quantum chemistry will enable us not only to calculate stabilization energies but also to evaluate bond dissociation energies (BDE) and heats of formation of new radicals.

The analysis of experimental data concerning radical transfer reactions has demonstrated the prominent role of exothermicity and polar effects among other factors. An interesting empirical rule has been proposed which correlates the electrophilicity (or nucleophilicity) of attacking and leaving radicals to the activation energy. Elaborate calculations using ab initio methods of quantum chemistry could help us to interpret the empirical rules and also to predict the exothermicity of reactions not yet experimentally studied.

Among many interesting reactions of free-radicals we have only considered the dimerization which is one of the termination pathways in chain reactions. The reason of this choice lies in the increasing importance of dimerization as a synthetic route (13). Two main problems remain unsolved concerning these reactions. How can we interpret the influence of substituents on the reaction rates and do these reactions have an activation energy? These questions concern obviously the quantum chemist as long as the experimental data (reaction rate and Arrhenius parameters) remain lacking or very imprecise.

## II. THEORETICAL CONTRIBUTION.

### II.1. Historical development.

The rationalization of free-radical properties has been the subject of many theoretical investigations based on simple models and more or less elaborate quantum mechanical calculations. Obviously we cannot describe in detail all these interesting contributions. We shall just draw up a tentative list of the main subjects which have been envisaged and point out the methods used for studying them :

| Subject | Method |
|---|---|
| Electronic structure ; | Linnett's double quartet theory (31) and elaborate calculations of spin properties (32); |
| Configuration and conformation ; | Geometry optimizations by ab initio methods of quantum chemistry (33); |
| Free-radical reactions ; | Qualitative approach based on the three-center three electron model (34) and explicit calculations of reaction paths (35). |

This list is certainly not exhaustive and other fundamental contributions need to be mentioned (36).

In this work, we shall try to rationalize the static and dynamic properties of carbon-centered radicals in terms of their electronic structure and thermodynamical stability as obtained by ab initio calculations. All the problems which have been envisaged in part I will be analyzed here from a theoretical point of view.

II.2. Computational method.

Since free-radicals are open shell systems, we have used the unrestricted Hartree-Fock (UHF) method (37) and all computations were carried out using the GAUSSIAN-70 series of programs (38). The main limitation of the UHF method is that the computed wave functions are not eigenfunctions of the spin operator $S^2$. However the UHF treatment allows us to obtain better spin density values than the RHF method. Furthermore, if we except the highly unsaturated radicals, spin contamination was found to be very small ($\langle S^2 \rangle \simeq 0.75$).

Geometry optimizations were performed using the minimal STO-3G basis set (39) and in some cases the split-valence 4-31G basis set (40). In order to obtain more reliable results for thermochemical studies, total energies were computed at the 4-31G level using the geometries obtained at the STO-3G level.
The molecular orbitals of each radical have been localized using the Boys criterion (41) which consists in maximizing the distances between the charge centroids of the orbitals. So, one obtains not only the positions of these charge centroids but also the second moments of the localized orbitals. According to Csizmadia (42) these second moments measure the size of the corresponding orbitals. In the case of open shell systems, the charge centroids have the same meaning as the Linnett symbols for α and β electrons (o and x). In this way the localization procedure allows us to reexamine and to generalize the Linnett's theory. The list of the carbon-centered radicals (C XYZ) studied in this work is given in table II.1.

II.3. Ab initio results.

1° Geometry optimization.

The energy and geometrical parameters of $CH_3$ and monosubstituted derivaties were obtained from the literature (43-44). Only the most important parameters of the di- and trisubstituted radicals were optimized using the same procedure as Pople and al. (45). We have also respected the symmetry constraints. The only parameters a priori choosen are the bond lengths C-H, N-H and O-H in the substituents $CH_3$, $NH_2$ and OH and the angles HCH in the $CH_3$ groups. The following values were adopted.

$$r(C\text{-}H) = 1.09\,Å;\; r(N\text{-}H) = 1.014\,Å;\; r(O\text{-}H) = 0.99\,Å;\; H\hat{C}H = 109.5°$$

The deviation from planarity is measured by θ, the angle between the C-Z' bond axis and the C X'Y' plane. In monosubstituted radicals, X' and Y' = H and Z' is the heavy atom of Z attached to the α carbon atom. In disubstituted radicals, X' and Y' are the heavy atoms of X and Y attached to $C_\alpha$ and Z' = H. Finally in $CH_3$, $(CH_3)_3C$

and $CF_3$, X', Y' and Z' = H, C or F. It is to be noted that in planar radical ($\theta = 0°$), the hydrogen atoms of the substituents are not necessarily in the plane of the heavy atoms. For non planar radicals ($\theta \neq 0$), the total energy of the planar configuration ($\theta = 0$) was calculated at the STO-3G level keeping the other parameters fixed at their optimized value corresponding to the most stable structure.

Table II.1. Carbon-centered radicals (C XYZ) .

| Radical | X | Y | Z |
|---|---|---|---|
| $CH_3$ | H | H | H |
| $CH_3CH_2$ | H | H | $CH_3$ |
| $(CH_3)_2CH$ | H | $CH_3$ | $CH_3$ |
| $(CH_3)_3C$ | $CH_3$ | $CH_3$ | $CH_3$ |
| $CH_2F$ | H | H | F |
| $CHF_2$ | H | F | F |
| $CF_3$ | F | F | F |
| $CH_2CN$ | H | H | CN |
| $CH_2NH_2$ | H | H | $NH_2$ |
| $CH_2OH$ | H | H | OH |
| $CH(CN)_2$ | H | CN | CN |
| $CH(CN)(NH_2)$ | H | CN | $NH_2$ |
| CH(CN)(OH) | H | CN | OH |
| $CH(OH)_2$ | H | OH | OH |

The energy difference $E(\theta) - E(0)$ is the inversion barrier of the system. For planar radicals ($\theta = 0$) the total energy was calculated for an arbitrary value of $\theta$ ($\theta = 10°$), at the STO-3G level, the other parameters being not re-optimized. So, we can estimate the shape of the potential energy curve around $\theta = 0$.

The optimized parameters, the total energies at the STO-3G level and the corresponding energies at the 4-31G level are collected in table II.2 with the energy differences $|E(\theta) - E(0)|$. The geometry of alkyl radicals was also fully optimized at the 4-31G level. The energies obtained are given in table II.2. The most stable geometries of the carbon-centered radicals studied in this work are represented in figure II.1.

Table II.2. Optimized parameters and energies of carbon-centered radicals.

| ·Radical | Parameters (STO-3G) (Å or °) | θ (°) |
|---|---|---|
| $CH_3$ | [43] | 15.2<br>0 |
| $CH_3$-$CH_2$ | [43] | 22.9<br>0 |
| $(CH_3)_2CH$ | CH = 1.084; CC = 1.522<br>CCC = 119.3; HCCC = 180.0 | 26.9<br>0 |
| $(CH_3)_3C$ | CC = 1.529; CCC = 117.6 | 26.6<br>0 |
| $CH_2F$ | [43] | 32.1<br>0 |
| $CHF_2$ | CH = 1.105; CF = 1.354<br>FCF = 113.2 | 40.1<br>0 |
| $CF_3$ | CF = 1.351; FCF = 111.6 | 49.2<br>0 |
| $CH_2CN$ | [44] | 0<br>10.0 |
| $CH_2NH_2$ | [43] | 36.3<br>0 |
| $CH_2OH$ | [43] | 33.4<br>0 |
| $CH(CN)_2$ | CH = 1.090; CC = 1.416<br>CN = 1.218; HCC = 119.1 | 0<br>10.0 |
| $CHCNNH_2$ | CH = 1.085; CC = 1.401<br>CN = 1.440; C≡N = 1.217<br>HCN = 118.0; CCH = 120.0; HNH = 107.4<br>Angle between CN and NH = 56.5<br>$NH_2$ symmetric/CN bond axis | 10.0<br>0 |
| CHCNOH | CH = 1.085; CC = 1.400<br>CO = 1.398; C≡N = 1.219<br>HCO = 115.7; CCH = 120.7<br>COH = 105.6; HOCH = 180.0 | 0<br>10.0 |
| $CH(OH)_2$ | CH = 1.080; CO = 1.400<br>OCO = 118.0; COH = 105.0<br>HOCH = 180.0 | 43.0<br>0 |

Table II.2. Contin.

| Energies (a.u.) STO-3G | 4-31G | 4-31G (optimized) | \|E(θ) - E(0)\| STO-3G (kcal/mole) |
|---|---|---|---|
| -39.07701<br>-39.07670 | -39.50392 | -39.50497<br>(θ = 0.0) | 0.19 |
| -77.66300<br>-77.66227 | -78.48527 | -78.48618<br>(θ = 8.4) | 0.46 |
| -116.24849<br>-116.24737 | -117.46593 | -117.46680<br>(θ = 0.0) | 0.70 |
| -154.83418<br>-154.83199 | -156.44788 | -156.44852<br>(θ = 16.0) | 1.37 |
| -136.53503<br>-136.53237 | -138.22470 | - | 1.67 |
| -234.00392<br>-233.99466 | -236.95808 | - | 5.81 |
| -331.48065<br>-331.45488 | -335.69449 | - | 16.17 |
| -129.65076<br>-129.65039 | -131.10962 | - | 0.23 |
| -93.39869<br>-93.38907 | -94.44437 | - | 6.03 |
| -112.91611<br>-112.91206 | -114.24123 | - | 2.54 |
| -220.21248<br>-220.21185 | -222.69686 | - | 0.40 |
| -183.96915<br>-183.96871 | -186.04937 | - | 0.28 |
| -203.48784<br>-203.48735 | -205.84043 | - | 0.30 |
| -186.76240<br>-186.75066 | -188.98832 | - | 7.36 |

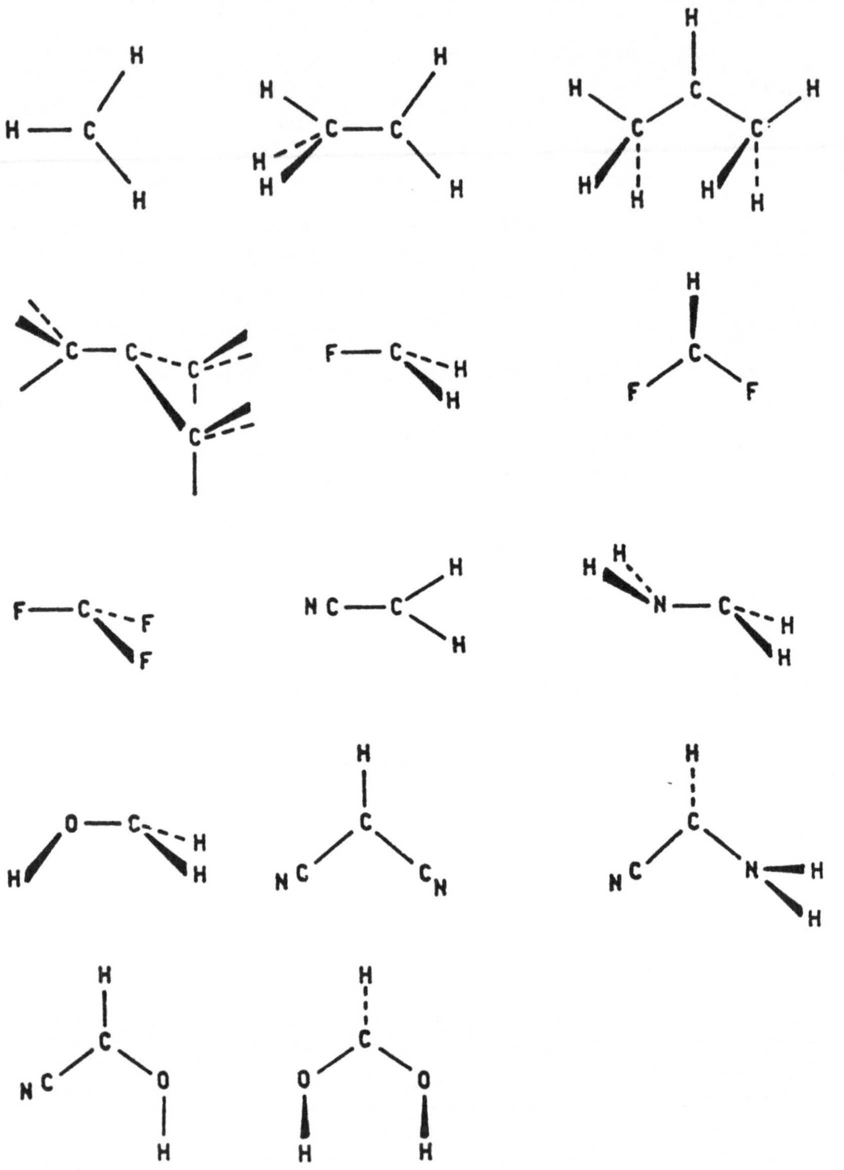


Figure II.1. Optimized geometry of carbon-centered radicals.

First of all, it is to be noted that the out-of-plane angle θ is over-estimated, at the STO-3G level. The first three alkyl radicals are found to be planar ($|E(8.4) - E(0)| = 0.07$ kcal/mole for $(CH_3)_2CH$) when all the parameters are optimized using the 4-31G basis set. On the other hand, the inversion barrier of $CH_2F$

is 0.5 kcal/mole with the 4-31G basis set to be compared with 1.67 kcal/mole at the STO-3G level. Then we shall assume that a substituted alkyl radical is planar if the difference $|E(\theta) - E(0)|$ obtained using the minimal basis set is of the order of 1 kcal/mole. In its most stable configuration the tert-butyl radical is found to be not planar at the two basis set levels. The potential energy curve as a function of $\theta$ is represented in figure II.2. The energy barrier for the inversion is presumably over-estimated. The theoretical results concerning the geometry of alkyl radicals are in qualitative agreement with the experimental ones (see I.2).

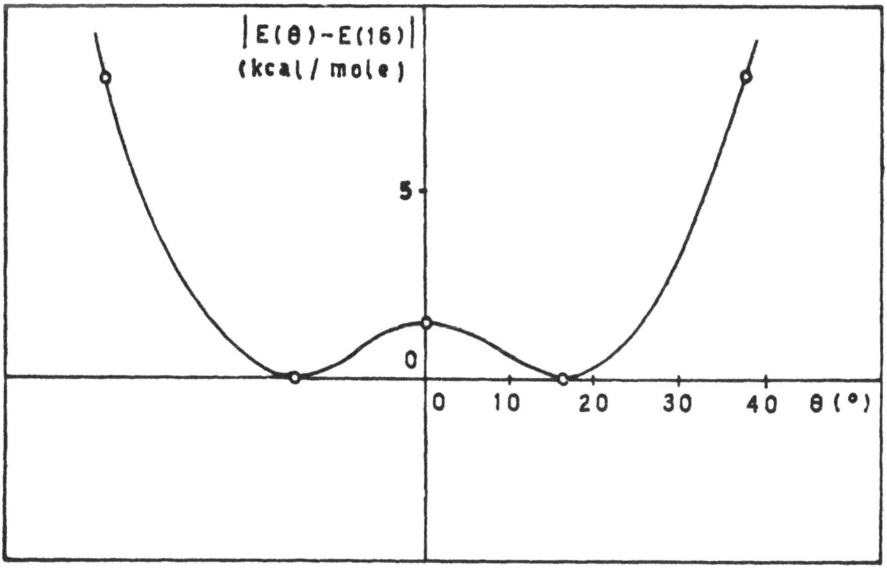


Figure II.2. Potential curve of tert-butyl radical (4-31G level).

The main following conclusions can be deduced from table II.2.

- Monosubstituted alkyl radicals can be divided in two classes : non planar species where Z = F, OH or $NH_2$ and the planar cyanomethyl radical whose potential curve around $\theta = 0$ is very flat. The inversion barriers for non planar systems are in the following order :

  $$F < OH < NH_2$$

  The C-Z' bond length is significantly shorter than that computed for the parent molecule as shown in table II.3 and the C-H bond lengths are only slightly shorter in the radical. In $CH_2CN$, the C≡N bond length is much larger than that computed for the corresponding molecule. These conclusions already deduced from previous theoretical calculations

Table II.3. Bond lengths in radicals and molecules (Å).

| Radical | C-X | C-H | C≡N |
|---|---|---|---|
| $CH_3$ | - | 1.080 | - |
| $CH_4$ | - | 1.083 | - |
| $CH_3CH_2$ | 1.516 | 1.083 | - |
| $CH_3CH_3$ | 1.538 | 1.086 | - |
| $(CH_3)_2CH$ | 1.522 | 1.084 | - |
| $(CH_3)_2CH_2$ | 1.547 | 1.088 | - |
| $(CH_2)_3C$ | 1.529 | - | - |
| $(CH_3)_3CH$ | 1.550 | 1.089 | - |
| $CH_2F$ | 1.350 | 1.091 | - |
| $CH_3F$ | 1.384 | 1.097 | - |
| $CHF_2$ | 1.354 | 1.105 | - |
| $CH_2F_2$ | 1.388 | 1.109 | - |
| $CF_3$ | 1.351 | - | - |
| $CHF_3$ | 1.371 | 1.118 | - |
| $CH_2CN$ | 1.400 | 1.083 | 1.221 |
| $CH_3CN$ | 1.488 | 1.088 | 1.158 |
| $CH_2NH_2$ | 1.444 | 1.085 | - |
| $CH_3NH_2$ | 1.486 | 1.090 | - |
| $CH_2OH$ | 1.394 | 1.087 | - |
| $CH_3OH$ | 1.439 | 1.095 | - |
| $CH(CN)_2$ | 1.416 | 1.090 | 1.218 |
| $CH_2(CN)_2$ | 1.494 | 1.095 | 1.158 |
| $CHCNNH_2$ | 1.440 (CN) | 1.085 | 1.217 |
| | 1.401 (CC) | - | - |
| $CH_2CNNH_2$ | 1.460 (CN) | 1.093 | 1.158 |
| | 1.508 (CC) | - | - |
| CHCNOH | 1.398 (CO) | 1.085 | 1.219 |
| | 1.400 (CC) | - | - |
| $CH_2CNOH$ | 1.442 (CO) | 1.098 | 1.158 |
| | 1.506 (CC) | - | - |
| $CH(OH)_2$ | 1.400 | 1.080 | - |
| $CH_2(OH)_2$ | 1.430 | 1.101 | - |

(44) are in agreement with those of ESR experiments according to which "atoms or substituents having electron pairs induce bending at the radical center". The influence of substituents on the geometry of substituted alkyl radicals is clearly related to their ± M effect : π-donor groups or atoms favor a pyramidal structure and π-acceptor groups, a planar one. Here again $CH_3$ behave as weak π-donor from this structural point of view.

- The angle θ and the inversion barrier increase with an increasing number of π-donor groups.
- Alkyl radicals substituted by two or three π-acceptor groups remain planar and their energy tend to increase more rapidly with increasing values of θ.
- cd-Substituted radicals are also planar and their potential curve around θ = 0 is very flat. The modifications of their bond lengths are approximately the same than that of the corresponding monosubstituted radicals (see table II.3). There is no structural evidence of an interaction between the two groups. Table II.4 summarizes the substituent effects on the geometry of alkyl radicals.

Table II.4. Substituent effects on the geometry of C XYZ radicals.

| Substituent | Structure | Energy barrier |
|---|---|---|
| X,π-donor | Pyramidal | F < OH < $NH_2$ |
| X,Y,Z,π-donor | Pyramidal | $CH_2X$ < CHXY < CXYZ |
| X,π-acceptor | Planar | Very small |
| X,Y,π-acceptor | Planar | Very small |
| X,π-acceptor, Y,π-donor | Planar | Very small |

2° Electronic structure of free-radicals.

The electronic structure of radicals can be analyzed in terms of densities, populations and properties of localized orbitals.
In the orbital approximation, the electron density of a radical is written :

$$\rho(M) = \sum_j n_j \phi_j^2(M) \quad \text{(II.1)}$$

where $n_j$ is the occupation number of the corresponding orbital. The UHF treatment allows us to separate the contributions of "α orbitals" and "β orbitals" :

$$\rho(M) = \sum_j^{n_\alpha} \phi_j^2(M) + \sum_j^{n_\beta} \phi_j^2(M) = \rho^\alpha(M) + \rho^\beta(M) \qquad (II.2)$$

where $n_\alpha$ and $n_\beta$ are the numbers of electrons of each spin.
We shall put : $n_\alpha - n_\beta = 1$ .
We can also define the spin density as the difference between $\rho^\alpha(M)$ and $\rho^\beta(M)$, at every point M :

$$\rho_S(M) = \sum_j^{n_\alpha} \phi_j^2(M) - \sum_j^{n_\beta} \phi_j^2(M) = \rho^\alpha(M) - \rho^\beta(M) \qquad (II.3)$$

The spin density is either positive, negative or equal to zero. Its value at a given nucleus, $\rho_S(N)$, allows us to calculate the corresponding hyperfine splitting constant a(N) using the equation (I.14).
In the framework of the LCAO-MO method, the total number of electrons of a system is calculated by the relation :

$$n = \int_E \rho(M)\, dv = \sum_j n_j \int \phi_j^2(M)\, dv$$

$$= \sum_A \sum_{p\in A} \sum_j n_j \sum_q C_{jp} C_{jq} S_{pq} \qquad (II.4)$$

or :

$$n = \sum_A \sum_{p\in A} P_p = \sum_A P_A \qquad (II.5)$$

where $P_p$ is the population of an atomic orbital and $P_A$, the population of atom A according to Mulliken's definitions. The sum over j can be separated in contributions of α and β orbitals. Equation (II.4) becomes, putting $n_j = 1$ :

$$n = \sum_A \sum_{p\in A} \sum_j^{n_\alpha} \sum_q C_{jp} C_{jq} S_{pq} + \sum_A \sum_{p\in A} \sum_j^{n_\beta} \sum_q C_{jp} C_{jq} S_{pq} \qquad (II.6)$$

or :

$$n = \sum_A \sum_{p\in A} (P_p^\alpha + P_p^\beta) = \sum_A (P_A^\alpha + P_A^\beta) \qquad (II.7)$$

where $P_p^\alpha$, $P_p^\beta$ and $P_A^\alpha$, $P_A^\beta$ are the numbers of electrons of a given spin (α or β) respectively associated to an atomic orbital or to an atom. The orbital spin density is defined by :

$$P_S^p = P_p^\alpha - P_p^\beta \qquad (II.8)$$

and the atomic spin density by :

$$P_S^A = P_A^\alpha - P_A^\beta \qquad (II.9)$$

$P_S^A$ is either positive, negative or equal to zero.

We have also the relations :

$$\int_E \rho_S(M)\,dv = 1 \tag{II.10}$$

and

$$\sum_A P_A^S = 1 \tag{II.11}$$

is the system contains an odd number of electrons.
The $\pi$ orbital spin densities are related to the proton hyperfine splitting constants by McConnell type equations (see part I).
We shall describe the electronic structure of some typical radicals in terms of populations, spin densities and properties of localized orbitals.

A. Small free-radicals.

A.1. Systems with 9 valence electrons.

- Cyanide radical CN; spin properties calculated at the 4-31G level :

$$P_S^C = 2.235$$

$$P_S^N = -1.235$$

$$\rho_S(C) = 1.0665$$

$$\rho_S(N) = -0.2051$$

The spin density is represented in figure II.3. The distribution of the charge centroids is given below :

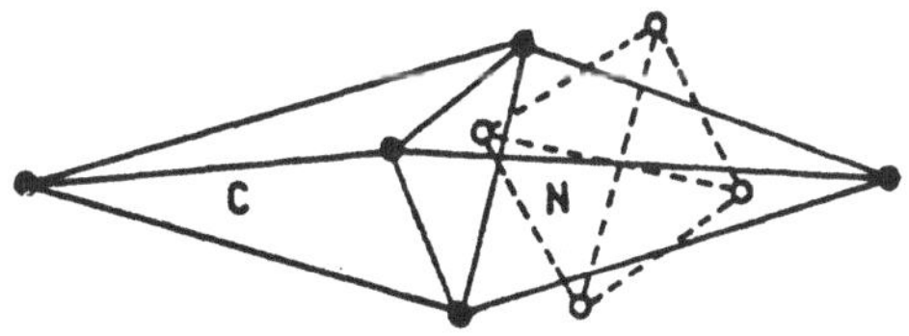


All these results suggest the following representation of CN radical :

• C ≡ N | or • C = N |

Figure II.3 allows us to understand the large carbon 13 hyperfine splitting constant observed in the ESR spectrum of this radical :

$$a(^{13}C) = 210\ G$$

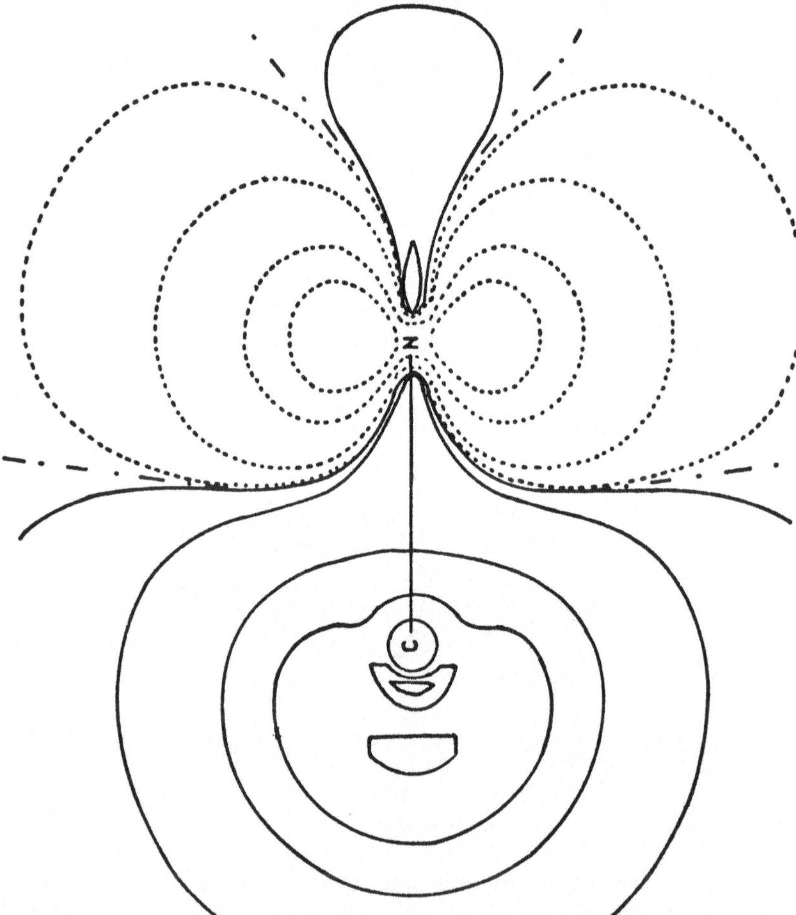


Figure II.3. Spin density of cyanide radical.

It is to be noted that the Lewis octet rule is not satisfied for the carbon atom. Cyanide radical is clearly an atom-centered radical of σ type.

- Nitrogen radical cation $N_2^+$; spin properties (4-31G) :

$$P_S^N = 0.500$$

$$\rho_S(N) = 0.0713$$

Charge centroids distribution :

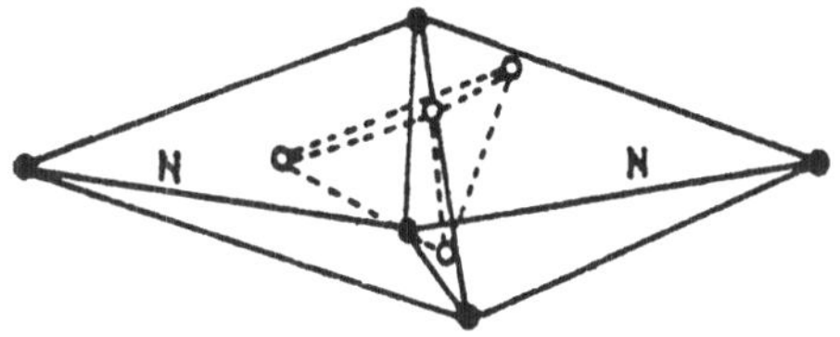


The spin density represented in figure II.4 shows the delocalization of the unpaired electron on the whole system.

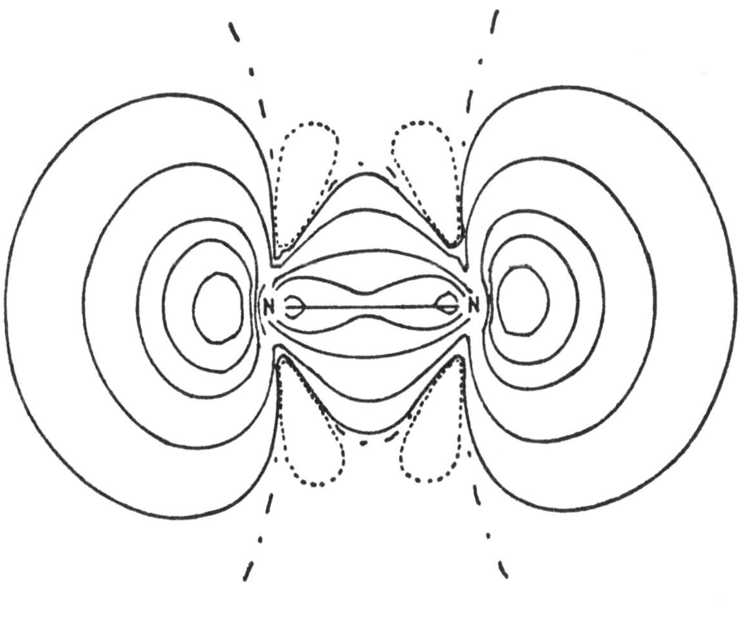


Figure II.4. Spin density of $N_2^+$ radical cation.

We can propose the following electronic structure :

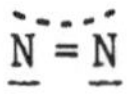

Here the octet rule is satisfied for the two atoms but the double quartet rule is not. $N_2^{+}$ is a delocalized radical.

A.2. System with 11 valence electrons.

- <u>The nitric oxide molecule NO</u>, spin properties (4-31G) :

$$P_S^N = 1.367$$

$$P_S^O = -0.367$$

Charge centroids distribution :

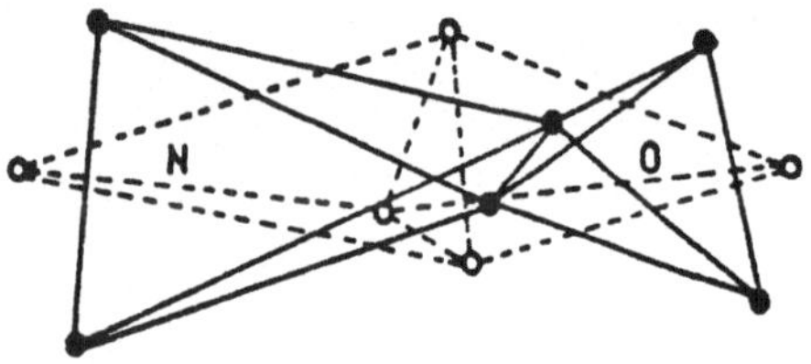


As shown in figure II.5 the unpaired electron is delocalized on the whole molecule.

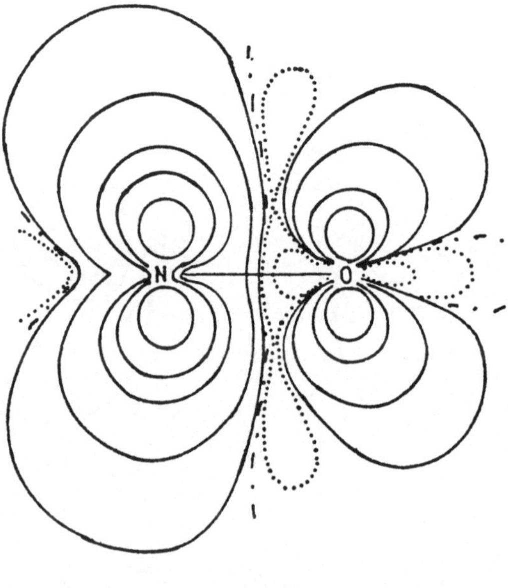


Figure II.5. Spin density of NO molecule.

Electronic structure :

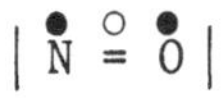


The octet and double quartet rules are satisfied for the two atoms of this delocalized open shell molecule.

A.3. System with 13 valence electrons.

- The FO radical; spin properties (4-31G) :

$$P_S^F = -0.093$$

$$P_S^O = 1.093$$

$$\rho_S(F) = 0.0726$$

$$\rho_S(O) = 0.1905$$

Charge centroids distribution.

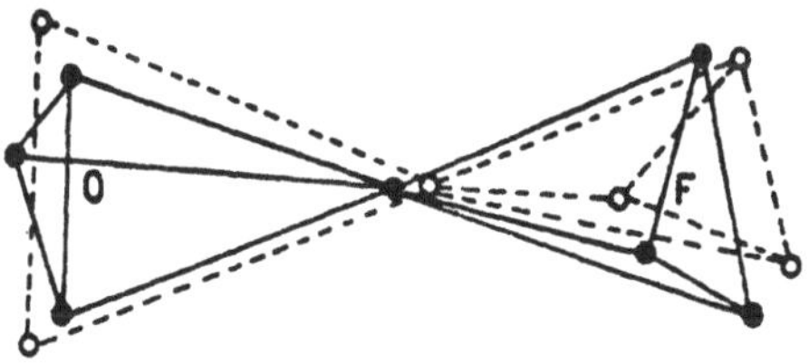


FO and CN have similar spin densities (figure II.6).

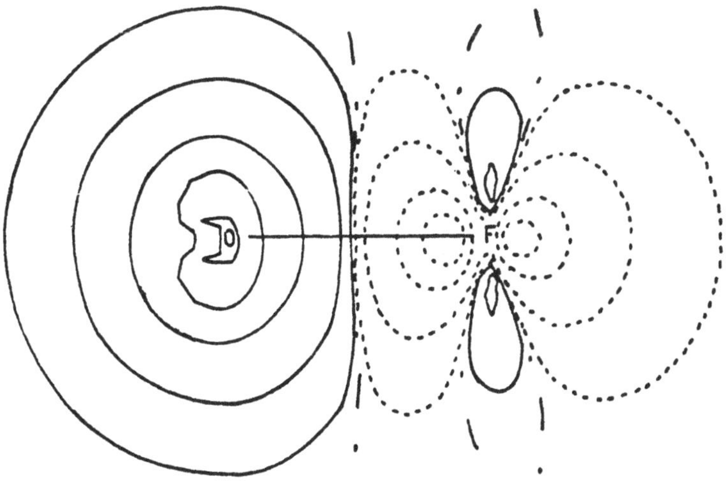


Figure II.6. Spin density of FO radical.

FO is also an atom centered radical for which the following electronic structure can be proposed :

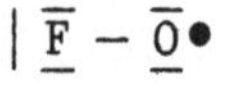

The oxygen atom has only seven electrons in its valence shell.

B. Model radicals.

B.1. Hydroxyl radical, prototype of RO radicals.

Spin properties (4-31G) :

$$P_S^O = 1.062$$

$$P_S^H = -0.062$$

$$\rho_S(O) = 0.1941$$

$$\rho_S(H) = -0.0298$$

OH is clearly a $\pi$ oxygen-centered radical (see figure II.7). As shown by the electronic structure, the octet rule is not satisfied for the oxygen atom :

OH is very similar to $NH_2$ and $CH_3$ (see below).

B.2. $NH_2$, prototype of $R_2N$ radicals.

Spin properties (4-31G) :

$$P_S^N = 1.152$$

$$P_S^H = -0.076$$

$$\rho_S(N) = 0.2086$$

$$\rho_S(H) = -0.0297$$

$NH_2$ is a $\pi$ nitrogen-centered radical which can be represented by the formula :

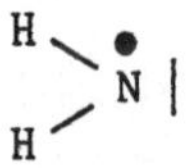

B.3. $CH_3$, prototype of alkyl radicals (see section B).

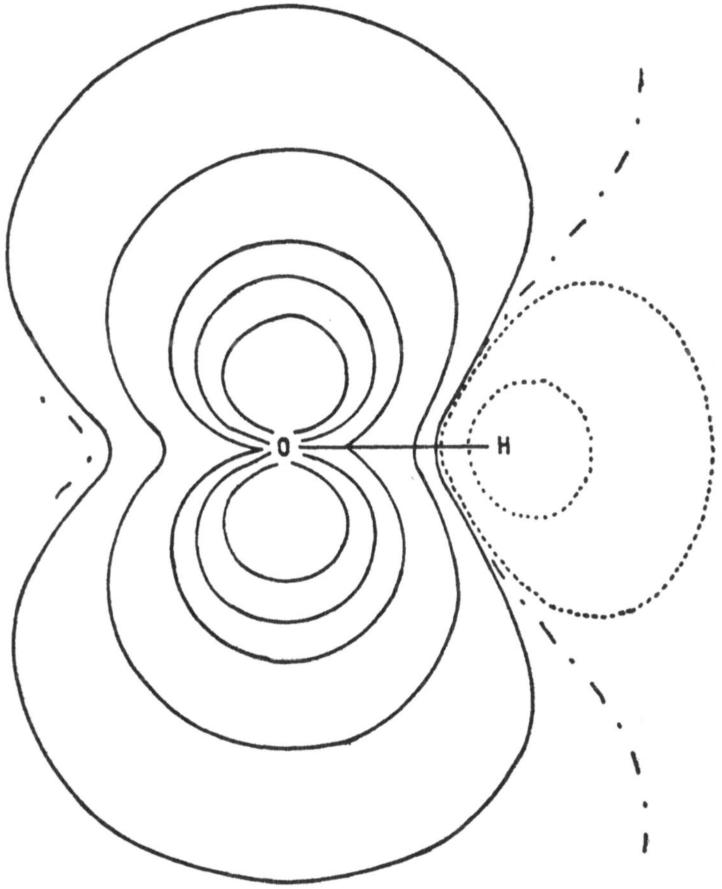


Figure II.7. Spin density of hydroxyl radical.

Spin properties (4-31G) :

$$P_S^C = 1.324$$

$$P_S^H = -0.108$$

$$\rho_S(C) = 0.2293$$

$$\rho_S(H) = 0.0335$$

Methyl radical is a π carbon-centered radical :

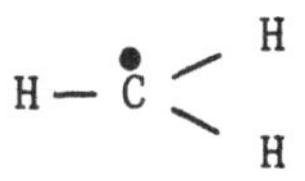

B.4. $H_2NO$, model of the nitroxide radicals.

Spin properties :

$P_S^O = 0.837$ $\quad P_S^\pi(O) = 0.796$

$P_S^N = 0.205$ $\quad P_S^\pi(N) = 0.191$

$P_S^H = -0.021$

$\rho_S(O) = 0.2403$

$\rho_S(N) = 0.0840$

$\rho_S(H) = -0.086$

Charge centroids distribution :

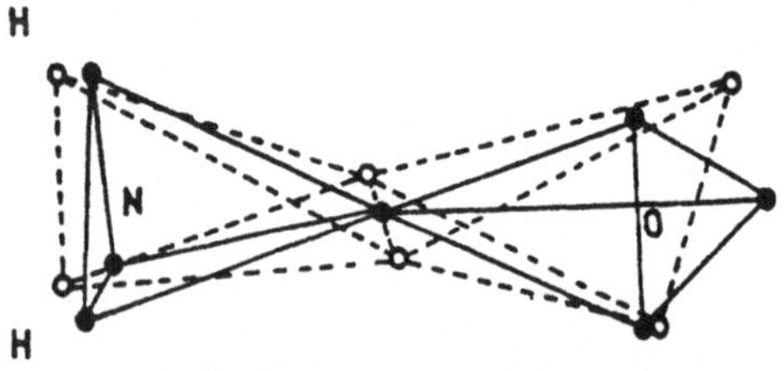


The NO bond is a "three-electron" bond. These results suggest the following representation of nitroxide radicals :

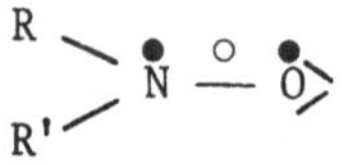


The delocalization of the unpaired electron is clearly shown by the spin density of figure II.8.
$H_2NO$ radical is very similar to $NH_2NH$ (see below).

B.5. $NH_2NH$ prototype of hydrazyl radicals.

Spin properties (4-31G) :

$P_S^{N_1} = 0.964$ $\quad P_S^\pi(N_1) = 0.838$

$P_S^{N_2} = 0.142$ $\quad P_S^\pi(N_2) = 0.153$

$P_S^{H_1} = -0.019$

$P_S^{H_2} = -0.017$

$$P_S^{H_3} = -0.069$$

$$\rho_S(N_1) = 0.2468$$

$$\rho_S(N_2) = 0.0624$$

$$\rho_S(H_1) = -0.0084$$

$$\rho_S(H_2) = -0.0069$$

$$\rho_S(H_3) = -0.0278$$

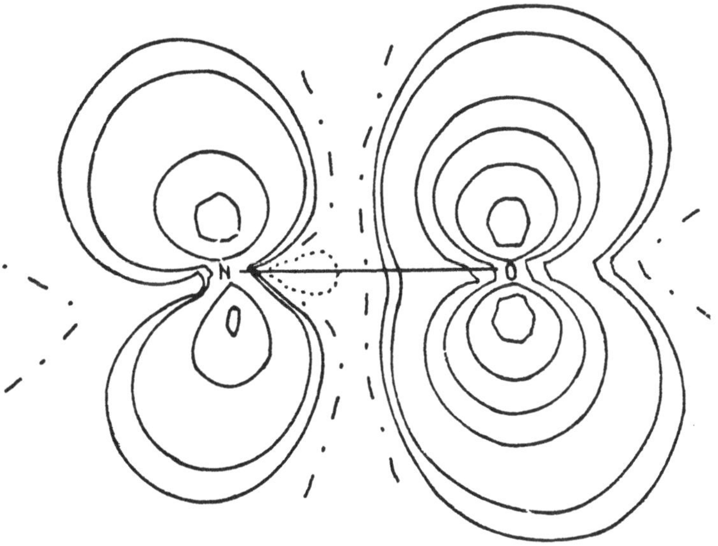


Figure II.8. Spindensity of $H_2NO$ (perpendicular plane).

$NH_2NH$ has the same charge centroids distribution as $H_2NO$. Its electronic structure is represented by the formula :

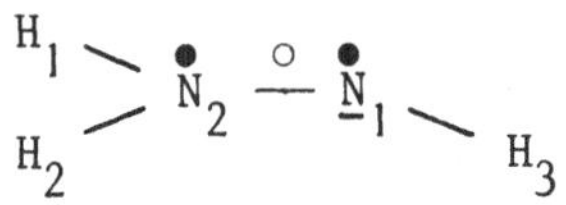


The spin density is given in figures II.9 and II.10.
In this delocalized radical the two nitrogen atoms have eight electrons in their valence shells.

C. Carbon-centered radicals C XYZ.

The electronic properties of carbon-centered radicals calculated

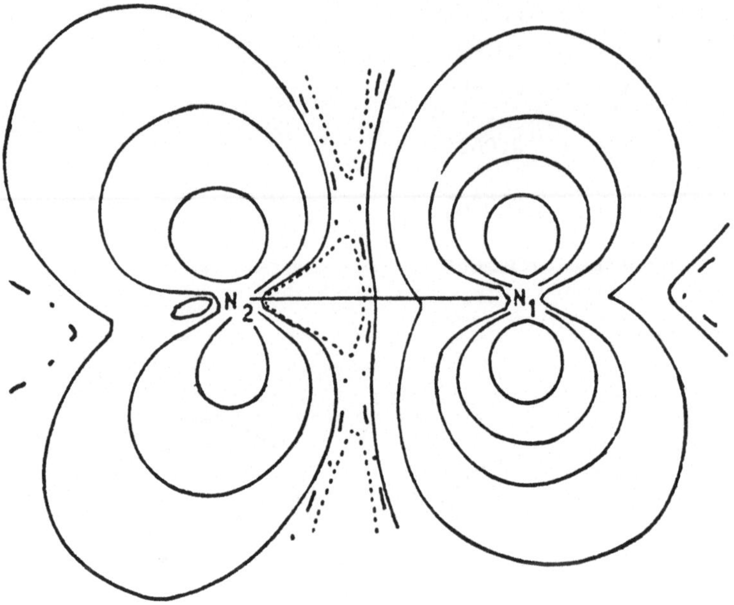


Figure II.9. Spin density of $NH_2NH$ (perpendicular plane).

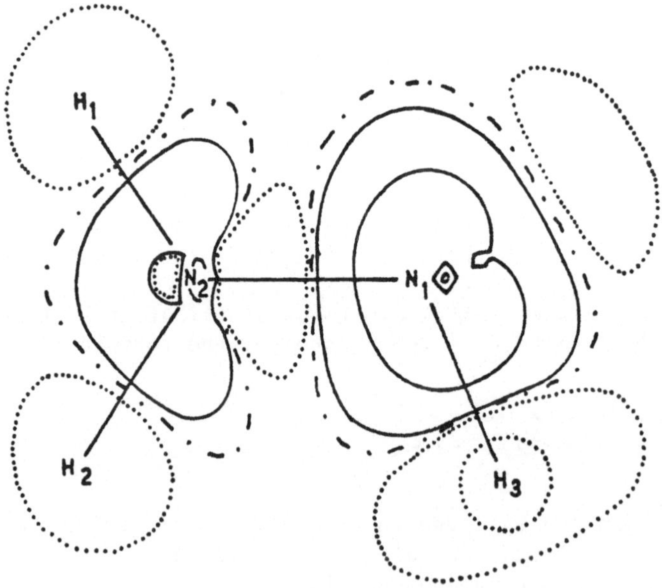


Figure II.10. Spin density of $NH_2NH$ (molecular plane).

at the STO-3G level are collected in table II.5. They are respectively: atom populations: $P_A^\alpha + P_B^\beta$, atomic spin densities: $P_A^\alpha - P_A^\beta$, $\pi$ orbital spin densities of planar radicals: $P_S^\pi$, spin densities at the nuclei: $\rho_S(N)$, distances between the charge centroid of the unpaired electron and the $\alpha$ carbon: $d_C$, the second moment of the localized orbital of the unpaired electron: $< r^2 >$ and the mean value of $S^2$: $< S^2 >$. Net charges of $CH_2$ (CH,C) and X(Y,Z) moities and, for planar radicals, $\sigma$ and $\pi$ electron transfers are given in figure II.11. The position and the shift of the charge centroid of the unpaired electron are indicated respectively by a star and an arrow. The spin densities of $CH_3$, $CH_2OH$, $CH_2CN$ and CH(CN)OH, in some particular planes, are represented in figure II.12.

Some interesting conclusions can be deduced from table II.5 and figure II.11 :

- All the substituents are electron withdrawing but the net charges are essentially due to $\sigma$ transfers.
- $CH_3$, F, OH and $NH_2$ behave as $\pi$-donor.
- The conjugation between $\pi$-donor and $\pi$-captor groups enhances considerably the $\pi$ transfer.
- The two groups of substituents can be distinguished by their influence on the properties of localized orbitals. $\pi$-donor substituents remove the charge centroid of the unpaired electron from the $\alpha$ carbon atom and decrease the corresponding second moment. On the other hand, $\pi$-acceptor substituents draw this charge centroid towards the bond $C_\alpha$-captor and increase the second moment. According to these results, the unpaired electron of $\pi$ radicals is involved in the chemical bonds but, in $\sigma$ radicals, it is much more centered on the $\alpha$ carbon atom.

These conclusions are summarized in the following formulae which are in complete agreement with the results of geometry optimization and the spin densities of figure II.12.

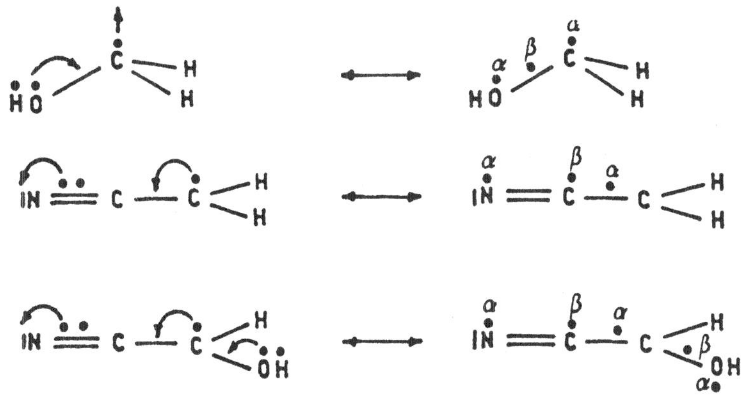

Table II.5. Electronic properties of C XYZ radicals.

| Radical | Atom | $P_A$ | $P_S^A$ | $P_S^\pi$ | $\rho_S$ (N) | $d_C$ | $\langle r^2 \rangle$ | $\langle S^2 \rangle$ |
|---|---|---|---|---|---|---|---|---|
| $CH_3$ | C | 6.170 | 1.265 | - | 0.279 | 0.567 | 1.485 | 0.76 |
| | H | 0.943 | -0.088 | | -0.031 | | | |
| $CH_3CH_2$ | C ($CH_2$) | 6.095 | 1.231 | - | 0.283 | 0.572 | 1.500 | 0.76 |
| | H ($CH_2$)* | 0.952 | -0.086 | | -0.030 | | | |
| | C ($CH_3$) | 6.190 | -0.145 | | -0.055 | | | |
| | H ($CH_3$)* | 0.937 | 0.029 | | 0.008 | | | |
| $(CH_3)_2CH$ | C (CH) | 6.023 | 1.199 | - | 0.296 | 0.600 | 1.504 | 0.76 |
| | H (CH) | 0.961 | -0.082 | | -0.029 | | | |
| | C ($CH_3$) | 6.190 | -0.137 | | -0.051 | | | |
| | H ($CH_3$)* | 0.940 | 0.026 | | 0.008 | | | |
| $(CH_3)_3C$ | C | 5.953 | 1.169 | - | 0.284 | 0.598 | 1.524 | 0.77 |
| | C ($CH_3$) | 6.192 | -0.133 | | -0.052 | | | |
| | H ($CH_3$)* | 0.941 | 0.026 | | 0.012 | | | |
| $CH_2F$ | C | 5.981 | 1.115 | - | 0.301 | 0.669 | 1.439 | 0.76 |
| | H | 0.943 | -0.071 | | -0.024 | | | |
| | F | 9.133 | 0.027 | | 0.043 | | | |

Table II.5. Cont.

| | | | | | | | | |
|---|---|---|---|---|---|---|---|---|
| $CHF_2$ | C | 5.792 | 0.967 | - | 0.346 | 0.757 | 1.397 | 0.76 |
| | H | 0.939 | -0.038 | | -0.009 | | | |
| | F | 9.135 | 0.036 | | 0.037 | | | |
| $CF_3$ | C | 5.585 | 0.808 | - | 0.432 | 0.821 | 1.368 | 0.76 |
| | F | 9.138 | 0.064 | | 0.023 | | | |
| $CH_2CN$ | C | 6.094 | 1.196 | 0.852 | 0.258 | 0.320 | 1.708 | 1.29 |
| | H | 0.908 | -0.107 | | -0.040 | | | |
| | C (CN) | 5.949 | -1.317 | -0.527 | -0.298 | | | |
| | N | 7.141 | 1.334 | 0.676 | 0.123 | | | |
| $CH_2NH_2$ | C | 6.037 | 1.099 | - | 0.292 | 0.678 | 1.437 | 0.76 |
| | H | 0.950 | -0.068 | | -0.023 | | | |
| | N | 7.386 | 0.021 | | -0.001 | | | |
| | H $(NH_2)^*$ | 0.839 | 0.004 | | 0.003 | | | |
| $CH_2OH$ | C | 6.016 | 1.086 | - | 0.296 | 0.680 | 1.433 | 0.76 |
| | $H^*$ | 0.950 | -0.067 | | -0.023 | | | |
| | O | 8.277 | 0.054 | - | 0.022 | | | |
| | H (OH) | 0.807 | -0.005 | | 0.001 | | | |
| $CH(CN)_2$ | C | 6.016 | 1.170 | 0.759 | 0.284 | 0.258 | 1.803 | 1.84 |
| | H | 0.882 | -0.120 | | -0.047 | | | |
| | C (CN) | 5.932 | -1.339 | -0.531 | -0.321 | | | |
| | N | 7.119 | 1.314 | 0.651 | 0.120 | | | |

Table II.5. Cont.

| | | | | | | | | |
|---|---|---|---|---|---|---|---|---|
| $CHCNNH_2$ | C | 5.962 | 1.073 | 0.757 | 0.241 | 0.733 | 1.646 | 1.28 |
| | H | 0.913 | -0.096 | - | -0.036 | | | |
| | C (CN) | 5.962 | -1.287 | -0.506 | -0.291 | | | |
| | N (CN) | 7.383 | 1.311 | 0.654 | 0.121 | | | |
| | N ($NH_2$) | 7.146 | -0.017 | 0.044 | -0.001 | | | |
| | H ($NH_2$)* | 0.817 | 0.008 | | 0.005 | | | |
| CHCNOH | C | 5.943 | 1.059 | 0.756 | 0.243 | 0.980 | 1.589 | 1.28 |
| | H | 0.904 | -0.095 | | -0.036 | | | |
| | C (CN) | 5.974 | -1.297 | -0.507 | -0.292 | | | |
| | N | 7.136 | 1.330 | 0.663 | 0.123 | | | |
| | O | 8.254 | 0.005 | 0.089 | 0.030 | | | |
| | H (OH) | 0.789 | -0.001 | | 0.000 | | | |
| $CH(OH)_2$ | C | 5.858 | 0.949 | - | 0.330 | 0.757 | 1.394 | 0.76 |
| | H | 0.937 | -0.054 | | -0.017 | | | |
| | O | 8.284 | 0.053 | - | 0.003 | | | |
| | H (OH) | 0.819 | -0.001 | | 0.002 | | | |

* Mean values

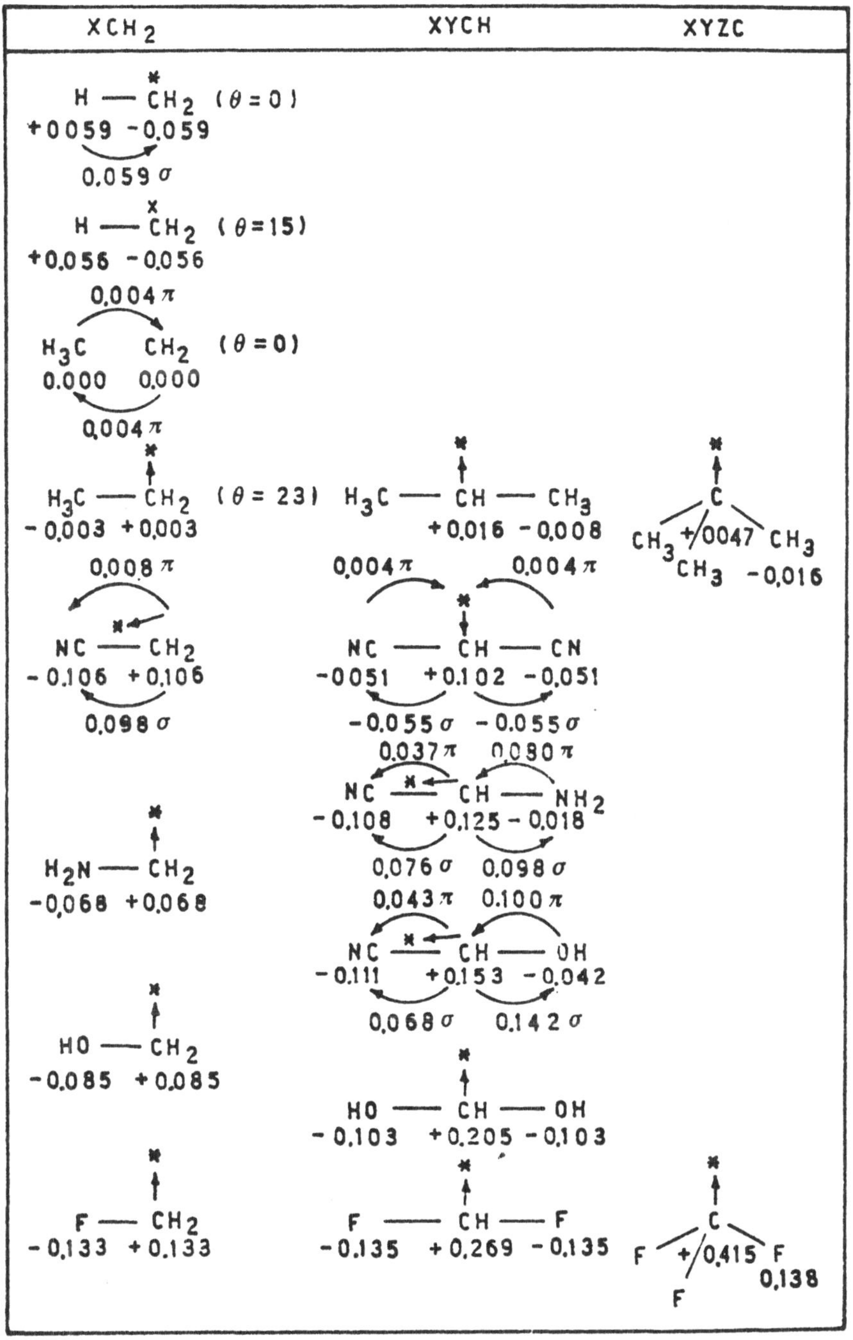


Figure II.11. Net charges; charge centroid; σ and π electron transfers.

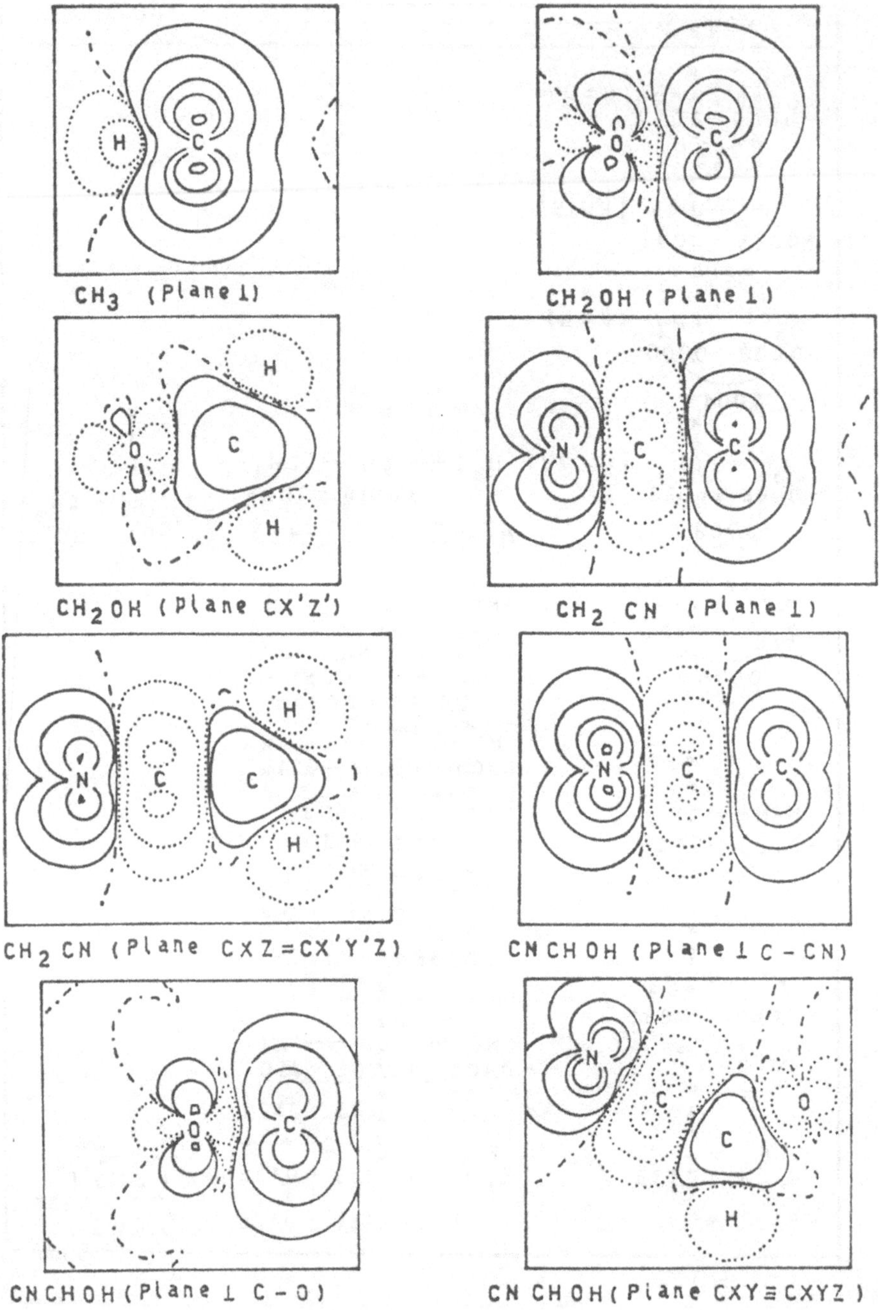


Figure II.12. Spin densities of some C XYZ radicals: continuous lines: positive values; dotted lines: negative values.

The spin properties collected in table II.5 are not very reliable because UHF wave functions do not represent pure doublet spin states. Furthermore the basis sets used in this work are very inadequate near the nuclei because they do not satisfy the cusp conditions. Consequently they fail to predict spin densities at the nuclei, $\rho_S(N)$, and the corresponding hyperfine splitting constants, a(N). However, the sign of the spin density seems to be correctly predicted. Indeed, this sign is generally in perfect agreement with the predictions based on the models of spin polarization and hyperconjugation. So, the $\alpha$ hydrogen atoms of an alkyl radical are always found to lie in regions of negative spin density and $\beta$ hydrogen atoms, in regions of positive spin density as shown in figure II.13.

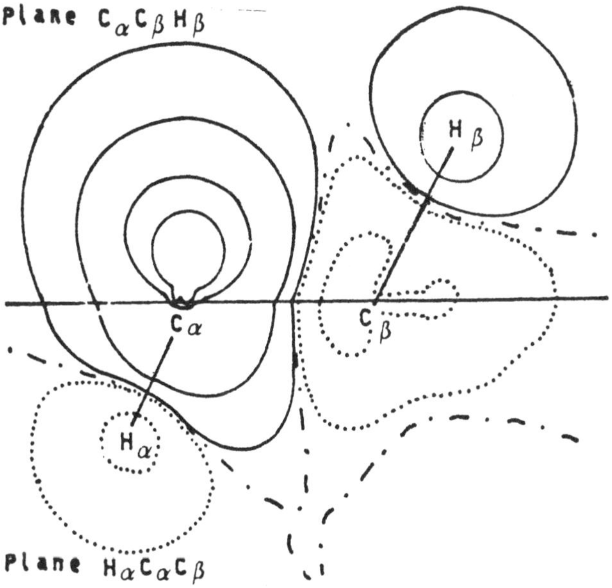


Figure II.13. Spin density of ethyl radical in two perpendicular planes.

In spite of their deficiency, we shall compare the calculated spin properties to the experimental ones.
Atomic spin densities can not be measured but the informations they yield on the spin-withdrawing influence of substituents are comparable to those deduced from Fischer's parameters: all the substituents are able to delocalize the unpaired electron of an alkyl radical; $P_S^{C\alpha}(C\,XYZ) < P_S^{C\alpha}(CH_3)$. The same conclusion is obtained from the analysis of $\pi$ spin densities. Moreover, there is a good correlation between $P_S^{\pi}(C_\alpha)$ and the corresponding $H_\alpha$ hyperfine splitting constants as shown in table II.6. The McConnell relation :

$$a(H_\alpha) = Q_\alpha P_S^\pi(C_\alpha) \qquad (II.12)$$

is approximately satisfied.

Table II.6. $\pi$ spin densities of C XYZ radicals.

| R | $P_S^\pi(C_\alpha)$ | $a(H_\alpha)$,G | $Q_\alpha$,G |
|---|---|---|---|
| $CH_3$ | 0.955 | 23.04 (47) | 24.13 |
| $CH_3CH_2$ | 0.927 | 22.38 (47) | 24.14 |
| $CH_2CN$ | 0.852 | 20.88 (48) | 24.51 |
| $CH(CN)_2$ | 0.759 | 19.18 (49) | 25.37 |

Spin densities at the nuclei are roughly correlated to the hyperfine splitting constants. Table II.7 collects $\rho_S(C_\alpha)$ and $a(^{13}C_\alpha)$ for some substituted alkyl radicals. These two quantities increase with the out of plane angle $\theta$ (or the $sp^3$ character of $\alpha$ carbon atom).

Table II.7. Comparison of $\rho_S(^{13}C_\alpha)$ and $a(^{13}C_\alpha)$.

| R | $\rho_S(C_\alpha)$ | $a(^{13}C_\alpha)$,G | | $\rho_S(C_\alpha)$exp [a] |
|---|---|---|---|---|
| $CH_3$ | 0.279 | 38.34 | (47) | 0.096 |
| $CH_3CH_2$ | 0.283 | 39.07 | (47) | 0.097 |
| $CH_2F$ | 0.301 | 54.80 | (51) | 0.137 |
| $CHF_2$ | 0.346 | 148.80 | (51) | 0.371 |
| $CF_3$ | 0.432 | 271.60 | (51) | 0.677 |
| $CH_2CN$ | 0.258 | 30.00-38.30 | (50) | 0.075-0.095 |
| $CH(CN)_2$ | 0.284 | 29.18 | (49) | 0.073 |
| $CH_2OH$ | 0.296 | 47.40 | (52-53) | 0.110 |
| CN | 1.065 | 210.00 | (54) | 0.524 |

[a] Calculated using equation (I.14)

It is also to be mentioned that $\rho_S(H_\alpha)$ is approximately proportional to $P_S^{\pi}(C_\alpha)$ as predicted by the spin polarization model (see table II.8).

Table II.8. Comparison of $\rho_S(H_\alpha)$ and $P_S^{\pi}(C_\alpha)$.

| R | $\rho_S(H_\alpha)$ | $P_S^{\pi}(C_\alpha)$ | $\rho_S/P_S^{\pi}$ |
|---|---|---|---|
| $CH_2CN$ | -0.040 | 0.852 | 0.047 |
| $CH(CN)_2$ | -0.047 | 0.759 | 0.062 |
| CH(CN)(OH) | -0.036 | 0.756 | 0.048 |
| $CH(CN)(NH_2)$ | -0.036 | 0.757 | 0.048 |

We must conclude that the most interesting theoretical results concerning the electronic structure of carbon-centered radicals are the spin density $\rho_S(M)$ and the properties of the unpaired electron localized orbital.

3° Thermochemistry of free-radicals.

It is generally assumed that heats of reaction of isodesmic transformations are correctly predicted at the SCF level, using 4-31G basis set. Thus, our theoretical results allow us to calculate the stabilization energies and the heat of formation of carbon-centered radicals. The dissociation energy of the C-H bond of the parent molecule can also be obtained. ZPE and temperature corrections will not be carried out in this preliminary study. So, experimental standard heats of reaction and the corresponding theoretical $\Delta E$ values will be directly compared.

We shall illustrate our procedure by considering the case of cyanomethyl radical. The intrinsic stability of this species is the $\Delta E$ of the bond separation reaction :

$$CH_2CN + 2\,CH_4 \longrightarrow CH_3 + CH_3CH_3 + HCN \qquad [1]$$

$\Delta E_1$ = 23.96 kcal/mole, at the 4-31G level; A.S. = 23.96 kcal/mole. The relative stability, or resonance energy as defined by Szwarc, is the $\Delta E$ of the isodesmic reaction :

$$CH_2CN + CH_4 \longrightarrow CH_3 + CH_3CN \qquad [2]$$

$\Delta E_2$ = 11.71 kcal/mole, at the 4-31G level; RE = 11.71 kcal/mole. We can write :

$$\Delta E_2 = BDE(CH_4) - BDE(CH_3CN) \qquad (II.13)$$

and also :

Table II.9. Intrinsic stabilization energies (kcal/mole).

| Radical | Bond separation reaction | ΔE(4-31G) | $\Delta H^o(exp)$[a] |
|---|---|---|---|
| $CH_3CH_2$ | $CH_3CH_2 + CH_4 \rightarrow CH_3 + CH_3CH_3$ | 3.32 | 6.36 |
| $CH_3CHCH_3$ | $CH_3CHCH_3 + 2\,CH_4 \rightarrow CH_3 + 2\,CH_3CH_3$ | 6.20 | 11.42 |
| $(CH_3)_3C$ | $(CH_3)_3C + 3\,CH_4 \rightarrow CH_3 + 3\,CH_3CH_3$ | 9.90 | 19.28 |
| $CH_2F$ | $CH_2F + CH_4 \rightarrow CH_3 + CH_3F$ | 2.31 | 3.30 |
| $CHF_2$ | $CHF_2 + 2\,CH_4 \rightarrow CH_3 + 2\,CH_3F$ | 12.52 | 14.78 |
| $CF_3$ | $CF_3 + 3\,CH_4 \rightarrow CH_3 + 3\,CH_3F$ | 24.63 | 29.60 |
| $CH_2CN$ | $CH_2CN + 2\,CH_4 \rightarrow CH_3 + CH_3CH_3 + HCN$ | 23.96 | 23.66 |
| $CH_2NH_2$ | $CH_2NH_2 + CH_4 \rightarrow CH_3 + CH_3NH_2$ | 9.55 | 9.70 |
| $CH_2OH$ | $CH_2OH + CH_4 \rightarrow CH_3 + CH_3OH$ | 6.21 | 10.33 |
| $CH(CN)_2$ | $CH(CN)_2 + 4\,CH_4 \rightarrow CH_3 + 2\,CH_3CH_3 + 2\,HCN$ | 36.35 | - |
| $CHCNNH_2$ | $CHCNNH_2 + 3\,CH_4 \rightarrow CH_3 + CH_3CH_3 + CH_3NH_2 + HCN$ | 33.07 | - |
| $CHCNOH$ | $CHCNOH + 3\,CH_4 \rightarrow CH_3 + CH_3CH_3 + CH_3OH + HCN$ | 26.10 | - |
| $CH(OH)_2$ | $CH(OH)_2 + 2\,CH_4 \rightarrow CH_3 + 2\,CH_3OH$ | 18.57 | - |

[a] Calculated using the data of table II.13.

Table II.10. Resonance energies and C-H bond dissociation energies (kcal/mole).

| Radical | Isodesmic reaction | ΔE(4-31G) | $\Delta H_1^o$ [a] | BDE(4-31G) | BDE(exp) |
|---|---|---|---|---|---|
| $CH_3$ | - | - | - | - | 104.30 (55) |
| $CH_3CH_2$ | $CH_3CH_2 + CH_4 \rightarrow CH_3 + CH_3CH_3$ | 3.32 | 6.3 | 100.98 | 97.94 (55) |
| $(CH_3)_2CH$ | $(CH_3)_2CH + CH_4 \rightarrow CH_3 + (CH_3)_2CH_2$ | 5.22 | 9.2 | 99.08 | 95.13 (55) |
| $(CH_3)_3C$ | $(CH_3)_3C + CH_4 \rightarrow CH_3 + (CH_3)_3CH$ | 7.31 | 11.8 | 96.99 | 92.52 (55) |
| $CH_2F$ | $CH_2F + CH_4 \rightarrow CH_3 + CH_3F$ | 2.31 | 3.3 | 101.99 | 101.00 (56) |
| $CHF_2$ | $CHF_2 + CH_4 \rightarrow CH_3 + CH_2F_2$ | 2.24 | 2.3 | 102.06 | 101.95 (56) |
| $CF_3$ | $CF_3 + CH_4 \rightarrow CH_3 + CHF_3$ | -2.43 | -2.2 | 106.73 | 106.40 (56) |
| $CH_2CN$ | $CH_2CN + CH_4 \rightarrow CH_3 + CH_3CN$ | 11.71 | 11.3 | 92.59 | 93.00 (57) |
| $CH_2NH_2$ | $CH_2NH_2 + CH_4 \rightarrow CH_3 + CH_3NH_2$ | 9.55 | 9.7 | 94.75 | 94.60 (58) |
| $CH_2OH$ | $CH_2OH + CH_4 \rightarrow CH_3 + CH_3OH$ | 6.22 | 10.3 | 98.08 | 93.97 (59) |
| $CH(CN)_2$ | $CH(CN)_2 + CH_4 \rightarrow CH_3 + CH_2(CN)_2$ | 21.70 | - | 82.60 | - |
| $CHCNNH_2$ | $CHCNNH_2 + CH_4 \rightarrow CH_3 + CNCH_2NH_2$ | 18.11 | - | 86.19 | - |
| CHCNOH | $CHCNOH + CH_4 \rightarrow CH_3 + CNCH_2OH$ | 16.21 | - | 88.09 | - |
| $CH(OH)_2$ | $CH(OH)_2 + CH_4 \rightarrow CH_3 + CH_2(OH)_2$ | 3.53 | - | 100.77 | - |

[a] Calculated from the data of table II.13.

$$\Delta E_1 = \Delta E_f(CH_3) + \Delta E_f(C_2H_6) + \Delta E_f(HCN) - 2\,\Delta E_f(CH_4) - \Delta E_f(CH_2CN) \quad (II.14)$$

Introducing respectively in equations (II.13) and (II.14) the experimental BDE of methane and the standard heats of formation of $CH_3$, $C_2H_6$, HCN, and $CH_4$, one obtains semi-empirical values of $\Delta E_f$ $(CH_2CN)$ and BDE$(CH_3CN)$ :

$$\Delta E_f(CH_2CN) = 58.20 \text{ kcal/mole}$$

$$BDE(CH_3CN) = 92.59 \text{ kcal/mole} .$$

The intrinsic stabilization energies of carbon-centered radicals are given in table II.9 with the corresponding bond separation reactions. They are compared with the available experimental results. The resonance energies and the C-H bond dissociation energies are collected in table II.10 and compared with the corresponding experimental data. The total energies of molecules used in the calculations are given in table II.11. The 4-31G results have been obtained using geometries optimized at the STO-3G level.

Table II.11. Total energies of molecules (a.u.).

| Molecule | STO-3G | 4-31G |
|---|---|---|
| $CH_4$ | -39.72686 (43) | -40.13976 (43) |
| $CH_3CH_3$ | -78.30618 (43) | -79.11582 (43) |
| $CH_3CH_2CH_3$ | -116.88580 | -118.09345 |
| $(CH_3)_3CH$ | -155.46594 | -157.07207 |
| $CH_3F$ | -137.16906 (43) | -138.85686 (43) |
| $CH_2F_2$ | -234.62730 | -237.59035 |
| $CHF_3$ | -332.10007 | -336.33421 |
| $CH_3CN$ | -130.27150 | -131.72679 |
| $CH_3NH_2$ | -94.03286 (43) | -95.06498 (43) |
| $CH_3OH$ | -113.54919 (43) | -114.86716 (43) |
| $CH_2(CN)_2$ | -220.80616 | -223.29811 |
| $CH_2(OH)_2$ | -187.38833 | -189.61854 |
| $CNCH_2NH_2$ | -184.56930 | -186.65634 |
| $CNCH_2OH$ | -204.09302 | -206.45043 |
| HCN | -91.67521 (43) | -92.73120 (43) |

We can also obtain semi-empirical heats of formation of molecules by considering the bond separation reactions of table II.12. The corresponding ΔE's measure the intrinsic stabilizations of these compounds. Finally, theoretical and experimental heats of formation of radicals and parent molecules are compared in table II.13.

In general, the theoretical results do not differ strongly from the experimental ones which are seldom very approximate. If we except the case of $CF_3$, all the radicals studied here have a positive "resonance energy". According to the criterion of intrinsic stabilization energy, all the substituents stabilize carbon-centered radicals. Whatever the criterion used may be, the stabilization of monosubstituted species increases in the following order:

$$F < CH_3 < OH < NH_2 < CN$$

according to the theoretical results and, in the order :

$$F < CH_3 < NH_2 \simeq OH < CN$$

if experimental data are used.

The intrinsic stability of carbon-centered radicals increases always with an increasing number of substituents. The effects of methyl groups are approximately additive :

| | A.S.(Th) | A.S.(Exp) |
|---|---|---|
| $CH_3CH_2$ | 3.32 | 6.36 |
| $(CH_3)_2CH$ | 6.20 | 11.42 |
| $(CH_3)_3C$ | 9.90 | 19.28 |

In the case of disubstituted radicals the following order of intrinsic stability is observed :

$$CH_3,CH_3 < F,F < OH,OH < CN,OH < CN,NH_2 < CN,CN$$

However, if we consider the resonance energy which measures the delocalization of the unpaired electron, the order of increasing stabilization is now :

$$F,F < OH,OH < CH_3,CH_3 < CN,OH < CN,NH_2 < CN,CN$$

It is interesting to note that capto-dative substituted radicals have approximately the same resonance energy as allyl and benzyl radicals (see table I.7).

4° Radical transfer reactions.

Some informations on radical transfer reactions :

Table II.12. Intrinsic stabilization of molecules (kcal/mole).

| Molecule | Bond separation reaction | ΔE(4-31G) | $\Delta H^{o}(exp)$ [a] |
|---|---|---|---|
| $CH_3CH_2CH_3$ | $CH_3CH_2CH_3 + CH_4 \rightarrow 2\,CH_3CH_3$ | 0.98 | 2.25 |
| $(CH_3)_2CH$ | $(CH_3)_2CH + 2\,CH_4 \rightarrow 3\,CH_3CH_3$ | 2.59 | 7.50 |
| $CH_2F_2$ | $CH_2F_2 + CH_4 \rightarrow 2\,CH_3F$ | 10.28 | 12.43 |
| $CHF_3$ | $CHF_3 + 2\,CH_4 \rightarrow 3\,CH_3F$ | 27.07 | 31.70 |
| $CH_3CN$ | $CH_3CN + CH_4 \rightarrow CH_3CH_3 + HCN$ | 12.25 | 12.36 |
| $CH_2(CN)_2$ | $CH_2(CN)_2 + 3\,CH_4 \rightarrow 2\,CH_3CH_3 + 2\,HCN$ | 14.65 | 14.32 |
| $CNCH_2NH_2$ | $CNCH_2NH_2 + 2\,CH_4 \rightarrow CH_3CH_3 + CH_3NH_2 + HCN$ | 14.97 | 14.58 |
| $CNCH_2OH$ | $CNCH_2OH + 2\,CH_4 \rightarrow CH_3CH_3 + CH_3OH + HCN$ | 9.89 | 14.33 |
| $CH_2(OH)_2$ | $CH_2(OH)_2 + CH_4 \rightarrow 2\,CH_3OH$ | 15.04 | 14.76 |

[a] Calculated from the data of table II.13.

Table II.13. Heats of formation of radicals and molecules (kcal/mole).

| Molecule | $\Delta E_f$ STO-3G | $\Delta E_f$ 4-31G | $\Delta H_f^o$(exp) | Radical | $\Delta E_f$ STO-3G | $\Delta E_f$ 4-31G | $\Delta H_f^o$(exp) |
|---|---|---|---|---|---|---|---|
| $CH_4$ | - | - | -17.90 (60) | $CH_3$ | - | - | 34.30 (55) |
| $C_2H_6$ | - | - | -20.24 (60) | $C_2H_5$ | 27.78 | 28.64 | 25.60 (55) |
| $(CH_3)_2CH_2$ | -22.76 | -23.56 | -24.83 (60) | $(CH_3)_2CH$ | 21.57 | 23.42 | 18.20 (55) |
| $(CH_3)_3CH$ | -25.62 | -27.51 | -32.42 (60) | $(CH_3)_3C$ | 15.23 | 17.38 | 8.00 (55) |
| $CH_3F$ | - | - | -56.80 (61) | $CH_2F$ | -14.52 | -6.91 | -7.90 (56) |
| $CH_2F_2$ | -105.76 | -105.98 | -108.13 (60) | $CHF_2$ | -70.17 | -56.02 | -58.28 (56) |
| $CHF_3$ | -163.84 | -161.67 | -166.30 (60) | $CF_3$ | -130.73 | -107.03 | -112.00 (56) |
| $CH_3CN$ | 19.31 | 17.71 | 17.60 (57) | $CH_2CN$ | 53.25 | 58.20 | 58.50 (57) |
| $CH_3NH_2$ | - | - | -5.50 (60) | $CH_2NH_2$ | 36.86 | 37.15 | 37.00 (58) |
| $CH_3OH$ | - | - | -48.07 (60) | $CH_2OH$ | -6.39 | -2.08 | -6.20 (59) |
| $CH_2(CN)_2$ | 62.79 | 63.17 | 63.5 (60) | $CH(CN)_2$ | 79.76 | 93.67 | - |
| $CNCH_2NH_2$ | 36.86 | 27.39 | 27.78 (62) | $CHCNNH_2$ | 57.88 | 61.49 | - |
| $CNCH_2OH$ | -10.34 | -10.10 | -14.54 (62) | CHCNOH | 13.83 | 25.89 | - |
| $CH_2(OH)_2$ | -88.78 | -93.28 | -93.00 (63) | $CH(OH)_2$ | -51.59 | -44.64 | - |
| HCN | - | - | 32.30 (64) | $CH{=}CH{-}CH_2$ | - | - | 39.40 (65) |
| $CH_2{=}CH_2$ | - | - | 12.45 (60) | $C_6H_5CH_2$ | - | - | 47.80 (65) |
| | | | | $CCl_3$ | - | - | 19.00 (66) |

$$RH + M \longrightarrow R + MH \qquad [3]$$

can be deduced from our theoretical results. We shall take here $M \equiv CH_3$. The heat of reaction is given by the relation :

$$\Delta H_3^o \simeq \Delta E_3 = BDH(RH) - BDH(CH_4) \qquad (II.15)$$

Using this equation and the data of table II.10, we obtain the results collected in table II.14. We give also the experimental activation energies which are not always very accurately determined. If we except the case where $RH \equiv CHF_3$, all the radical transfer reactions of table II.14 are exothermic. One observes a qualitative agreement between the theoretical and experimental results.

As we have seen before several factors can be considered for interpreting and predicting the activation energies of hydrogen abstraction reactions. The most important seem to be the polar effects and the exothermicity of the reaction. If we except the case of thermoneutral transformations, these two factors must always be considered simultaneously. The Evans-Polanyi relationship which must be used with caution holds quite well for methyl reacting with alkanes. Polar effects do not favor these hydrogen abstraction reactions because the attacking and leaving radicals are both nucleophilic. The reactions involving fluoroalkanes are favoured by polar effects but have a small exothermicity; here the Evans-Polanyi relation does not hold. Polar effects should also favor the reactions of methyl with mono- and disubstituted alkanes $CH_3X$ and $CH_2XY$ (where $X, Y \equiv CN, NH_2, OH$). Evans-Polanyi equation will be satisfied if the exothermicities of the reactions are not tooclose. Thus, we can anticipate that the activation energies for methyl reacting with disubstituted alkanes will decrease in the following order :

$$OH,OH > CN,OH > CN,NH_2 > CN,CN$$

If the attacking radical is electrophilic, this order could be modified but the activation energy relative to $CH_2(OH)_2$ would presumably remain the highest.
It is also interesting to compare the net charges of $H_\alpha$ and $C_\alpha$ atoms in the different molecules (see table II.15). There is apparently a correlation between the net charge of the $\alpha$ carbon atom and the activation energy of the hydrogen abstraction reaction, for narrow series of compounds (alkanes and the series: $CH_3CN$, $CH_3NH_2$ and $CH_3OH$): $E_a$ decreases when $Q(C_\alpha)$ becomes less negative. A similar correlation is observed between $E_a$ and the sum $[Q(C_\alpha) + Q(H_\alpha)]$. These correlations are obviously dependent on the polarity of the attacking radical.
In practice it is very difficult to estimate the relative importance of the various factors which are assumed to be responsable

Table II.14. Heats of reaction and activation energies of hydrogen abstraction reactions (kcal/mole).

| Reaction | $\Delta E$(4-31G) | $\Delta H^{o}$(exp) | $E_a$(exp) |
|---|---|---|---|
| $CH_4 + CH_3 \rightarrow CH_3 + CH_4$ | 0.0 | 0.0 | 14.1 (67) |
| $CH_3CH_3 + CH_3 \rightarrow CH_3CH_2 + CH_4$ | -3.32 | -6.3 | 11.1 (67) |
| $(CH_3)_2CH_2 + CH_3 \rightarrow (CH_3)_2CH + CH_4$ | -5.22 | -9.0 | 10.0 (67) |
| $(CH_3)_3CH + CH_3 \rightarrow (CH_3)_3C + CH_4$ | -7.31 | -11.8 | 7.8 (67) |
| $CH_3F + CH_3 \rightarrow CH_2F + CH_4$ | -2.31 | -3.3 | 11.40 (68) |
| $CH_2F_2 + CH_3 \rightarrow CHF_2 + CH_4$ | -2.24 | -2.3 | 10.20 (68) |
| $CHF_3 + CH_3 \rightarrow CF_3 + CH_4$ | 2.43 | 2.2 | 10.20 (69) |
| $CH_3CN + CH_3 \rightarrow CH_2CN + CH_4$ | -11.71 | -11.3 | 10.00 (69) |
| $CH_3NH_2 + CH_3 \rightarrow CH_2NH_2 + CH_4$ | -9.55 | -9.7 | 8.70 (70) |
| $CH_3OH + CH_3 \rightarrow CH_2OH + CH_4$ | -6.22 | -10.3 | 9.30 (67) |
| $CH_2(CN)_2 + CH_3 \rightarrow CH(CN)_2 + CH_4$ | -21.70 | - | - |
| $CNCH_2NH_2 + CH_3 \rightarrow CNCHNH_2 + CH_4$ | -18.11 | - | - |
| $CNCH_2OH + CH_3 \rightarrow CNCHOH + CH_4$ | -16.21 | - | - |
| $CH_2(OH)_2 + CH_3 \rightarrow CH(OH)_2 + CH_4$ | -3.53 | - | - |

Table II.15. Net charges of $H_\alpha$ and $C_\alpha$ atoms.

| Molecule | $Q(C_\alpha)$ | $Q(H_\alpha)$ |
|---|---|---|
| $CH_4$ | -0.263 | 0.066 |
| $CH_3CH_3$ | -0.174 | 0.056<br>0.059 (x 2) |
| $(CH_3)_2CH_2$ | -0.174 | 0.057 |
| $(CH_3)_3CH$ | -0.013 | 0.046 |
| $CH_2F_2$ | 0.160 | 0.066 |
| $CHF_3$ | 0.361 | 0.069 |
| $CH_3CN$ | -0.181 | 0.102 |
| $CH_3NH_2$ | -0.092 | 0.047<br>0.067 (x 2) |
| $CH_3OH$ | -0.070 | 0.072<br>0.055 (x 2) |
| $CH_2(CN)_2$ | -0.095 | 0.130 |
| $CH_2(CN)(NH_2)$ | 0.005 | 0.103 |
| $CH_2(CN)(OH)$ | 0.009 | 0.103<br>0.087 (x 2) |
| $CH_2(OH)_2$ | 0.109 | 0.061 |

for the activation energy of radical transfer reactions. In some cases, one of them is important enough to permit qualitative predictions. For example, in the series of disubstituted alkyl radicals, the exothermicities are sufficiently different to allow the prediction that the species : $CH(CN)_2$, CH(CN)(OH) and CH(CN)($NH_2$) will be easier to obtain than $CH(OH)_2$, under the same experimental conditions.

Only elaborate quantum chemical calculations would enable us to solve completely this problem, as every chemical reactivity problem. Nome has been performed at the present time. However, some theoretical studies deserve to be mentioned (71). For our part we have performed ab initio calculations at the SCF-CI level on the reaction (72) :

$$CH_4 + OH \longrightarrow CH_3 + H_2O \qquad [4]$$

This work is still in progress but a few interesting results have yet been ontained. The geometries of the reactants, the transition state and the products, optimized at the 6-31G level are given in table II.16. The transition state has a tight structure in which the C...H and H...O bonds are both about 0.26 Å larger than their normal single bond length. The distribution of the charge centroids (figure II.13), the spin density (figure II.14) and the net charges of the atoms (figure II.13) suggest to describe the electronic structure of the transition state by the following formula :

$$[\, H_3C\bullet \; \overset{\delta+}{H}\circ \; \overset{\delta-}{\bullet OH} \,]^{\neq}$$

The calculated activation entropy is consistent with a tight transition state :

$$\Delta S^{\neq} = -\,22.37 \text{ e.u.}$$

These results must be refined in many aspect before to be compared with the corresponding experimental ones (73).

Table II.16. Optimized geometries (6-31G; Å or degree).

| Parameter | $CH_4$ | OH | ≠ | $CH_3$ | $H_2O$ |
|---|---|---|---|---|---|
| C-H | 1.082 | - | 0.960 | 1.071 | - |
| $H\hat{C}H$ | 109.5 | - | - | 120.0 | - |
| O-H | - | 0.967 | 1.080 | - | 0.950 |
| $H\hat{O}H$ | - | - | 104.6 | - | 111.5 |
| C...H | - | - | 1.340 | - | - |
| H...O | - | - | 1.210 | - | - |
| θ | - | - | 14.20 | - | - |
| $C\widehat{...H...}O$ | - | - | 186.1 | - | - |

5° The recombination reaction.

Our theoretical results provide also some informations on the thermochemistry of the dimerization reaction :

$$R + R \longrightarrow R - R \qquad [5]$$

For estimating the heat of reaction and, at the same time, the C-C bond dissociation energy of R-R, we have optimized the geometry of the dimers at the STO-3G level.

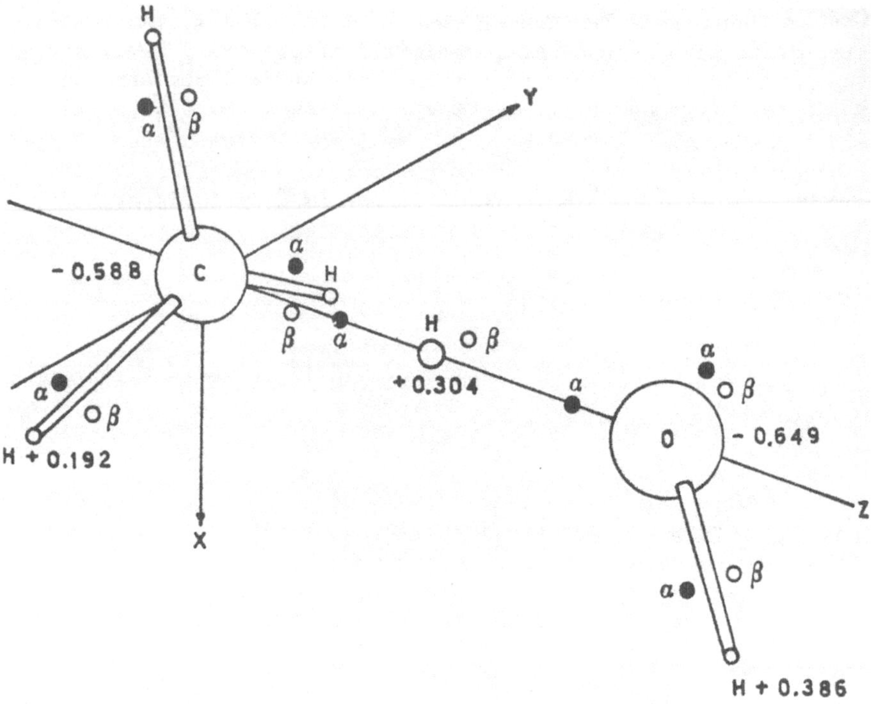


Figure II.13. Charge centroids and net charges (6-31G; ≠).

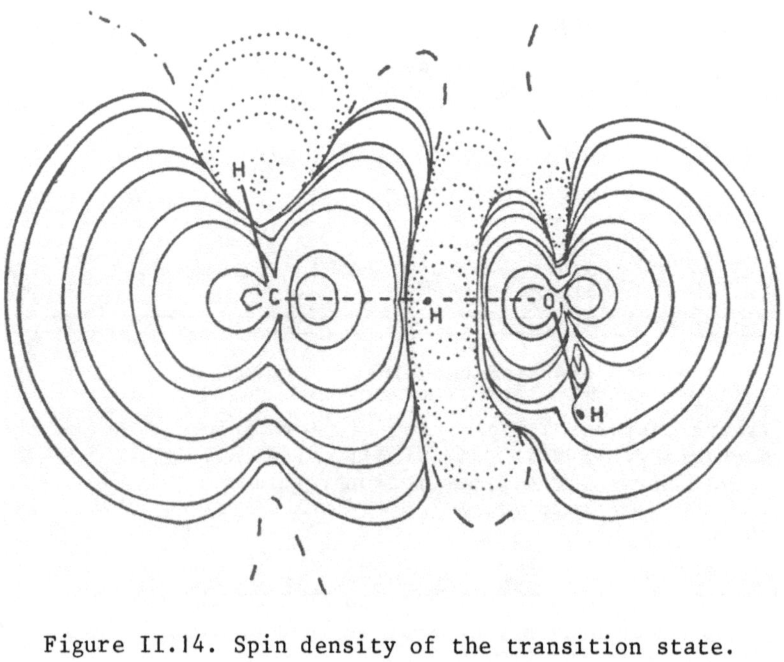


Figure II.14. Spin density of the transition state.

The C-C bond lengths and the total energies calculated at the two basis set levels are given in table II.17.

Table II.17. C-C bond lengths (Å) and total energies (a.u.).

| Dimer | C-C | E(STO-3G) | E(4-31G) |
|---|---|---|---|
| $(CH_3)_2$ | 1.538 | -78.30618 | -79.11582 |
| $(CH_3CH_2)_2$ | 1.549 | -155.46594 | -157.07020 |
| $[(CH_3)_2CH]_2$ | 1.582 | -232.61923 | -235.01929 |
| $[(CH_3)_3C]_2$ | 1.636 | -309.76611 | - [a] |
| $(CH_2F)_2$ | 1.564 | -273.19820 | -276.56425 |
| $(CHF_2)_2$ | 1.590 | -468.11821 | -474.02868 |
| $(CF_3)_2$ | 1.624 | -663.05947 | - [a] |
| $(CH_2OH)_2$ | 1.555 | -225.95978 | -228.58959 |
| $(CH_2CN)_2$ | 1.560 | -259.39335 | -262.28733 |
| $[CH(CN)_2]_2$ | 1.584 | -440.45422 | - [a] |
| $[CH(OH)_2]_2$ | 1.574 | -373.64109 | -378.09078 |
| $[CH(CN)(OH)]_2$ | 1.577 | -407.04307 | - [a] |

[a] Not yet calculated.

The other structural parameters are not reported because they have approximately the standard STO-3G values. The (STO-3G and 4-31G) heats of formation are estimated by using the bond separation reactions of table II.18. In the same table, we compare them to the available experimental data or to the results obtained by assuming the model of additive group contributions. Finally,the heats of dimerization and the corresponding BDE(C-C) calculated at the 4-31G level are collected in table II.19.
The C-C bond lengths are generally much larger than the normal C-C single bond. This result could be an artefact of the methodology used in this work. However, X-rays studies have demonstrated the existence of abnormally large C-C central bonds (1.638 Å and even 1.7 Å) in substituted ethane compounds (74). Furthermore, as shown in figure II.15, our results are not too far from the experimental ones in the case of a capto-dative substituted dimer (75). There seems to exist a correlation between BDE(C-C) and the C-C bond length in the alkane series but the theoretical results and the experimental data need to be refined before generalizing

Table II.18. Bond separation reactions and heats of formation of the dimers (kcal/mole).

| Dimer | Bond separation reaction | ΔE 3G | ΔE 4-31G | $\Delta E_f$ 3G | $\Delta E_f$ 4-31G | $\Delta E_f$ Exp |
|---|---|---|---|---|---|---|
| $(CH_3CH_2)_2$ | $CH_3-CH_2-CH_2-CH_3 + 2\,CH_4 \rightarrow 3\,CH_3-CH_3$ | 0.70 | 1.42 | -25.62 | -26.34 | -30.36 (60) |
| $[(CH_3)_2CH]_2$ | $(CH_3)_2CH-CH(CH_3)_2 + 4\,CH_4 \rightarrow 5\,CH_3-CH_3$ | -2.65 | -0.48 | -26.95 | -29.12 | -42.6 (60) |
| $[(CH_3)_3C]_2$ | $(CH_3)_3C-C(CH_3)_3 + 6\,CH_4 \rightarrow 7\,CH_3-CH_3$ | -10.03 | - | -24.25 | - | -54.0 (60) |
| $(CH_2F)_2$ | $FCH_2-CH_2F + 2\,CH_4 \rightarrow 2\,CH_3F + CH_3-CH_3$ | 4.78 | 8.93 | -102.82 | -106.97 | -105.4 (76) |
| $(CHF_2)_2$ | $F_2CH-CHF_2 + 4\,CH_4 \rightarrow 4\,CH_3F + CH_3-CH_3$ | 27.12 | 27.89 | -202.96 | -203.73 | -207.9 (76) |
| $(CF_3)_2$ | $F_3C-CF_3 + 6\,CH_4 \rightarrow 6\,CH_3F + CH_3-CH_3$ | 62.78 | - | -316.42 | - | -320.9 (77) |
| $(CH_2NH_2)_2$ | $H_2NCH_2-CH_2NH_2 + 2\,CH_4 \rightarrow 2\,CH_3NH_2 + CH_3-CH_3$ | - | - | - | - | -4.07 (78) |
| $(CH_2OH)_2$ | $HOCH_2-CH_2OH + 2\,CH_4 \rightarrow 2\,CH_3OH + CH_3-CH_3$ | 5.61 | 11.90 | -86.19 | -92.48 | -93.9 (60) |
| $(CH_2CN)_2$ | $NCCH_2-CH_2CN + 4\,CH_4 \rightarrow 2\,HCN + 3\,CH_3-CH_3$ | 19.97 | 22.90 | 55.51 | 52.58 | 46.32 (79) |
| $[CH(CN)_2]_2$ | $(NC)_2CH-CH(CN)_2 + 8\,CH_4 \rightarrow 4\,HCN + 5\,CH_3-CH_3$ | 23.43 | - | 147.77 | - | 132.3 (79) |
| $[CH(OH)_2]_2$ | $(HO)_2CH-CH(OH)_2 + 4\,CH_4 \rightarrow 4\,CH_3OH + CH_3-CH_3$ | 28.58 | 41.00 | -169.52 | -181.92 | -184.2 (80) |
| $[CH(CN)(OH)]_2$ | $[CH(CN)(OH)]_2 + 6\,CH_4 \rightarrow 2\,CH_3OH + 2\,HCN + 3\,CH_3-CH_3$ | 23.14 | - | -8.00 | - | -34.0 (79) |
| $[CH(CN)(NH_2)]_2$ | $[CH(CN)(NH_2)]_2 + 6\,CH_4 \rightarrow 2\,CH_3NH_2 + 2\,HCN + 3\,CH_3-CH_3$ | - | - | - | - | 62.12 (79) |

Table II.19. Heats of dimerization and BDE(C-C) (kcal/mole).

| Radical | ΔE(4-31G) | $\Delta H^o_o$ | BDE(4-31G) | $DH^o$(C-C) |
|---|---|---|---|---|
| $CH_3$ | - | -88.8 | - | 88.8 (81) |
| $CH_3CH_2$ | -83.62 | -81.6 | 83.62 | 81.6 (81) |
| $(CH_3)_2CH$ | -75.96 | -79.0 | 75.96 | 79.0 (81) |
| $(CH_3)_3C$ | - | -70.0 | - | 70.0 (31) |
| $CH_2F$ | -93.15 | -89.6 | 93.15 | 89.6 (76) |
| $CHF_2$ | -91.69 | -91.4 | 91.69 | 91.4 (76) |
| $CF_3$ | - | -96.9 | - | 96.9 (81) |
| $CH_2NH_2$ | - | -78.1 | - | 78.1 (81) |
| $CH_2OH$ | -88.32 | -81.5 | 88.32 | 81.5 (81) |
| $CH_2CN$ | -63.82 | -70.7 | 63.82 | 70.7 (81) |
| $CH(CN)_2$ | - | -55.0 | - | 55.0 (82) |
| $CH(OH)_2$ | -92.6 | -94.9 | 92.6 | 94.9 (82) |
| CH(CN)(OH) | - | -85.8 | - | 85.8 (82) |
| $CH(CN)(NH_2)$ | - | -60.9 | - | 60.9 (82) |

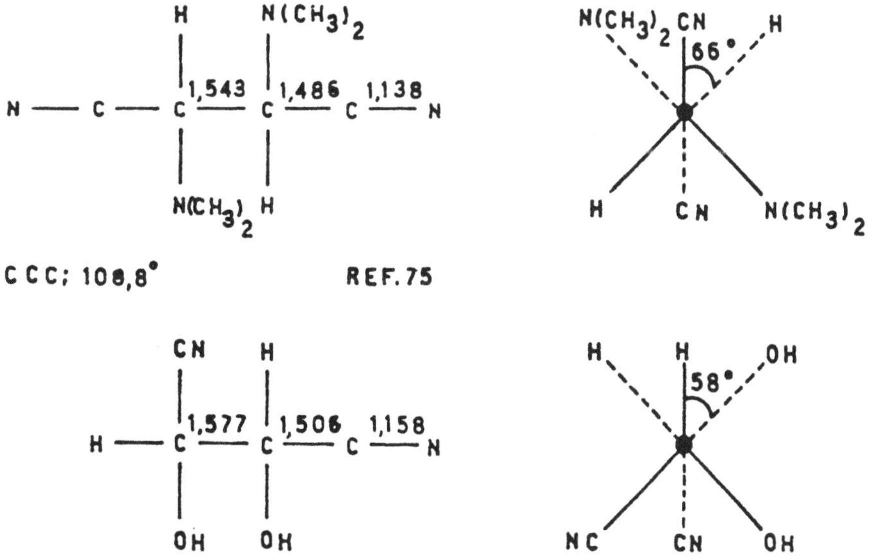


Figure II.15. Experimental structure and optimized geometry of dimers.

this correlation. The rather poor agreement between calculated and measured BDE(C-C) can also be explained by the imprecision of the two series of results. ZPE and thermal corrections and geometry optimizations at the 4-31G level would presumably improve this situation but some experimental results seem also to be wrong especially in the fluoro-dimers series.

In spite of the general imprecision of the data, some conclusions can be deduced from table II.19 :

- If we except the fluoro - and hydroxy - compounds, $DH^{o}$(C-C) decreases with an increasing number of substituents ;
- In the 1,2 disubstituted ethanes, $DH^{o}$(C-C) increases in the following order :

$$1,2\ CN < 1,2\ NH_2 < OH \simeq 1,2\ CH_3 < 1,2\ F \quad ;$$

- A different order is observed for the tetrasubstituted compounds :

$$1,1,2,2\ CN < 1,1,2,2\ F < 1,1,2,2\ CH_3 < 1,1,2,2\ OH$$

- For the dimer of a capto-dative substituted radical, we have :

$$1,1,2,2\ CN < 1,1\ CN,OH;\ 2,2,CN,OH < 1,1,2,2\ OH$$

- 1,1,2,2 tetracyanoethane has a particularly small $DH^{o}$(C-C).

All the dimerization reactions have a large exothermicity and correspondingly a large negative $\Delta G^{o}$. Thus, they are all thermodynamically favoured.

The lack of theoretical studies concerning the recombination reaction of free-radicals makes very difficult the quantitative interpretation of the kinetic results (see table I.10). The common assumption according to which there is no activation energy for the dimerization of radicals is to be taken with caution. Indeed several factors could be responsible for such an activation energy. We shall only notice the polar effects and, in the case of stabilized planar radicals, the pyramidalization barrier which is related to the relocalization of the unpaired electron on the $\alpha$ carbon atom. This hypothesis of a positive activation energy due to electronic factors seems to be confirmed by the persistent character of highly stabilized radicals, such as $C(CN)_2\ N(CH_3)_2$ (84) and NO, and by the transient character of non stabilized or destabilized radicals, such as respectively $CH_3$ and CN. In the absence of other factors (steric and (or) polar effects), the kinetic stability would be directly related to the thermodynamic stabilization of the radical.

The decreasing reactivity observed in the alkyl series could be interpreted in these terms :

| | $RE^{o}$ | $\log k_c$ |
|---|---|---|
| $CH_3$ | 0.00 | 10.5 |
| $CH_3CH_2$ | 6.36 | 9.89 |
| $(CH_3)_2CH$ | 9.17 | 9.71 |
| $(CH_3)_3C$ | 11.78 | 9.38 |

The lower reactivity of tertiobutyl radical is also probably due to steric factors.
Polar effects could be responsible of the relatively low reactivity of $CF_3$ (83) which, as a characteristic carbon-centered ($\sigma$) destabilized radical would have to recombine faster than $CH_3$. As emphasized by Griller and Ingold, the persistence of many radicals is mainly a consequence of steric factors which decrease the pre-exponential factor of the Arrhenius equation. Triphenylmethyl is a well-known example of this type of radicals.
The presumed activation energy could also be correlated to the exothermicity of the reaction and to the delocalization of the unpaired electron as measured by the second moment of the corresponding localized orbital.

Some of these factors can not be calculated and many of them are related one to the other. It is the reason why it is difficult to rationalize the experimental results and to predict the rate of the self-reaction of free-radicals. However the influence of each factor on the rate of recombination can be independently anticipated; $\log k_c$ will increase with :

(1) increasing localization of the unpaired electron ;
(2) decreasing resonance energy ;
(3) decreasing pyramidalization barrier ;
(4) decreasing polar effects ;
(5) decreasing steric effects .

Utilizing the correlations (1), (4) and (5), we can try to set up a tentative classification of free-radicals based on their "self-reactivity". This classification is given in table II.20 and illustrated by figure II.16.

## CONCLUSION.

This work is an illustration of the quantum chemistry contribution in the study of organic chemistry problems. Considering the peculiar example of the structure and reactivity of free-radicals, we have shown that ab initio calculations and experiments are quite complementary.

Table II.20. Tentative classification of free-radicals.

| Radical | $< r^2 >$ | Polar effects | Steric effects | $\log k_c$ |
|---|---|---|---|---|
| CN | Small | | | high (85) |
| $CH_3$ | 1.485 | | | 10.5 (86) |
| $CF_3$ | 1.368 | + | | 10.0 (86) |
| $CH(OH)_2$ | 1.394 | + | | - |
| $CHF_2$ | 1.397 | | | - |
| $CH_2OH$ | 1.433 | | | - |
| $CH_2NH_2$ | 1.437 | | | - |
| $CH_2F$ | 1.439 | | | - |
| $CH_3CH_2$ | 1.500 | | | 9.89 (86) |
| $(CH_3)_2CH$ | 1.504 | | | 9.71 (86) |
| $(CH_3)_3C$ | 1.524 | | + | 9.38 (86) |
| CH(CN)(OH) | 1.589 | | | - |
| $CH(CN)(NH_2)$ | 1.646 | | | - |
| $CH_2CN$ | 1.708 | | | - |
| $CH(CN)_2$ | 1.803 | + | | 9.1 (49)[a] |
| $C(CN)_3$ | high | + | | 6.7 (49)[a] |
| $C(CN)_2(NH_2)$ | high | | + | 0.0 |
| $(C_6H_5)_3C$ | small | | ++ | 0.0 |
| NO | high | | | 0.0 |

[a] In solution. There is some doubt that the dimer has been obtained in this reaction.

Theoretical results allow us :

1° to generalize qualitative models such as the Linnett's model of the electronic structure of molecules and the spin polarization and hyperconjugation models which are commonly used by the experimentalists for determining the sign of hyperfine splitting constants ;

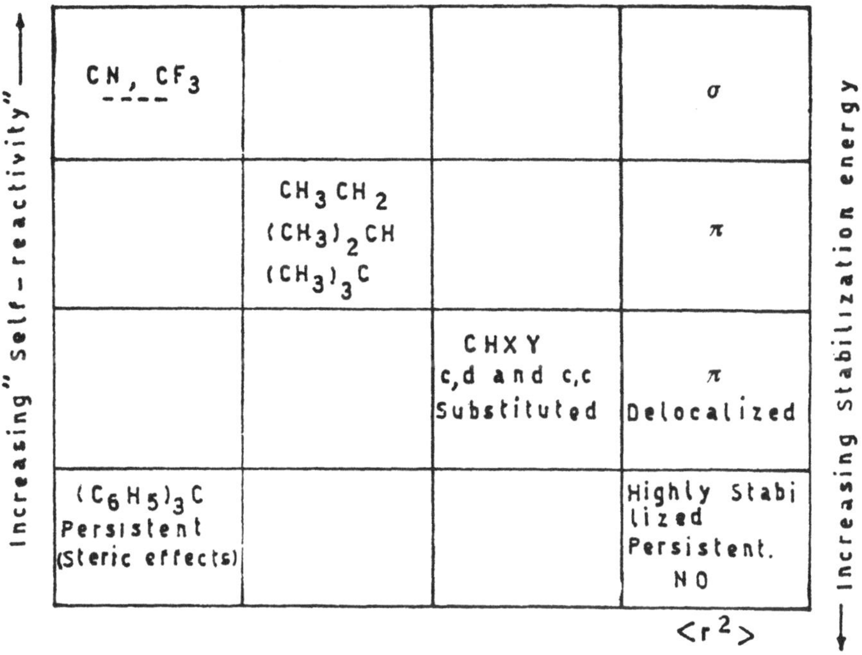


Figure II.16. Relative reactivity of free-radicals.

2° to analyze qualitative concepts (stability, polar effects ...) and empirical relations (McConnell and Evans-Polanyi equations) ;

3° to rationalize experimental data and to provide informations on properties which can not be directly measured (spin densities for example) ;

4° to predict, at least qualitatively, data not yet obtained experimentally (BDE, $\Delta H_f$ ...).

We shall summarize some interesting results of this preliminary study of the structure and reactivity of free-radicals.
We have shown that two main classes of unpaired electron systems can be distinguished : the atom-centered radicals where the Lewis octet rule (or Linnett's double quartet rule) is not satisfied and the delocalized radicals where the unpaired electron participates in the chemical bonds. NO, nitroxide and hydrazyl radicals are typical examples of the second class. Their persistent character is presumably due to the delocalization of the unpaired electron which is also responsible for the important stabilization of these systems. CN, $CH_3$, $NH_2$, OH and, in general σ radicals are representative of the first class. Their transient character, especially in their self-reaction, can be attributed to the localization of the unpaired electron which is related to a small stabilization energy.

Alkyl substituted radicals are intermediate species which, in limiting cases, can belong either to the first class or to the second one, according to the nature of the substituents. Alkyl radicals substituted by $\pi$ donor group(s) are typical non planar ($\sigma$) atom-centered radicals. Their stabilization energy is generally small and their inversion barrier, which increases with the number of substituents, can be very high (~ 7 kcal/mole in $CH(OH)_2$ and ~ 16 kcal/mole in $CF_3$). They are also characterized by a large $\rho_S(^{13}C_\alpha)$ and correspondingly a large $^{13}C_\alpha$ hyperfine splitting constant. Some of them (such as $CH(OH)_2$) are presumably difficult to obtain by hydrogen abstraction from the parent molecule.

Alkyl radicals substituted by $\pi$ acceptor group(s) or simultaneously by $\pi$ donor and $\pi$ acceptor groups are planar ($\pi$) delocalized radicals, the stabilization energy of which can be very high. However, their pyramidalization barrier is small and they can very easily adopt the non planar configuration of $\sigma$ radicals. For this reason, they should also dimerize as rapidly as $CH_3$ if other factors are not to be considered. Nevertheless alkyl substituted radicals could become persistent if their stabilization energy due to the delocalization of the unpaired electron is very large. The radical $C(CN)_2\,N(CH_3)_2$ is probably such a persistent system.

REFERENCES.

(1a) Kochi J.K. (ed.) "Free-Radicals", (1973) Vols. 1 and 2, Wiley, New York.
b) Huyser E.S. (Ed.) "Methods in Free-Radical Chemistry", (1969) Vols. 1 and 2; (1972) Vol. 3; (1973) Vol. 4; (1974) Vol. 5, M. Dekker.
c) Williams G.H. (Ed.) "Advances in Free-Radical Chemistry", (1965) Vol. 1, (1969) Vol. 3; (1972) Vol. 4, Logos Press; (1975) Vol. 5, Elek Science, London.
d) Waters W.A. (Ed.) "Free-Radical Reactions".
e) Pryor W.A., "Free-Radicals", (1976) McGraw Hill, New York.
f) Huang R.L., Goh S.H., Ong, S.H., "The Chemistry of Free-Radicals", (1974) Edward Arnold.
g) Nonhebel D.C., Tedder J.M., Walton J.C., "Radicals", (1979) Cambridge University Press.
h) Forrester A.R., Hay J.M., Thomson R.H., "Organic Chemistry of Stable Free-Radicals", (1968) Academic Press.
i) Hay J.M., "Reactive Free-Radicals", (1974) Academic Press.
j) Stirling C.M.J., "Radicals in Organic Chemistry", (1965) Oldbourne Press, London.
k) Rüchardt C., "Relations between Structure and Reactivity in

Free-Radical Chemistry",
Angew. Chem. Int. Edit. (1970), 9, 830.
1) O'Neal H.E., Benson S.W., "Thermochemistry of Free-Radicals" in "Free-Radicals", Ed. Kochi J.K., (1973) Vol. 2, Wiley.
(2a) Wertz J.E., Bolton J.R., "Electron Spin Resonance: Elementary Theory and Practical Applications", (1972) McGraw Hill.
b) Knowles P.F., Marsh D., Rattle H.W.E., "Magnetic Resonance of Biomolecules: an Introduction to the Theory and Practice of NMR and ESR in Biological Systems", (1976) J. Wiley.
c) Sales K.D., "The Theory of Isotopic Hyperfine Splitting Constants for Organic Free-Radicals", in "Advances in Free-Radicals Chemistry", Ed. Williams G.H., (1969) Vol. 3, Logos Press.
(3a) Gilbert B.C., "Organic Radicals: Structure" in "Electron Spin Resonance", (1977) Vol. 4, Chemical Society, Special Publication, London.
b) Kaplan L., "The Structure and Stereochemistry of Free-Radicals" in "Free-Radicals", Ed. Kochi J.K., (1973) Vol. 2 - Chapter 18, Wiley, New York.
c) Kochi J.K., "Configurations and Conformations of Transient Alkyl Radicals in Solution by Electron Spin Resonance Spectroscopy", in "Advances in Free-Radical Chemistry", (1975) Vol. 5, Elek Science, London, Ed. Williams G.H..
d) Norman R.O.C., "Structures of Organic Radicals", (1970) Chemistry in Britain, 6, 66.
e) Fischer H. "Structure of Free-Radicals by E.S.R. Spectroscopy" in "Free-Radicals", Ed. Kochi J.K., (1973) Vol. 2 - Chapter 19, Wiley, New York.
f) Beckwith A.L.J., "Structure, Reactivity and Rearrangement", in "Free-Radical Reactions", Ed. Waters W.A., (1973) Vol. 10, Butterworths.
g) Krusic P.J., Bingham A.G., J. Am. Chem. Soc., (1976), 98, 1, 230.
(4) Sullivan P.D., Menger E.M., "Temperature-dependent Splitting Constants in the ESR Spectra of Organic Free-Radicals", in "Advances in Magnetic Resonance", Ed. Waugh J.S., (1977) Vol. 9, Academic Press.
(5a) Griller D., Marriott P.R., PRESTON K.F., J. Chem. Phys., (1979), 71, 9, 3703.
b) Griller D., Preston K.F., J. Am. Chem. Soc., (1979), 101, 8, 1975.
c) Griller D., Krusic P.J., Fischer H., J. Am. Chem. Soc., (1978), 100, 21, 6750.
(6a) Fessenden R.W., J. Phys. Chem., (1967), 71, 74.
b) Fischer H., "Structure of Free-Radicals by ESR Spectroscopy", in "Free-Radicals", Ed. Kochi J.K.,

(1973) Vol. 2 - Chapter 18, Wiley.
c) Krusic J.P., Meakin P., J. Am. Chem. Soc., (1976), 98, 228.
d) Schlüter K., Berndt A., Tetrahedron Lett., (1979), 11, 929.
(7) Krusic P.J., Meakin P., J. Am. Chem. Soc., (1976), 98, 228.
(8) McConnell H.M., J. Chem. Phys., (1956), 24, 764.
(9a) Fessenden R.W., Ogawa S., J. Am. Chem. Soc., (1964), 86, 3591.
b) Bolton J.R., Mol. Phys., (1963), 6, 219.
c) Carter M.K., Vincow G., J. Chem. Phys., (1967), 47, 292.
d) Carrington A., Smith I.C.P., Mol. Phys., (1963), 7, 99;
Vincow G., Morrell M.I., Volland W.V., Dauben,Jr. H.J., Hunter F.R., J. Am. Chem. Soc., (1965), 87, 3527.
e) Katz T.J., Strauss H.I., J. Chem. Phys., (1960), 32, 1873.
(10a) Chestnut D.B., J. Chem. Phys., (1958), 29, 43.
b) Lazdins D., Karplus M., J. Chem. Phys., (1966), 44, 1600.
c) Fessenden R.W., Schuler R.H., J. Chem. Phys., (1963), 39, 2147.
(11) Fischer H., Z. Naturforsch., (1965), 20A, 428.
(12) Fischer H., Z. Naturforsch., (1964), 19A, 866.
(13) Viehe H.G., Merenyi R., Stella L., Janousek Z., Angew. Chem. Int. Ed. Engl., (1979), 18, 917.
(14) Stella L., Merenyi R., Janousek Z., Viehe H.G., Tordo P., Munoz A., J. Chem. Chem., (1980), 84, 304.
(15) Hehre W.J., Ditchfield R., Radom L., Pople J.A., J. Am. Chem. Soc., (1970), 92 16, 4796.
(16) Szwarc M., J. Chem. Phys., (1948), 16, 128.
(17) Benson S.W., J. Chem. Educ., (1965), 42, 502.
(18) Rodgers A.S., Wu M.C.R., Kuitu L., J. Phys. Chem., (1972), 76, 918.
(19a) Krusic P.J., Meakin P., Smart B.E., J. Am. Chem. Soc.,(1974), 96, 6211.
b) Sustman R., Trill H., J. Am. Chem. Soc., (1974), 96, 4343.
c) Crawford R.J., Hamelin J., Strehlke B., J. Am. Chem. Soc., (1971), 93, 3810.

d) Conradi M.S., Zeldes H., Livingston R., J. Phys. Chem., (1979), 83, 2160.
e) Chen K.S., Edge D.J., Kochi J.K., J. Am. Chem. Soc., (1973), 95, 7036.
f) Krusic P.J., Meakin P., Jesson J.P., J. Phys. Chem., (1971), 75, 3438.

(20) Griller D., Ingold K.U., Account Chem. Res., (1976), 9, 13.

(21a) Laurie S.H., J. Chem. Educ., (1972), 49, 746.
b) Fleming I., "Frontier Orbitals and Organic Chemical Reactions", (1976), p. 20-21, J. Wiley.

(22a) Trotman-Dickenson A.F., Milne G.S., "Tables of Bimolecular Gas Reactions", NSRDS-NBS 9, (1967).
b) Ratajczak E., Trotman-Dickenson A.F., "Supplementary Table of Bimolecular Gas-Reactions", (1969), University of Wales.
c) Kerr J.A., Ratajczak E., "Second Supplementary Tables of Bimolecular Gas Reactions", (1972), University of Birmingham.
d) Gray, P., Herod A.A., Jones A., Chem. Rev., (1971), 71, 247.
e) Kerr J.A., in "Free-Radicals", Ed. Kochi J.K., (1973), Vol. 1 - Chapter 1, Wiley.
f) Ingold K.U., in "Free-Radicals", Ed. Kochi J.K., (1973), Vol. 1 - Chapter 2, Wiley.
g) Trotman-Dickenson A.F., in "Advances in Free-Radical Chemistry", Ed. Williams G.H., (1965), Vol. 1 - Chapter 1.
h) Howard J.A., in "Advances in Free-Radical Chemistry", Ed. Williams G.H., (1972), Vol. 4 - Chapter 2.
i) Danen W.C., in "Methods in Free-Radical Chemistry, Ed. Huyser E.S., (1974), Vol. 5 - Chapter 1, M. Dekker.
j) Alfassi Z.B., Benson S.W., Internat. J. Chem. Kinetics, (1973), 5, 879.
k) Zavitsas A.A., Melikian A.A., J. Am. Chem. Soc., (1975), 97, 2757; Zavitsas A.A., J. Am. Chem. Soc., (1972), 94, 2779.
l) Norman R.O.C., Pure and Appl. Chem., (1979), 51, 1009.
m) Macken K.V., Sidebottom H.W., Internat. J. Chem. Kinetics, (1979), 11, 511.
n) Howard C.J., Evenson K.M., J. Chem. Phys., (1976), 64, 197.
o) Wilson W.E., J. Phys. Chem. Ref. Data, (1972), 1 2, 535.

p) Bell T.N., Perkins P.G.,
J. Phys. Chem., (1977), 81, 2012.
(23) Evans M.G., Polanyi M.,
Trans. Faraday Soc., (1938) 34, 11.
(24a) Davis W.H., Pryor W.A.,
J. Am. Chem. Soc., (1977), 99, 6365;
Pryor W.A., Davis W.H.,
J. Am. Chem. Soc., (1973), 95, 4754.
b) Shinohara H., Imamura A., Masuda T., Kondo M.,
Bull. Chem. Soc. Jap., (1979), 52, 974;
Bull. Chem. Soc. Jap., (1979), 52, 2801;
Bull. Chem. Soc. Jap., (1979), 52, 3265.
c) Walling C., "Free-Radicals in Solution",
(1957), Wiley.
d) Pryor W.A., "Free-Radicals",
(1966), McGraw-Hill, New York.
e) Russel G.A., in "Free-Radicals", Ed. Kochi J.K.,
(1973), Vol. 1, Wiley.
f) Henderson R.W.,
J. Am. Chem. Soc., (1975), 97, 213.
g) Minisci F., Mondelli R., Gardini G.P., Porta O.,
Tetrahedron, (1972), 28, 2403;
Clerici A., Minisci F., Porta O.,
J. Chem. Soc., Perkin II, (1974), 1699;
Clerici A., Minisci F., Porta O.,
Tetrahedron, (1973), 29, 2775.
h) Fossey J., Lefort D.,
Tetrahedron, (1980), 36, 1023.
i) Fleming I., "Frontier Orbitals and Organic Chemical Reactions", (1976), Chapter 5, J. Wiley.
j) Hay J.M., "Reactive free-Radicals",
(1974), Academic Press.
k) Zavitsas A.A., Pinto J.A.,
J. Am. Chem. Soc., (1972), 94, 7390.
l) Screttas C.G.,
J. Org. Chem., (1979), 44, 1471.
(25) Alfassi Z.B., Benson S.W.,
Internat. J. Chem. Kinetics, (1973), 5, 879.
(26a) Rosenstock H.M., Draxl K., Steiner B.W., Herron J.T.,
"Energetics of Gaseous Ions",
J. Phys. and Chem. Ref. Data, (1977), 6, Suppl. n°1.
b) Bartmess J.E., Scott J.A., McIver R.T.,
J. Am. Chem. Soc., (1979), 101, 6046.
c) Tanaka K., Mackay G.I., Payzant J.D., Bohme D.K.,
Canad. J. Chem., (1976), 54, 1643.
d) Page F.M., in "Handbook of Chemistry and Physics",
E 55, 53 nd ed. (1972-1973); and "Handbook of Chemistry and Physics" E 55, E 56, E 62 - E 68, 52 nd (1971-1972).
e) Janousek B.K., Zimmerman A.H., Reed K.J., Brauman J.I.,
J. Am. Chem. Soc., (1978), 100, 6142.

f) Zimmerman A.H., Brauman J.I.,
J. Am. Chem. Soc., (1977), 99, 3565;
J. Am. Chem. Soc., (1977), 99, 7203.
g) Ellison G.B., Engelking P.C., Lineberger W.C.,
J. Am. Chem. Soc., (1978), 100, 2556.
h) Richardson J.H., Stephenson L.M., Brauman J.I.,
Chem. Phys. Lett., (1975), 30, 17;
J. Chem. Phys., (1975), 63, 74.
i) Nalley S.J., Compton R.N., Schweinler H.C., Anderson V.E.,
J. Chem. Phys., (1973), 59, 4125.
j) Berry R.S.,
Chem. Rev., (1969), 69, 533.
k) Berkowitz J., Chupka W.A., Walter T.A.,
J. Chem. Phys., (1969), 50, 1497.
l) Smyth K.C., Brauman J.I.,
J. Chem. Phys., (1972), 56, 4620.
m) Houle F.A., Beauchamp J.L.,
J. Am. Chem. Soc., (1979), 101, 4067;
J. Am. Chem. Soc., (1978), 100, 3290.
n) Dekock R.L., Barbachyn M.R.,
J. Am. Chem. Soc., (1979), 101, 6516.
o) Lossing F.P.,
J. Am. Chem. Soc., (1977), 99, 7526.
(27) Krech R.H., McFadden D.L.,
J. Am. Chem. Soc., (1977), 99, 8402.
(28) Johnston H.S., Parr C.,
J. Am. Chem. Soc., (1963), 85, 2544.
(29a) Gilliom R.D.,
J. Chem. Phys., (1976), 65, 5027;
J. Am. Chem. Soc., (1977), 99, 8399.
b) Jordan R.M., Kaufman F.,
J. Chem. Phys., (1975), 63, 1691.
c) Arthur N.L., Donchz K.F., McDonell J.A.,
J. Chem. Phys., (1975), 62, 1585;
J. Chem. Soc. Faraday Trans.I (1975), 71, 2431;
J. Chem. Soc. Faraday Trans.I (1975), 71, 2442.
d) Berces T., Dombi J.,
Internat. J. Chem. Kinetics, (1980), 12, 123;
Internat. J. Chem. Kinetics, (1980), 12, 183.
e) Zavitsas A.A., Melikian A.A.,
J. Am. Chem. Soc., (1975), 97, 2757.
(30a) Parkes D.A., Quinn C.P.,
J. Chem. Soc., Faraday I, (1976), 72, 1952.
b) Adachi H., Basco N., James D.G.L.,
Chem. Phys. Lett., (1978), 59, 502.
c) James F.C., Simons J.P.,
Internat. J. Chem. Kinetics, (1974), 6, 887.
d) Van Den Berg H.E., Callear A.B., Nostram R.J.,
Chem. Phys. Lett., (1969), 4, 101;
Trans. Faraday Soc., (1979), 66, 2681.

e) Hochanadel C.J., Sworski T.J., Ogren P.J., J. Phys. Chem., (1980), 84, 231.
f) Rossi M., Golden D.M., Internat. J. Chem. Kinetics, (1979), 11, 775.
g) Hiatt R., Benson S.W., Internat. J. Chem. Kinetics, (1972), 4, 479.
h) Rossi M., King K.D., Golden D.M., J. Am. Chem. Soc., (1979), 101, 1223.
i) Benson S.W., "Thermochemical Kinetics", 2nd Ed., Wiley (1976).
j) De Maré G.R., Huybrechts G.H., Trans. Faraday Soc., (1968), 64, 1311; Chem. Phys. Lett., (1967), 1, 64.
k) Matheson I.A., Sidebottom H.W., Tedder J.M., Internat. J. Chem. Kinetics, (1974), 6, 493.
(31a) Linnett J.W., "The Electronic Structure of Molecules", (1966), Methuen.
b) Linnett J.W., J. Am. Chem. Soc., (1961), 83, 2643.
c) Firestone R.A., J. Org. Chem., (1969), 34, 2621; Tetrahedron Lett., (1968), 8, 971.
(32a) Chipman D.M., J. Chem. Phys., (1979), 71, 761.
b) Snyder L.C., Amos T., J. Chem. Phys., (1964), 41, 1773; J. Chem. Phys., (1965), 42, 3670.
c) Hamano H., Kondo H., Bull. Chem. Soc. Jap., (1979), 52, 1255.
d) Salotto A.W., Burnelle L., J. Chem. Phys., (1970), 53, 333.
e) Bernardi F., Epiotis N.D., Cherry W., Schlegel H.B., Whangbo M-H., Wolff S., J. Am. Chem. Soc., (1976), 98, 469.
f) Bernardi F., Cherry W., Shaik S., Epiotis N.D., J. Am. Chem. Soc., (1978), 100, 1352.
g) Bernardi F., Camaggi C.M., Tiecco M., J. Chem. Soc., Perkin II, (1974), 518.
h) Hinchliffe A., Chem. Phys. Lett., (1971), 11, 131.
i) Cremaschi P., Gamba A., Morosi G., Simonetta M., Theoret. Chim. Acta, (1976), 41, 177.
j) Pacansky J., Dupuis M., J. Chem. Phys., (1978), 68, 4276; J. Chem. Phys., (1979), 71, 2095; Pacansky J., Coufal H., J. Chem. Phys., (1980), 72, 3298.
(33a) Lathan W.A., Curtiss L.A., Hehre W.J., Lisle J.B., Pople J.A., "Molecular Orbital Structures for Small Organic Molecules and Cations", in Progr. Phys. Org. Chem., (1974), 11, 175.

b) Schaefer III H.F. (ed.) "Applications of Electronic Structure Theory", in Modern Theoretical Chemistry (1977).
c) Lathan W.A., Hehre W.J., Pople J.A., J. Am. Chem. Soc., (1971), 93, 808.
d) Radom L., Lathan W.A., Hehre W.J., Pople J.A., J. Am. Chem. Soc., (1971), 93, 5339.
e) Radom L., Hariharan P.C., Pople J.A., Schleyer P.v.R., J. Am. Chem. Soc., (1973), 95, 6531.
f) Lathan W.A., Hehre W.J., Curtiss L.A., Pople J.A., J. Am. Chem. Soc., (1971), 93, 6377.
g) Radom L., Hehre W.J., Pople J.A., J. Chem. Soc., (A) (1971), 2299.
h) Radom L., Pople J.A., Schleyer P.v.R., J. Am. Chem. Soc., (1972), 94, 5935.
i) Ditchfield R., Hehre W.J., Pople J.A., J. Chem. Phys., (1971), 54, 724.
j) Newton M.D., Lathan W.A., Hehre W.J., Pople J.A., J. Chem. Phys., (1970), 52, 4064.
k) Radom L., Hehre W.J., Pople J.A., J. Am. Chem. Soc., (1971), 93, 289.
l) Radom L., Hehre W.J., Pople J.A., J. Am. Chem. Soc., (1972), 94, 2371.
m) Yarkony D.R., Schaefer III H.F., Rothenberg S., J. Am. Chem. Soc., (1974), 96, 656.

(34a) Baird N.C., Gupta R.R., Taylor K.F., J. Am. Chem. Soc., (1979), 101, 4531.
b) Baird N.C., Taylor K.F., Canad. J. Chem., (1980), 58, 733.
c) Baird N.C., Tetrahedron, (1979), 35, 289.
d) Harcourt R.D., Roso W., Canad. J. Chem., (1978), 56, 1093.
e) Gillespie R.J., J. Chem. Educ., (1970), 47, 18; J. Chem. Educ., (1963), 40, 295.

(35) Niblaeus K., Roos B.O., Siegbahn P.E.M., Chem. Phys., (1977), 26, 59.

(36a) Bernardi F., Epiotis N.D., "Progress in Theoretical Organic Chemistry", Ed. I.G. Csizmadia, (1977), Vol. 2, p. 47, Elsevier.
b) Carsky P., Zahradnik R., J. Am. Chem. Soc., (1972), 94, 5603.
c) Carsky P., Zahradnik R., Acc. Chem. Res., (1976), 9, 407.
d) Bonacic-Koutecky V., Koutecky J., Salem L., J. Am. Chem. Soc., (1977), 99, 842.
e) Claxton T.A., Platt E., Symons M.C.R., Mol. Phys., (1976), 32, 1321.
f) Surratt G.T., Goddard III W.A., Chem. Phys., (1977), 23, 39.

g) Adams G.F., Bent G.D., Purvis G.D., Bartlett R.J., J. Chem. Phys., (1979), 71, 3697.
h) Konishi H., Morokuma K., J. Am. Chem. Soc., (1972), 94, 5603.
i) Toriyama K., Iwasaki M., J. Am. Chem. Soc., (1979), 101, 2516.
j) Ihrig A.M., Jones P.R., Jung I.N., Lloyd R.V., Marshall J.L., Wood D.E., J. Am. Chem. Soc., (1975), 97, 4477.

(37) Pople J.A., Nesbet R.K., J. Chem. Phys., (1954), 22, 571.

(38) Hehre W.J., Lathan W.A., Ditchfield R., Pople J.A., Q.C.P.E. (1973), 11, 236.

(39) Hehre W.J., Stewart R.F., Pople J.A., J. Chem. Phys., (1969), 51, 2657.

(40) Ditchfield R.D., Hehre W.J., Pople J.A., J. Chem. Phys., (1969), 54, 724.

(41a) Boys S.F., Rev. Mod. Phys., (1960), 32, 296.
b) Foster J.M., Boys S.F., Rev. Mod. Phys., (1960), 32, 300.

(42a) Daudel R., Kapuy E., Kozmutza C., Goddard J.D., Csizmadia I.G., Chem. Phys. Lett., (1976), 44, 197.
b) Csizmadia I.G., "Progress in Theoretical Organic Chemistry", Ed. Csizmadia I.G., (1976), Vol. 1, Elsevier.

(43a) Lathan W.A., Curtiss L.A., Hehre W.J., Lisle J.B., Pople J.A., Progr. Phys. Org. Chem., (1974), 11, 175.
b) Lathan W.A., Hehre W.J., Pople J.A., J. Am. Chem. Soc., (1971), 93, 808.

(44) Bernardi F., Epiotis N.D., Cherry W., Schlegel H.B., M.-H. Whangbo, Wolff S., J. Am. Chem. Soc., (1976), 98, 469.

(45) Newton M.D., Lathan W.A., Hehre W.J., Pople J.A., J. Chem. Phys., (1970, 52, 4064.

(46) Peeters D., Deplus A., Private communication.

(47) Fessenden R.W., J. Phys. Chem., (1967), 71, 74.

(48) Smith P., Kaba R.A., Smith I.C., Pearson J.T., Wood P.B., J. Magn. Res., (1975), 18, 254.

(49) Kaba R.A., Ingold K.U., J. Am. Chem. Soc., (1976), 98, 523.

(50) Extrapolated value from ref. (49).

(51) Fessenden R.W., Schuler R.H., J. Chem. Phys., (1965), 43, 2704.

(52) Krusic P.J., Meakin P., Tesson J.P., J. Phys. Chem., (1971), 75, 3438.

(53) Dobbs A.J., Gilbert B.C., Norman R.O.C., J. Chem. Soc. (A), (1971), 124.
(54) Landolt-Börnstein, (1977), Volume 9 - Part a, Springer-Verlag.
(55) O'Neal H.E., Benson S.W., "Thermochemistry of Free-Radicals", in "Free-Radicals", Ed. Kochi J.K., (1973), Vol. 2 - Chapter 17, Wiley.
(56) Howard C.J., Evenson K.M., J. Chem. Phys., (1976), 64, 197.
(57) King K.D., Goddard R.D., Internat. J. Chem. Kinetics, (1975), 7, 837.
(58) Colussi A.J., Benson S.W., Internat. J. Chem. Kinetics, (1977), 9, 307.
(59) Golden D.M., Benson S.W., Chem. Rev., (1969), 125.
(60) Cox J.D., Pilcher G., "Thermochemistry of Organic and Organometallic Compounds" (1970), Academic Press.
(61) Rodgers A.S., Chao J., Wilhoit R.C., Zwolinski B.J., J. Phys. Chem. Ref. Data, (1974), 3, 117.
(62) Estimated values for $CNCH_2NH_2$ and $HOCH_2NH_2$ from incremental method of J.D. Cox, G. Pilcher, "Thermochemistry of Organic and Organometallic Compounds", (1970), Academic Press.
(63) Benson S.W., Angew. Chem. Internat. Ed. Engl., (1978), 17, 812.
(64) Benson S.W., Cruickshank F.R., Golden D.M., Haugen G.R., O'Neal H.E., Rodgers A.S., Shaw R., Walsh R., Chem. Rev., (1969), 69, 279.
(65) Rossi M., Golden D.M., J. Am. Chem. Soc., (1979), 101, 1230.
(66) Matheson A., Sidebottom H.W., Tedder J.M., Internat. J. Chem. Kinetics, (1974, 6, 493.
(67) Gilliom R.D., J. Am. Chem. Soc., (1977), 99, 8399.
(68) Pritchard G.O., Bryant J.T., Thommarson R.L., J. Phys. Chem., (1965), 69, 664.
(69) Wynen M.H.J., J. Chem. Phys., (1954), 22, 1074.
(70) Gray P., Chem. Rev., (1971), 71, 247.
(71a) Fukui K., Kato S., Fujimoto H., J. Am. Chem. Soc., (1975), 97, 1.
b) Nagase S., Morokuma K., J. Am. Chem. Soc., (1978), 100, 1666.
(72) Villaveces J.L., Sana M., Leroy G., In preparation.
(73) Zellner R., J. Phys. Chem., (1979), 83, 18.

(74) Littke W., Drück U.,
Angew. Chem. Internat. Ed. Engl., (1979), 18, 406.
(75) Declerq J.P., Germain G., Van Meerssche M.,
Private communication.
(76) Calculated from G.E. Millward, E. Tschuikow-Roux,
J. Phys. Chem., (1972), 76, 292 and
Table II.13 and Table II.18
(77) Chen S.S., Rodgers A.S., Chao J., Wilhoit R.C., Zwolinski B.J.,
J. Phys. Chem. Ref. Data, (1975), 4, 441.
(78) Good W.D., Moore R.T.,
J. Chem. Eng. Data, (1970), 15, 150.
(79) Estimated from group contribution method of Cox and
Pilcher (ref. 60).
(80) Eigenmann H.K., Golden D.M., Benson S.W.,
J. Phys. Chem.,(1973), 77, 1687.
(81) Calculated from Table II.13 and Table II.18
(82) Calculated from theoretical $\Delta E_f$ (Table II.13) and experimental $\Delta H_f$ (Table II.18)
(83) Tschuikow-Roux E.,
J. Phys. Chem., (1965), 69, 1075.
(84a) de Vries L.,
J. Am. Chem. Soc., (1977), 99, 1982;
J. Am. Chem. Soc., (1978), 100, 926;
J. Org. Chem., (1973), 38, 2604.
b) Heimer N.E.,
J. Org. Chem., (1977), 42, 3767.
c) Itoh M., Nagakura S.,
Tetrahedron Lett., (1965), 8, 417;
Kosower E.M., Poziomek E.J.,
J. Am. Chem. Soc., (1964), 86, 5515.
(85) Paul D.E., Dalby F.W.,
J. Chem. Phys., (1962), 37, 592.
(86) Rate constants in gas phase of Table I.10

# TRIPLET OXIRANES : APPLICATION OF QUANTUM MECHANICAL METHODS TO THE STUDY OF THE REACTIONS OF TRIPLET ISOMERIC OXIRANES.

George R. De Maré

Laboratoire de Chimie Physique Moléculaire,
Faculté des Sciences CP 160, Université Libre
de Bruxelles, 50 av. F.D. Roosevelt,
B-1050 Brussels, Belgium.

The thermal and photochemical reactions of simple oxiranes are reviewed. Molecular unrestricted Hartree-Fock calculations with geometry optimization have been carried out on eight triplet $C_3H_6O$ isomers considered to be possible intermediates in the $O(^3P)$ + propylene and in the Hg $6(^3P_1)$ sensitization of methyloxirane. The computed thermodynamic stabilities reveal that four of these species are available in the former while all eight are accessible in the latter reaction. The isomer $^{\bullet}CH(CH_3)-CH_2O^{\bullet}$ (MO1) is more stable than $^{\bullet}OCH(CH_3)-^{\bullet}CH_2$ (MO2) giving a satisfactory explanation for the observation that propionaldehyde is the major carbonyl product in the $O(^3P)$ + propylene reaction. The energy surfaces $E(\Theta_1,\Theta_2)$ for MO1 and MO2, and the energy hypersurfaces $E(\Theta_1,\Theta_2,\Theta_3)$ for $^{\bullet}CH(CH_3)-O-^{\bullet}CH_2$ (MO3), acetone and propionaldehyde were generated and the surfaces analyzed for the location and relative energies of the critical points (minima, saddle points and maxima). The overall stereochemical finding was that MO1, MO2, and MO3 possess rather flexible structures. For acetone and propionaldehyde, the barriers to inversion at the carbonyl group are 2.7 and 4.2 kcal/mol, respectively.

## 1. INTRODUCTION

We have been interested in the reactions, in particular the photochemistry and triplet mercury photosensitization, of the simple oxiranes (olefinic oxides, see



*I. G. Csizmadia and R. Daudel (eds.), Computational Theoretical Organic Chemistry, 335–369.*

diagram) for a number of years. The oxiranes are known

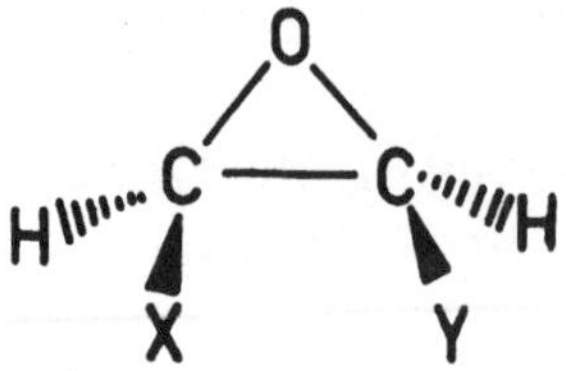


(a) Oxirane : X = Y = H.

(b) Methyloxirane : X = H; Y = $CH_3$.

(c) Cis-1,2-dimethyloxirane : X = Y = $CH_3$.

air pollutants(1) and some of them are noted mutagenics (2). The ring-opened $T_1$ state(s) of the oxiranes are the probable intermediates in the Hg $6(^3P_1)$ (hereafter denoted as $^3$Hg) photosensitization of the oxiranes(3). It has been suggested that the primary product of the $O(^3P)$ + olefin reaction is a triplet biradical, ·C-C-O· (3), synonymous with the ring distorted $T_1$ state(s) of the corresponding oxirane(4). (Note that a planar hydrogen-bonded intermediate has been invoked by Klein and Scheer(5) for this latter reaction.) The $T_1$ species can undergo intersystem crossing to the ground $S_0$ state and be stabilized as oxirane, or undergo isomerization via H-atom or alkyl migration and be stabilized as a carbonyl compound.

The industrial and commercial importance of the oxiranes is reflected in the rank held by the two simplest members : oxirane is among the top 30 industrial compounds with more than two million metric tons produced annually while methyloxirane is 41st with an annual production near one million metric tons(6). In spite of their industrial and bio-environmental implications, the oxiranes have been the object of relatively few fundamental photochemical investigations. One of the reasons for this lies in the complexity of their reactions. For instance, the triplet mercury sensitization of oxirane affords at least ten primary products(3)! Some of these are much more reactive than oxirane under the photosensitization conditions and their reactions may mask those of the parent compound. Knowledge of the primary photophysical and photochemical steps of the oxiranes, and of their relative importance, is sought nevertheless, since it should help us to understand (i) why carefully controlled production chains of methyloxirane sometimes explode without any apparent reason, (ii) the role played by excited state species in the mutagenic activity of oxiranes and (iii) the mechanisms of oxygen atom attack on olefins to yield oxiranes, their carbonyl isomers and fragmentation products.

As mentioned above, the primary steps in the photochemistry of the oxiranes are difficult to elucidate experimentally because of the multitude of products and the occurrence of secondary reactions. Thus quantum mechanical investigations of the possible (or probable) intermediates, giving their relative thermodynamic stabilities, should be very useful in predicting the various experimental methods or conditions which should be used in future studies.

## 2. HISTORICAL SURVEY

### 2.1. Oxirane : experimental studies

2.1.1. Thermal reactions. Benson(7) has discussed the thermal decompositions of oxirane and methyloxirane. He proposed that the initial step is C-O bond cleavage. The decompositions then proceed via "hot" aldehyde molecules formed by H-atom migration in the radicals. The electronic state of the aldehydes (ground singlet or triplet state) or of the radicals, or whether intersystem crossing occurred, was not discussed. The "hot" aldehydes can be collisionally deactivated or decompose to give molecular and free radical products. In more complex oxiranes, alkyl group migration also occurs. For instance Blades(8) has shown that the thermal reactions of methyloxirane yields the isomers methylvinylether, propionaldehyde, acetone and propen-3-ol. The expectation that the energy of the lowest triplet and singlet states of the ring-opened oxiranes lie close together has recently been confirmed in a quantum mechanical study by Yamaguchi et al(9). More information is required to allow one to show whether the thermal reactions proceed via the singlet, triplet or both of these, biradical states. I shall not dwell on the thermal reactions of the oxiranes here but the results will be referred to as they are needed in the text.

2.1.2. Photochemical reactions. The $^3$Hg photosensitization of oxirane was first studied by Phibbs, Darwent and Steacie(PDS; 10). The major products they observed were $H_2$, CO, acetaldehyde and a polymer. Smaller amounts of $CH_4$, "$C_2$ hydrocarbons" and formaldehyde were also found. PDS proposed a mechanism involving principally the initial formation of an activated molecule which isomerized to acetaldehyde or decomposed to

$$\underset{CH_2-CH_2}{\overset{O}{\triangle}} \xrightarrow{^3Hg} (\underset{CH_2-CH_2}{\overset{O}{\triangle}})^{\textbf{x}} \begin{cases} \nearrow CH_3CHO \\ \\ \searrow CH_2 + H_2 + CO \end{cases} \qquad (I)$$

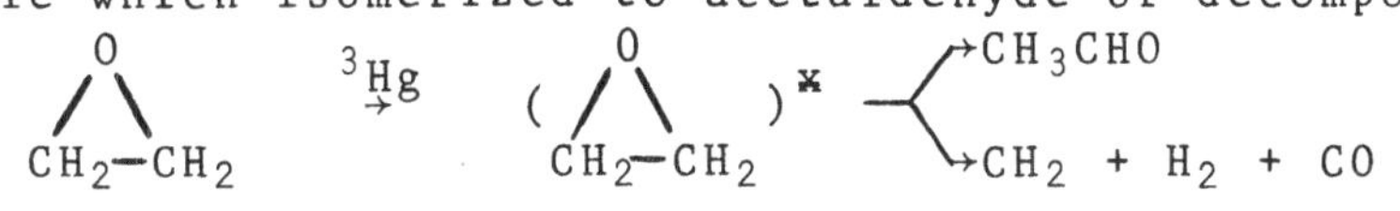

give methylene radicals and molecular products(eq. (I)).
The participation of a minor paraffinic type quenching
step, (II), was suggested by the observation that the

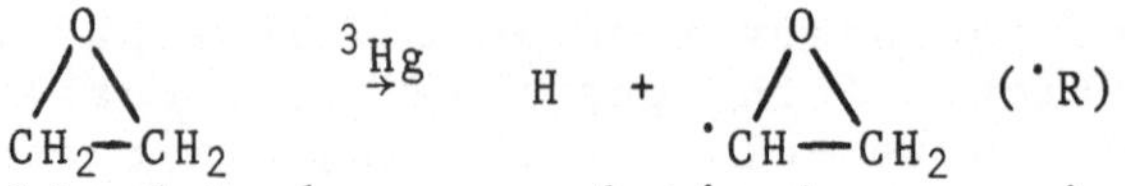

$$
\underset{CH_2-CH_2}{\overset{O}{\triangle}} \quad \overset{^3Hg}{\rightarrow} \quad H \;+\; \underset{\cdot CH-CH_2}{\overset{O}{\triangle}} \quad (\cdot R) \hspace{3cm} (II)
$$

yield of $H_2$ became relatively more important at higher
pressures. The oxirane radical, $\cdot R$, formed in step II
would then isomerize and abstract a hydrogen atom from
the substrate to give acetaldehyde, or decompose. The
limiting high pressure rates showed that only 0.30 mole
of oxirane isomerized or decomposed per Einstein absor-
bed, while the total quantum yield of disappearance was
0.66. According to these data, and depending on the
mechanism of the polymerization, there is a quantum
inefficiency of 0.3 to 0.7 in the photosensitization.

In 1955, Cvetanovic(11) found that the major products of the addition of oxygen atoms on ethylene were
remarkably similar to those observed by PDS(10) in the
$^3Hg$ sensitization of oxirane. This led him to reinves-
tigate this latter reaction. He showed that the "$C_2$
fraction" (-183 to -126°C) reported by the earlier wor-
kers(10) consisted mainly of ethane and propane(3(a)).
Traces of ethylene and small amounts of ketene and
$C_4H_{10}$ were also present. He concluded that at least
two, and probably more than two, primary steps occur.
He also reported that hydrogen atoms play an important
role in the decomposition, as well as $\cdot CH_3$, $\cdot CHO$, $\cdot CH_2CHO$,
ethyl and, to a lesser extent, methylene radicals(3(a)).

The addition of small amounts of free radical traps
such as olefins can often help to distinguish between
products formed in free radical reactions or in molecular decomposition processes. Cvetanovic(3(a)) found
that added ethylene suppressed hydrogen, methane and CO
formation. In contrast, propane, ketene and aldehyde
formation were increased compared to the uninhibited
reaction. The effect of added butene-1 was similar to
that of added ethylene with the exception that propane
formation was almost eliminated. $\Phi C_2H_4$ was higher than
in the uninhibited reaction indicating that ethylene is
formed in a primary process and that it disappears
through addition of H-atoms and/or methyl radicals to
it. Appreciable quantities of paraffinic aldehydes
(propanal to hexanal) and higher paraffins (up to octane)
were also formed. $H_2$ formation by the process III,

$$
(\underset{CH_2-CH_2}{\overset{O}{\triangle}})^x \quad \rightarrow \quad H_2 \;+\; CO \;+\; :CH_2, \hspace{3cm} (III)
$$

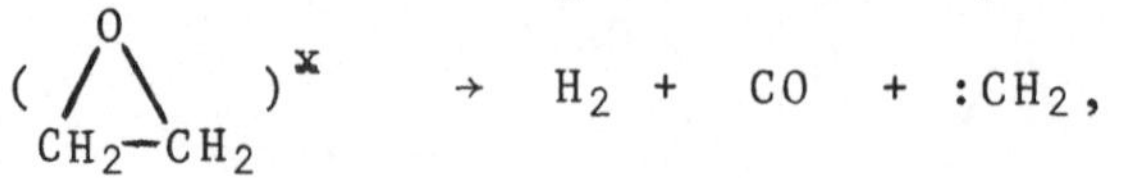

given by PDS(10) for the decomposition of excited oxirane molecules should not be suppressed by the addition of small quantities of ethylene or butene-1. Experiments with mixtures of normal and perdeuterated oxirane confirmed that only a small fraction, perhaps 10%, of the hydrogen is formed molecularly(3(a)). This is probably via steps III or IV :

$$\left(\overset{O}{\overbrace{CH_2-CH_2}}\right)^* \rightarrow H_2 + CH_2{=}C{=}O \qquad (IV)$$

Cvetanovic thus concluded that there is formation of hydrogen atoms in the photosensitization, probably via a mechanism of the type suggested by Gomer and Noyes(12) for the photolysis of oxirane :

$$\overset{O}{\overbrace{CH_2-CH_2}} \xrightarrow{h\nu} {}^{\bullet}CH_3 + {}^{\bullet}CHO \qquad (V)$$

followed by

$${}^{\bullet}CHO \rightarrow H + CO \qquad (VI)$$

$H_2$ is then formed by H-atom abstraction from the substrate, step VII, yielding the oxirane radical ${}^{\bullet}R$, which

$$H + \overset{O}{\overbrace{CH_2-CH_2}} \rightarrow H_2 + {}^{\bullet}R \qquad (VII)$$

undergoes H-atom migration and decomposes (step VIII)

$${}^{\bullet}R \rightarrow {}^{\bullet}CH_3 + CO \qquad (VIII)$$

or isomerizes(step IX). These steps were considered to

$${}^{\bullet}R \rightarrow {}^{\bullet}CH_2CHO \qquad (IX)$$

occur with equal probability, independent of temperature. The stable ${}^{\bullet}CH_2CHO$ radicals would then lead to acetaldehyde formation by H-atom abstraction, or to higher aldehydes by cross-combination with ${}^{\bullet}CH_3$ or another alkyl radical. Before pursueing the discussion of the $^3$Hg-sensitized reactions of oxirane, it is necessary to note that there is an apparent discrepancy between Gomer and Noyes' results on the UV photolysis(12) and Roquitte's flash photolysis results at about 210 nm(13). Indeed the former find aldehydes as the major products while the latter reports that no aldehydes are formed. In a recent study, Kawasaki, Ibuki, Iwasaki and Takazaki (KIIT; 14) have shown that the nature of the photolysis products of oxirane is wavelength (energy) dependent. This may partly explain the above discrepancy. We shall return to KIIT's(14) results later.

Cvetanovic(3(a)) gives two possible primary steps to account for ethylene formation :
- i) <u>a split into ethylene + an oxygen atom</u>. This would then be similar to the photolytic decomposition of oxirane(13,14) and of ethylene episulfide(15).
-ii) <u>a reaction involving CH$_2$ radicals</u>. This possibility is quite interesting since Gesser(16) reported that ethylene is formed in the reaction of CH$_2$ radicals with oxirane. Oxygen abstraction by hydrogen atoms or methyl radicals has not been reported(12,17) and is probably not important. Indeed, the reaction of ·CH$_3$ with oxirane to give CH$_3$O· and ethylene has a calculated activation energy of 40 kcal/mol(18).

Another possibility, which was not considered by Cvetanovic, is suggested by analogy with the thermal reactions of ethylene episulfide(19) :

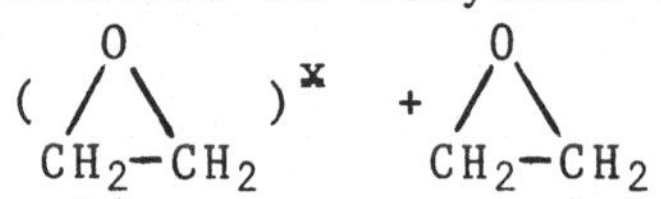

$$( \underset{CH_2-CH_2}{\overset{O}{\triangle}} )^x + \underset{CH_2-CH_2}{\overset{O}{\triangle}} \rightarrow 2\ C_2H_4 + O_2 \qquad (X)$$

In this case, one might expect the ethylene yield to increase with increasing pressure of oxirane.

The importance of secondary reactions in the early work is difficult to evaluate. PDS(10) reported that the quantum yield of acetaldehyde is 0.12. KIIT(14) have calculated, from Cvetanovic's results(3(a)) that $\Phi$(total aldehydes)$/\Phi CO$ = 0.78 at 100 Torr. If $\Phi CO$ = 0.3 then $\Phi$(total aldehydes) = 0.23. Some of the minor products probably arise from secondary reactions such as sensitization or direct photolysis of the acetaldehyde or of the other aldehydes formed. For example, acetaldehyde gives rise to ·CH$_3$ and ·CHO radicals in high yield in both the mercury photosensitization ($\Phi$=0.95; 20) and direct photolysis ($\Phi$=0.38; 21). It is therefore essential that reaction percentages be kept low to reduce the fraction of reaction products that is being consumed or is arising through secondary reactions.

In 1969 Neeley presented a Ph.D. dissertation on the mercury photosensitization of oxirane and methyloxirane(22). Unfortunately, the percentage conversion is too high in his experiments and his results do not further our understanding of the $^3$Hg sensitization of oxirane. Methyloxirane will be discussed in Section 2.2.

Recently, De Maré and Strausz(23) determined the quantum yields of H$_2$, CO, CH$_4$, C$_2$H$_4$, C$_2$H$_6$; and C$_3$H$_8$ in the $^3$Hg sensitization of oxirane as a function of substrate pressure at 28±2°C with an incident light intensity, $I_0 = 1.23 \times 10^{15}$ quanta/sec(see Figures 1 and 2).

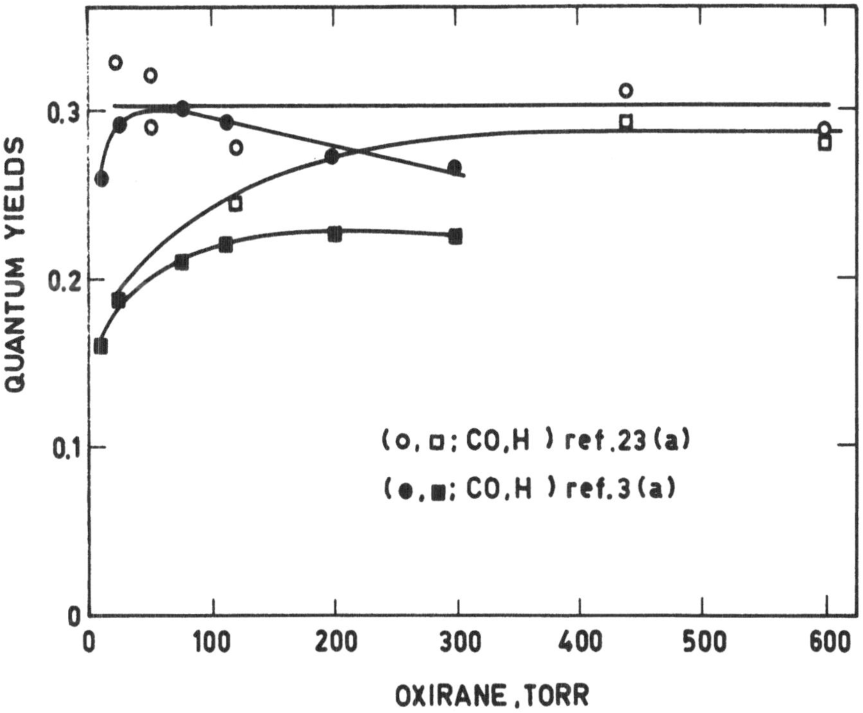


FIGURE 1. Major products of the $^3$Hg photosensitization.

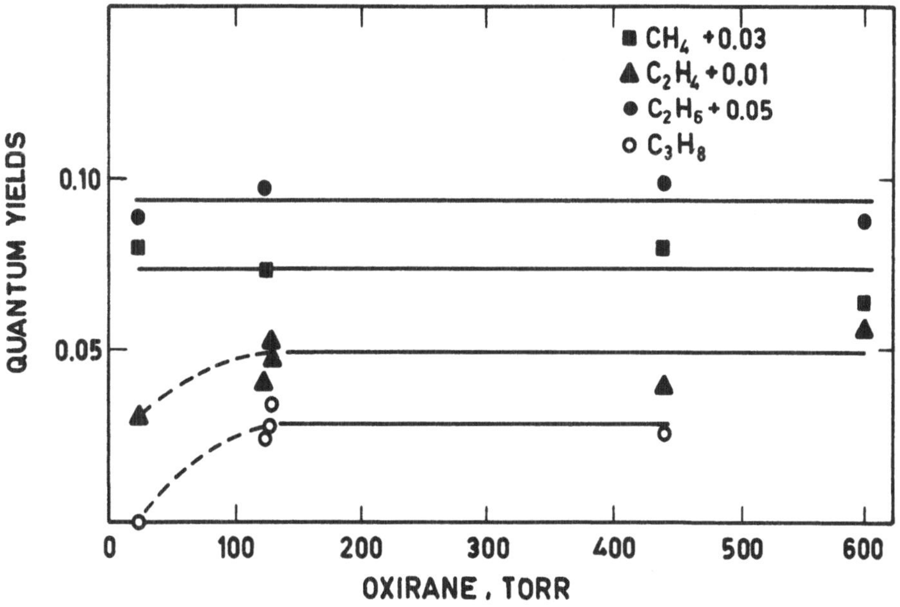


FIGURE 2. Minor products of the $^3$Hg photosensitization.

De Maré and Strausz(23) also identified formaldehyde, n-butane, ethanol, carbon dioxide and water as minor products. A specific search for oxygen failed to reveal its presence (even in trace quantities) among the products. This is not conclusive evidence that step X does not occur. Indeed, when a small amount of air was added to the starting oxirane, the mixture was $O_2$ depleted after the sensitization(23). The quantum yield of CO was 0.30±0.02, seemingly independent of pressure(Figure 1.). Thus the fall-off in the CO yield at high pressures, reported by Cvetanovic(3(a)), was not confirmed under the above experimental conditions. Also, within experimental error, $\Phi H_2 = \Phi CO$ for pressures >300 Torr. (These results are compatible with the finding that attack of $O(^3P)$ atoms on ethylene, studied in a crossed beam experiment, leads directly to methyl and formyl radicals as primary products(24).) De Maré and Strausz's(23) data require that one H-atom or molecule be generated for every CO molecule formed. The decrease in $\Phi H_2$ with decreasing pressure can be explained by the migration of H-atoms to the walls, where they recombine rather than abstracting a hydrogen atom from the substrate. The migration to the walls is enhanced by the high fraction of the absorption taking place in the first few millimeters of the cell. From the comparison of $\Phi H_2$ at high and low substrate pressure, the maximum yield of molecular $H_2$ in a primary step (III or IV) is calculated to be 0.13.

De Maré and Strausz(23) found $\Phi C_2H_4$ in the uninhibited reaction at 100 Torr oxirane, as high as that obtained by Cvetanovic in his butene-1 inhibited reactions. They therefore studied the effect of added cis-2-butene, which should not interfere with ethylene analysis or give rise to extraneous ethylene product. (To reduce the eventual importance of secondary reactions, they placed a 253.7 nm interference filter between the lamp and the cell. Although the filter reduced the incident light intensity from 12.3 to 2.15 x $10^{14}$ quanta/sec, this had little or no effect on $\Phi C_2H_4$.) The added 2-butene increased the ethylene yield dramatically : addition of only 0.97 Torr cis-2-butene to 125 Torr oxirane increased $\Phi C_2H_4$ nearly three-fold to 0.12 while the yield of propane was reduced by a factor of eight. Increasing the cis-2-butene pressure to 3.57 Torr further reduces the propane yield ($\Phi C_3H_8 = 0.001_3$) and causes a decrease in $\Phi C_2H_4$ to $0.08_2$. This decrease is about twice that expected due to the increase in $^3$Hg quenching by the cis-2-butene(25). Thus cis-2-butene seems to be competing with the process(es) leading to ethylene formation. A possible explanation is step III followed by

$$:CH_2 \quad + $$ 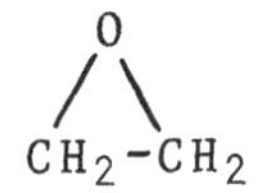 $$\rightarrow \quad CH_2O \quad + \quad C_2H_4 \qquad (XI)$$

and

$$:CH_2 \quad + \quad \underline{cis}\text{-2-butene} \quad \rightarrow \quad \text{products} \qquad (XII).$$

Such a mechanism would require that the quantum yield of molecular formation of hydrogen be at least 0.12 and that the decompositon be pressure independent. It would be very interesting to carry out competition experiments with cis-2-butene and look for cyclopropane products of step XII.

2.1.3. Biphotonic processes. The quantum inefficiency in the $^3$Hg sensitization of oxirane may be caused by quenching into the C-C bond, as observed in the $^3$Hg sensitization of spiropentane(26). The major fate of the triplet $\dot{C}H_2-O-\dot{C}H_2$ formed must then be intersystem crossing to the $S_0$ state and collisional deactivation to the ring closed oxirane. Under special experimental conditions, it should be possible to trap the biradical or to excite it further so that it will decompose. Since the quenching cross-section of oxirane for $^3$Hg is quite small compared to compounds with a double-bond(25,27), trapping the $\dot{C}H_2-O-\dot{C}H_2$ biradical with an olefin is not practical. In an attempt to observe reactions of the radical, De Maré and Strausz(23) increased the incident light intensity over 20-fold to $3.82 \times 10^{16}$ quanta/sec and determined the quantum yields of CO, $H_2$ and $CH_4$ (see Figure 3). The yields of these products are all

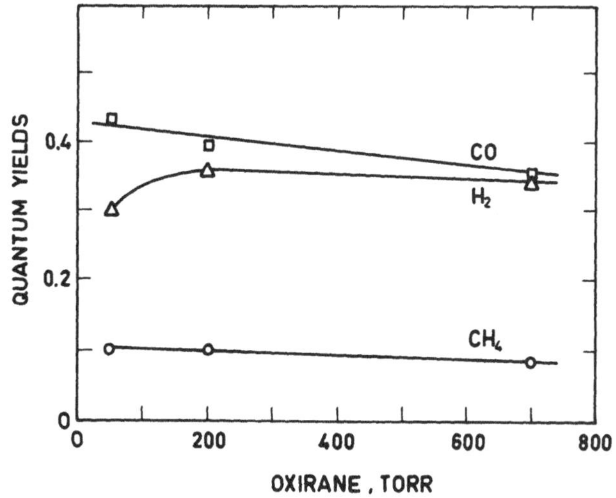


FIGURE 3. $^3$Hg photosensitization of oxirane.

much higher than those determined at the lower light intensities. Moreover, $\Phi CO$ and $\Phi CH_4$ decrease with increasing pressure, indicating possible deactivation of an excited species with a short lifetime. At the highest pressure, the quantum yields of $H_2$ and CO are again nearly equal. Using the quantum yields taken at 400 Torr oxirane pressure from smooth curves representing the experimental points and from the simple equation

$$\text{rate} = \text{constant} \times I_0^{\,n-1} \qquad \text{(XIII)}$$

values of n = 1.07, 1.06 and 1.24 were calculated for CO, $H_2$ and $CH_4$, respectively(23(b)). This clearly demonstrates the occurrence of a biphotonic process in the $^3$Hg sensitization of oxirane. The process may be either

$$^{\bullet}CH_2\text{-}O\text{-}^{\bullet}CH_2 \; + \; h\nu \; \rightarrow \; \text{products} \qquad \text{(XIV)}$$

or

$$^{\bullet}CH_2\text{-}O\text{-}^{\bullet}CH_2 \; + \; {}^3Hg \; \rightarrow \; \text{products} + Hg \qquad \text{(XV)}.$$

Experiments at a constant light intensity and variable concentrations of mercury vapour should allow one to distinguish between which of these two steps are responsible for the observed biphotonic product formation.

2.1.4. Photolysis of oxirane. As mentioned above, KIIT(14) studied the vacuum UV photolysis of oxirane at 147, 174.4 and 178.3-184.5 nm. They limited oxirane conversions to less than 2% to avoid secondary reactions. They found that the main process was decomposition into methyl and formyl radicals and that the mechanism depended on the exciting wavelength (energy). Thus decomposition into O-atoms and ethylene was seven times more important at the shortest wavelength. They concluded that five primary processes occur in the photolysis at 147 nm.

$$\text{oxirane} + h\nu \rightarrow \begin{cases} ^{\bullet}CH_3 + CHO(CO + H) & \text{(i)} \\ O + C_2H_4 & \text{(ii)} \\ H_2 + (CH_2CO) & \text{(iii)} \\ CH_3CHO & \text{(iv)} \\ :CH_2 + CH_2O & \text{(v)} \end{cases}$$

The relative importance of these processes was determined to be i:ii:iii:iv:v = 1.0:0.7:0.1:0.2:0.2 in the photolysis at 147 nm and 1.0:0.1:0.1:0.7:trace in the $^3$Hg photosensitization at 100 Torr. The latter results were obtained by reinterpretation of Cvetanovic's data(3(a)). At 174.4 nm, only steps i, ii, iii, and at 178.3-184.5 nm, only steps i and ii, are important, respectively(14). Since added NO does not affect the yield of acetaldehyde in the photolysis, KIIT(14) assume that all of it is formed by molecular isomerization and none via free radical reactions.

## 2.2. Methyl- and dimethyloxirane : experimental studies

2.2.1. Photolyses. The photolyses of these oxiranes do not appear to have been studied.

2.2.2. Methyloxirane. As noted in the introduction, both the $^3Hg$ sensitization of methyloxirane and the addition of $O(^3P)$ on propylene should yield triplet diradical intermediates. There are several reasons why these processes are more interesting than the corresponding $O(^3P) + C_2H_4$ or $^3Hg$ + oxirane reactions. First of all, in contrast to the $^{\bullet}CH_2-O-^{\bullet}CH_2$ biradical, MO3 can undergo H-atom migration to yield the triplet state of a stable product, methylvinylether. Also the addition of the oxygen atom on the $CH_2$ or $CHCH_3$ ends of the propylene double bond yields two different biradicals, MO1 and MO2, with different possibilies of product formation. The $C_3H_6O$ biradicals have more degrees of freedom for the distribution of the excess energy and they should be stabilized more easily than the $C_2H_4O$ biradicals. Thus, while $O(^3P) + C_2H_4$ yields only trace quantities of oxirane, $O(^3P) + C_3H_6$ yields methyloxirane as the major addition product under extremely varied experimental conditions (see Table I)(3,28-30). The second most important product is the isomer propionaldehyde. Trace quantities of acetone are also formed. We shall return to the discussion of the mechanism of O-atom addition to propylene in Section 3.

TABLE I. Fractional yields of the addition products of $O(^3P)$ on propylene.

| Products | A | B | C | D |
|---|---|---|---|---|
| Methyloxirane | 0.50 | 0.53 | 0.55 | 0.56 |
| Propionaldehyde | 0.45 | 0.46 | 0.42 | 0.38 |
| Acetone | 0.05 | 0.01 | 0.03 | 0.06 |

A.Gas phase, 298 K(28(a)); B.Liquid phase, 77 K(28(b)); C.Solid phase, 65-77 K(29); D.Solid phase, 90 K(30).

The only study, to our knowledge, on the $^3Hg$ sensitized reactions of methyloxirane is in the thesis by Neeley(22). Although the substrate conversions in his experiments are too high, it is obvious that the reaction mechanism is quite complex. Neeley identified ten saturated hydrocarbon products as well as propylene, ethylene, CO, $H_2$, acetaldehyde, 2-methylpropionaldehyde, butyraldehyde, methylethylketone, propionaldehyde and acetone. He explained the wide variety of products formed by assuming that primary processes involving an oxygen atom abstraction, a hydrogen atom abstraction and splitting into ethyl + formyl and methyl + $C_2H_3O$ radicals occur. He estimated that oxygen atom abstraction accounts for 8% of the reaction and the $CH_3$ + $C_2H_3O$ split for 19%. The other two reactions would then account for the remaining 73%, or the major part of the decomposition.

2.2.3. Dimethyloxirane. Cvetanovic and Doyle(31) reported the $^3$Hg photosensitization of trans-1,2-dimethyloxirane in 1957. As observed in the case of oxirane and methyloxirane, the reaction is quite complex. The main products are isobutanal, $C_2H_6$, CO,$C_2H_4$, $C_3H_8$, $CH_3CHO$, $C_2H_5CHO$, $CH_3COC_2H_5$, $CH_4$ and $H_2$. The yields of most products decreased with increasing pressure. The exceptions were ethane, methane and hydrogen which remained essentially pressure independent(31). The $^3$Hg sensitization of the cis-1,2-dimethyloxirane has not been reported. The addition of ground state oxygen atoms to cis- and trans-2-butene has been studied by Cvetanovic(see ref. 3(b), 28(a) and references cited therein).

## 2.3. Theoretical studies

2.3.1. Oxirane. Quantum mechanical studies on the triplet states of oxirane have been mainly on the C-O bond opened triplet biradical, $^\cdot CH_2CH_2O^\cdot$, and on the ring-closed triplet species (either with or without distortion of the C-O bonds or C-C-O angles(4,9,32). The calculated thermodynamical stabilities show that the energy of the undistorted ring-closed triplet species lies well above that of the triplet $^\cdot CH_2CH_2O^\cdot$ biradical (4). This biradical has been of great interest since it is the most likely primary intermediate in the addition of O($^3$P) on ethylene(3). The isomeric biradical, $^\cdot CH_2$-O-$^\cdot CH_2$, has been the object of a MINDO/3 calculation(33). The lowest triplet state of this biradical is a likely intermediate in the $^3$Hg sensitization of oxirane where interaction with the C-C bond can occur in the primary process(26). The $^\cdot CH_2$-O-$^\cdot CH_2$ biradical is expected to be less stable than the $^\cdot CH_2CH_2O^\cdot$ isomer since C-C bonds are relatively stronger than C-O bonds by about 30 kcal/mol(34). It is interesting to note that the quantum mechanical studies(4) lead to the conclusion that for ring-closed oxirane, the lowest triplet species lies so high above the ground state that a certain deformation of the molecule must take place before triplet mercury can transfer its energy to it.

In two recent papers, Bigot, Sevin and Devaquet(35) simulated "the complete evolution of ethylene oxide by monomolecular processes". They obtained the potential energy curves of the various electronic states by ab initio SCF-CI calculations with the STO-3G basis(36) of the GAUSS 70 programs(37). The configuration interaction was limited to the first 100 mono- and biexcited configurations. The authors point out that the technique(small basis set, limited CI) does not lead to precise quantitative information but is well suited to give semiquantitative information of chemical significance. They propose a semiquantitative rationale for the overall thermal reactivity and the photochemical behaviour of oxirane in the first accessible singlet and triplet valence excited states(35). The quantitative nature of the results could also be improved by complete geometry optimization of the initial and transitional states. The importance of complete geometry optimization in calculating energy barriers has recently been made very clear by Yamaguchi et al(9).

Although these previous studies have laid an impressive groundwork for further work on oxirane, we have chosen not to pursue the quantum mechanical investigations on oxirane at this time. Indeed, studies on methyloxirane, because of its asymmetric structure, should yield more valuable information and compensate for the relatively important increase in computer time required for each computation.

## 3. QUANTUM MECHANICAL INVESTIGATION ON TRIPLET ISOMERIC METHYLOXIRANES

### 3.1. Introduction

Accumulated experimental observations indicate that the unsubstituted olefinic end ($=CH_2$) of propylene is more prone to attack by $O(^3P)$ than the substituted end ($CH_3-CH=$)(28). Assuming that the addition of $O(^3P)$ to an olefinic double bond is electrophilic, the product MO1 may be classified as the Markovnikov adduct while MO2 is the anti-Markovnikov product. Since the basis of Markovnikov's rule is the thermodynamic control of the electrophilic addition, it is reasonable to expect that computed thermodynamic stabilities might be very useful in elucidating some of the features of $O(^3P)$ addition to propylene as well as the $^3Hg$ sensitized decomposition of methyloxirane. In this latter reaction, as mentioned in Section 2.2.2 an additional triplet species, MO3, is probably formed through interaction with the C-C ring bond; this type of interaction has been observed in the $^3Hg$ photosensitization of spiropentane(26). As shown in Figure 4, ring-cleavage may be followed by

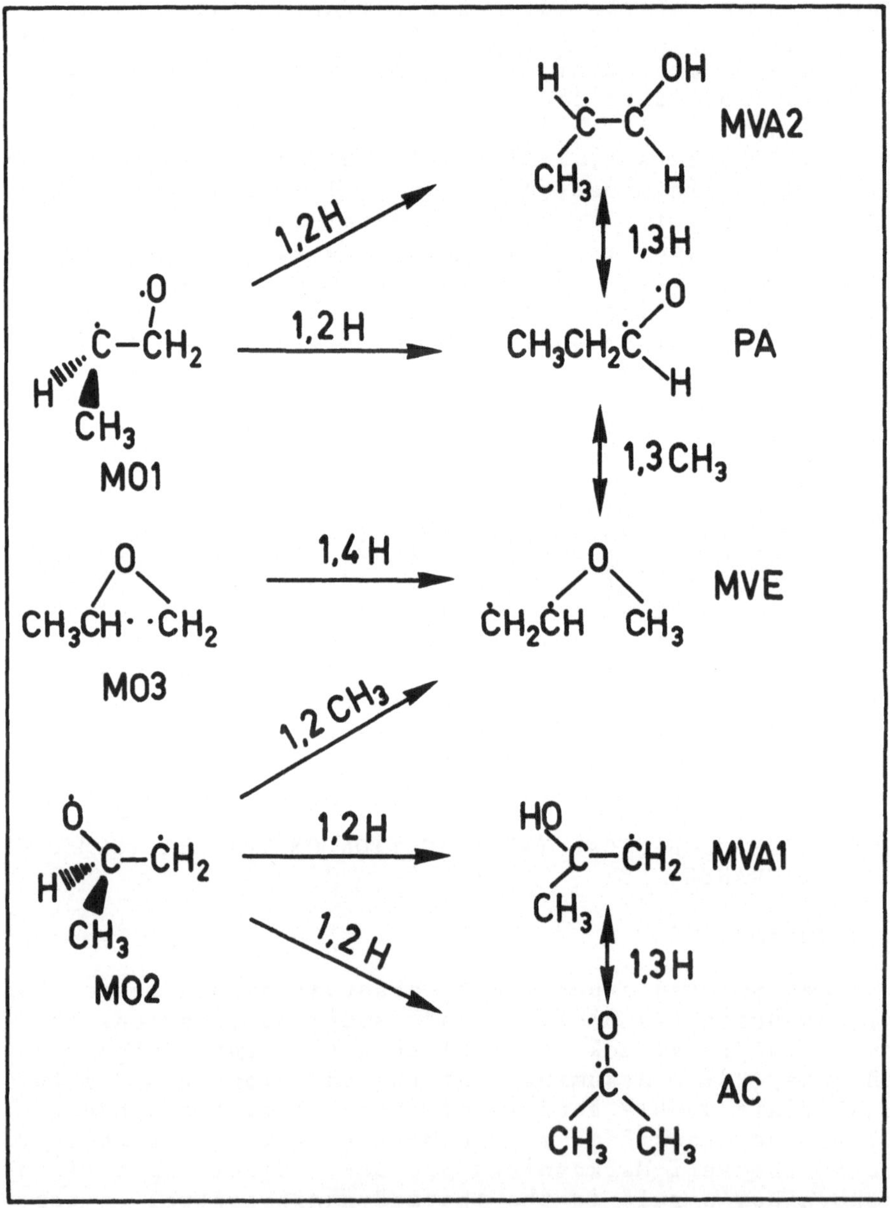


FIGURE 4. Possible H-atom and methyl group shifts in MO1, MO2, MO3 and their isomers (see ref. 38).

1,2-hydrogen (1,2H) shifts in MO1 to give triplet propionaldehyde (PA) or its enol, MVA2; 1,2H shifts in MO2 to yield triplet acetone (AC) or its enol, MVA1. A 1,2-methyl group (1,2$CH_3$) shift to the oxygen atom in MO2 or a 1,4H shift in MO3 leads to the formation of triplet methylvinylether, MVE. If one does not consider the eventual energetic barriers, all of these species could play a role in the $^3$Hg photosensitization of methyloxirane or the addition of O($^3$P) to propylene. The first objective of our study was therefore the calculation of the relative thermodynamic stabilities of these eight species at a chosen conformation to elucidate the question of which of them are energetically accessible in the O($^3$P) + propylene and $^3$Hg + methyloxirane reaction manifolds.

The second objective involves filling in the details of the rotational potential surfaces so that information is obtained about the barriers separating rotamers as well as the energies at the potential minima, saddle points and maxima. Some of the results on the generated energy surfaces or hypersurfaces for the species in Figure 4 will be analyzed in section 3.2.2. The data should be useful in determining some details of the dynamics of the unimolecular processes of these isomeric triplet $C_3H_6O$ species.

## 3.2. Computational method

A minimal contracted (STO-3G) basis set(36) was used for the SCF calculations within the unrestricted Hartree-Fock (UHF) formalism using the Gaussian 70 system(37) for the eight isomeric triplet species shown in Figure 4. Except for totally optimized structures (section 3.2.3), local $C_{2v}$ and $C_{3v}$ symmetry was imposed on the $CH_2$ and $CH_3$ groups, respectively. The energy of O($^3$P) was computed and the geometry of ground state propylene was optimized within the same formalism.

3.2.1. Geometry optimizations. In this first part, no torsion angles were optimized, they were fixed and studied later as rotational (hyper)surfaces. The geometry optimizations for all species were carried out by varying one parameter at a time. The bond lengths were optimized first and then the bond angles. After such a cycle of geometry optimization the heavy atom bond lengths and the angles were reoptimized.

3.2.2. Conformational studies. In MO1 and MO2 the torsions involved a methyl rotation ($\Theta_1$) and a rotation about the C-C bond ($\Theta_2$) which was part of the original three-membered ring. This led to the generation of an energy surface of the type $E = E(\Theta_1,\Theta_2)$. In MO3, in addition to methyl rotation, two nonequivalent C-O rotations ($\Theta_2,\Theta_3$) were considered. The AC conformational study involved two methyl group torsions ($\Theta_1,\Theta_2$) and the out-of-plane angle ($\Theta_3$). The PA study considered methyl rotation ($\Theta_1$), ethyl torsion ($\Theta_2$) and inversion at the carbonyl group ($\Theta_3$). Thus in MO3, AC and PA one obtains energy hypersurfaces $E = E(\Theta_1,\Theta_2,\Theta_3)$. These energy (hyper)surfaces represent rigid rotations because no further geometry optimization was attempted. The (0°,0°) and (0°,0°,0°) structures for the isomers and the definition of the $\Theta_3$ coordinate for AC and PA are shown in Figure 5. For MO2, $\Theta_1 = 0°$ corresponds to a methyl hydrogen eclipsed with the oxygen atom.

An analytical equation of the form

$$E(\Theta_1,\Theta_2) = \sum_{i=1}^{n} C_i \cdot f_i(\Theta_1) \cdot g_i(\Theta_2) \qquad \text{(XVI)}$$

was fitted with a stepwise least-squares regression procedure(39) to 24 computed SCF data points for MO2(n=7) and an equation of the same general form was similarly fitted to 18 symmetrically unique SCF data points for MO1 (n=10). An analytical equation of the form

$$E(\Theta_1,\Theta_2,\Theta_3) = \sum_{i=1}^{n} C_i \cdot f_i(\Theta_1) \cdot g_i(\Theta_2) \cdot h_i(\Theta_3) \qquad \text{(XVII)}$$

was fitted by the same procedure to sets of 68, 47 and 124 symmetrically unique SCF points for MO2, AC and PA, respectively. The average error for the surface fits was 0.020 (MO1, MO2, MO3), 0.028 (AC) and 0.022 (PA) kcal/mol, and all equation terms were significant at the 6.6% level, at least.

The fitted equations were analyzed for their critical points (minima, saddle points, and maxima) by solving the following system of gradient equations

$$\frac{\delta E}{\delta \Theta_1} = 0, \quad \frac{\delta E}{\delta \Theta_2} = 0, \quad \frac{\delta E}{\delta \Theta_3} = 0 \qquad \text{(XVIII)}$$

using the VA05AD optimization method(40). The critical points found were classified by determining the number of negative eigenvalues (the "order") of the Hessian (second derivatives) matrix.

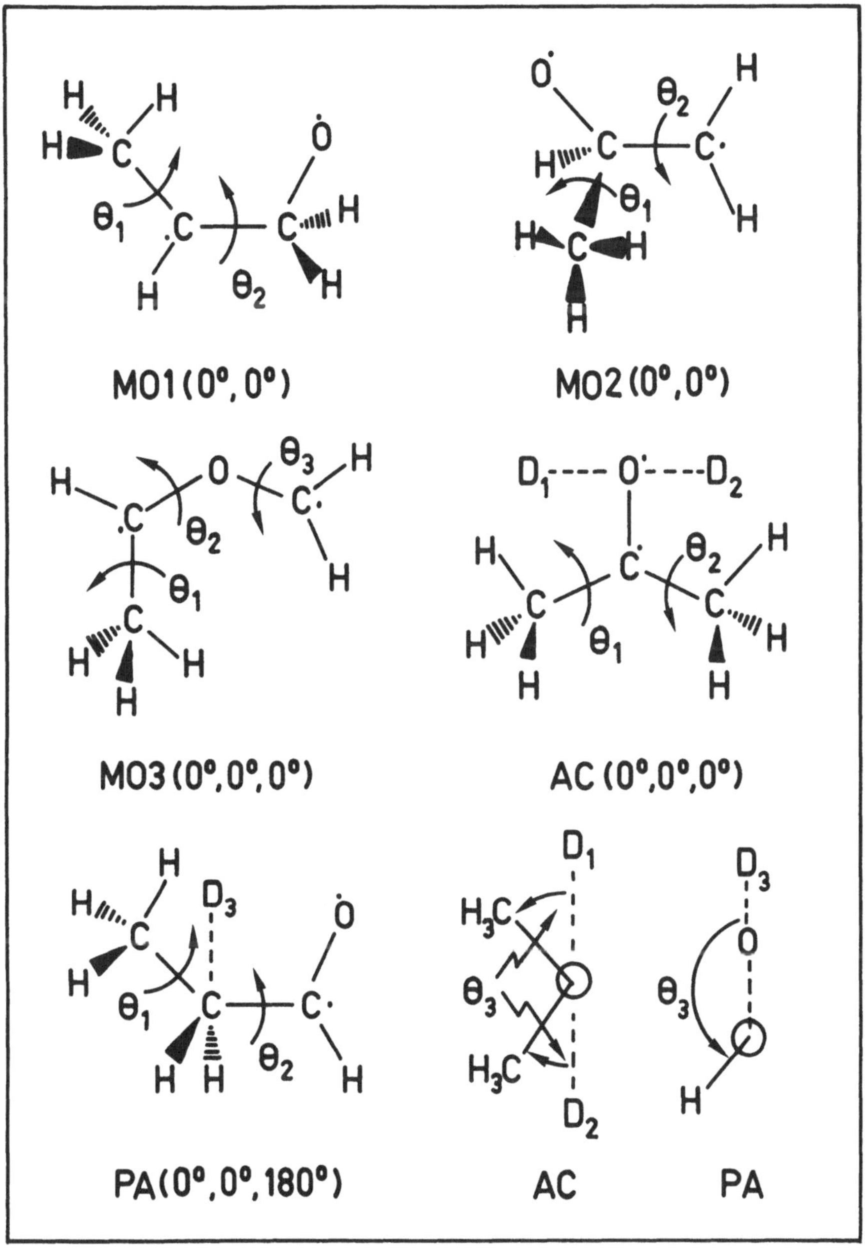


FIGURE 5. The reference structures for the three biradical structures MO1, MO2 and MO3 and for acetone and propionaldehyde.

3.2.3. Complete geometry optimization and critical points from direct optimization for comparison with surface fitting. For comparative purposes, the critical points of MO2 and also its optimized STO-3G geometry at the minimum energy were determined by either using the optimally conditioned minimization technique(41) or by solving the gradient equations $\delta E/\delta q_i=0$ with the VA05AD method(40). The former was utilized for minima and higher-order symmetry constrained critical points. The latter routine, which estimates the Hessian matrix by gradient differences during the first iteration, was utilized for saddle points lacking symmetry, and as a check on the order of some other critical points, by diagonalizing the Hessian to ascertain the number of negative eigenvalues. Both methods utilize the analytic gradients (42) of the SCF energy with respect to the internal geometrical parameters. These calculations were performed using the MONSTERGAUSS ab initio program(43).

### 3.3. Results and discussion

3.3.1. Geometry optimization. As an example of the optimized geometries, MO3 is shown in Figure 6. The optimized structures of the other triplet $C_3H_6O$ isomers

TABLE II. Total and relative energies of triplet $C_3H_6O$ isomers for selected geometries(38,45).

| Species and ($\Theta_1,\Theta_2$) or ($\Theta_1,\Theta_2,\Theta_3$) | Total energy, hartrees | Relative energy, kcal/mol |
|---|---|---|
| Methyloxirane + 112.7 kcal/mol | - | 42 |
| MVE (60°,180°,90°) | -189.43769 | 16.8 |
| MVA2 (60°,90°,0°) | -189.44225 | 13.9 |
| MO3 (60°,90°,90°) | -189.44319 | 13.3 |
| MVA1 (60°,90°,0°) | -189.44360 | 13.1 |
| $O(^3P)$ + Propylene | - 73.80415<br>-115.66027 | 0.0 |
| MO2 (60°,90°) | -189.48322 | -11.8 |
| MO1 (60°,90°) | -189.48635 | -13.8 |
| PA (60°,180°,138.3°) | -189.49488 | -19.1 |
| AC (60°,60°,0°) | -189.49835 | -21.3 |

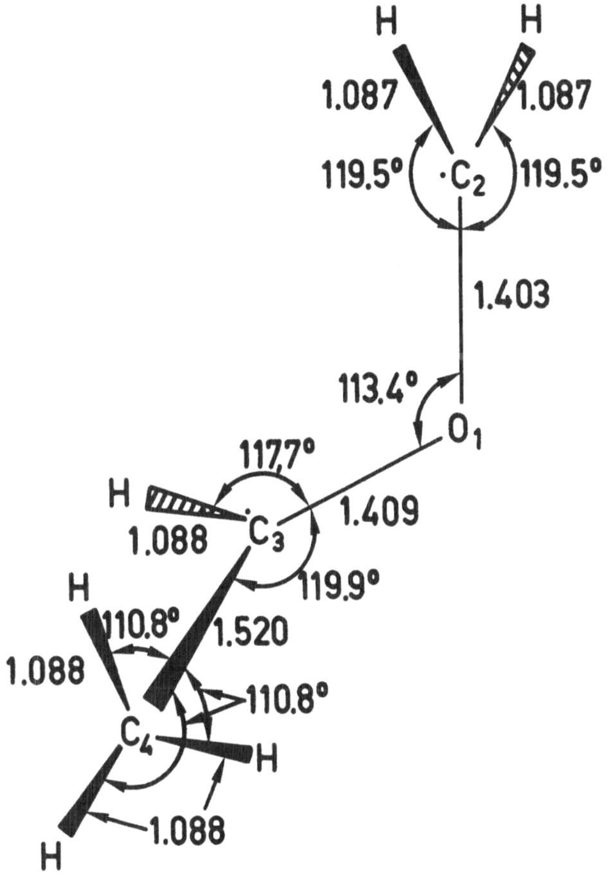


FIGURE 6. The STO-3G optimized structure for MO3. Local $C_{2v}$ symmetry was imposed on $C_2$ and $C_3$; $C_{3v}$ on $C_4$. The energy for the $(\Theta_1,\Theta_2,\Theta_3)=(60°,90°,90°)$ structure is -189.44319 hartree (38).

are reported in reference 38. For propylene with $C_{3v}$ symmetry imposed on the $CH_3$ group, the calculated energy is -115.66027 hartree, only slightly higher than that obtained (-115.66030 hartree(44)) for the geometry with the eclipsed methyl group H-atom nonequivalent to the other two methyl hydrogens. The computed total energies and relative thermodynamic stabilities of the selected $C_3H_6O$ isomers are given in Table II.

The results show (Table II) that the $T_1$ states of MO1, MO2, PA and AC are accessible in the $O(^3P)$ + propylene system. Transfer of the 112.7 kcal/mol of electronic energy from $^3$Hg to methyloxirane produces excited molecules with about 42 kcal/mol more energy than those produced by O($^3$P) + propylene(45). Thus in the $^3$Hg + methyloxirane system the $T_1$ states of MO3, MVE and of the two isomeric methylvinyl alcohols, MVA1 and MVA2, also become available. Optimization of the torsion angles and relaxation of the imposed geometric restrictions (local $C_{2v}$ and $C_{3v}$ for the $CH_2$ and $CH_3$ groups,resp.) may change the relative energies of the triplet species somewhat, but it should not change the above conclusions.

As shown in section 2.2.2 (Table I), the principal products of O($^3$P) + propylene at room temperature are methyloxirane, PA and AC, in the ratio 50:45:5. At lower lower temperatures methyloxirane is still the major product and (excluding the results at 90 K) the yield of AC is suppressed. Hirokami and Cvetanovic(28(b)) interpreted this as probably being due to a difference of about 0.3 kcal/mol in the activation energies of the anti-Markovnikov and Markovnikov additions. The present results give a 2.0 kcal/mol difference for the stabilities of the anti-Markovnikov and Markovnikov adducts. Thus only a fraction of this difference need be reflected in the activation energies of addition to account for the experimental results. It should also be mentioned that the H-migration in the adducts probably also has a small activation energy which, on steric grounds, could be slightly higher in the Markovnikov than in the anti-Markovnikov adduct.

As mentioned in the introduction, the $^3$Hg + oxirane reactions are more complex and less understood than the O($^3$P) + olefin systems. The formation of PA and AC from methyloxirane has been reported(22) but MVE has not been observed to date. Formation of MVE in its $T_1$ state from MO3($T_1$) by 1,4H shift or MO2($T_1$) by 1,2$CH_3$ migration to the O-atom is endothermic by 3.5 or 28.6 kcal/mol, resp.

TABLE III. Computed keto-enol tautomerization energies in singlet and triplet states.

| Molecule | ΔE, kcal/mol |
|---|---|
| Acetone($T_1$) | 34.4 |
| Propionaldehyde($T_1$) | 33.0 |
| Acetaldehyde($S_o$) | 18.4(ref. 46) |

Also, it can be concluded that H-atom migration to the O-atom in MO1 or MO2 is quite endothermic, requiring 25 or more kcal/mol. The relative energies of the keto-enol tautomers of triplet AC and PA and of ground state acetaldehyde(46) are given in Table III. Although the molecules are not the same, the results indicate that the energy differences between the keto-enol tautomers in the triplet state are about twice as large as those in the ground state. Thus passage from the triplet to the vibrationally excited ground state manifold may provide a plausible reaction path for the formation of enolic intermediates in the $^3$Hg sensitization of oxiranes.

3.3.2. Surface scans. The energy surface scans of triplet MO1, MO2 and MO3 have been reported(47) and those of triplet PA and AC will be published in the near future(48). Some examples of the surface scans and the major features of the stereochemistry will be given in the following discussion of the results.

MO1($T_1$). The parameters of eq. XVI, fitted to the 18 SCF data points of MO1, are given in Table IV. The equation representing the surface is plotted in Figure 7. A three-fold periodicity, characteristic of the local $C_{3v}$ symmetry, prevails along the coordinate of methyl rotation, $\Theta_1$. Rotation along the other C-C bond ($\Theta_2$)

TABLE IV. Parameters of the MO1 analytical surface equation.[a]

| Term | $f(\Theta_1)$ | $g(\Theta_2)$ | Coefficient (C) | Standard error of C |
|---|---|---|---|---|
| 1 | 1.0 | 1.0 | -189485.619 | 0.035 |
| 2 | $\cos 3\Theta_1$ | 1.0 | 0.224 | 0.017 |
| 3 | 1.0 | $\cos \Theta_2$ | 0.363 | 0.016 |
| 4 | $\cos 3\Theta_1$ | $\cos \Theta_2$ | 0.055 | 0.022 |
| 5 | 1.0 | $\cos 2\Theta_2$ | 0.538 | 0.016 |
| 6 | $\cos 3\Theta_1$ | $\cos 2\Theta_2$ | 0.086 | 0.022 |
| 7 | 1.0 | $\cos 3\Theta_2$ | 0.393 | 0.016 |
| 8 | $\cos 3\Theta_1$ | $\cos 3\Theta_2$ | 0.150 | 0.022 |
| 9 | 1.0 | $\cos 4\Theta_2$ | 0.085 | 0.012 |
| 10 | $\cos 3\Theta_1$ | $\cos 4\Theta_2$ | 0.040 | 0.017 |

[a] Each term is a product $C_i \cdot f_i(\Theta_1) \cdot g_i(\Theta_2)$, where the coefficients give the total molecular SCF energy in millihartrees.

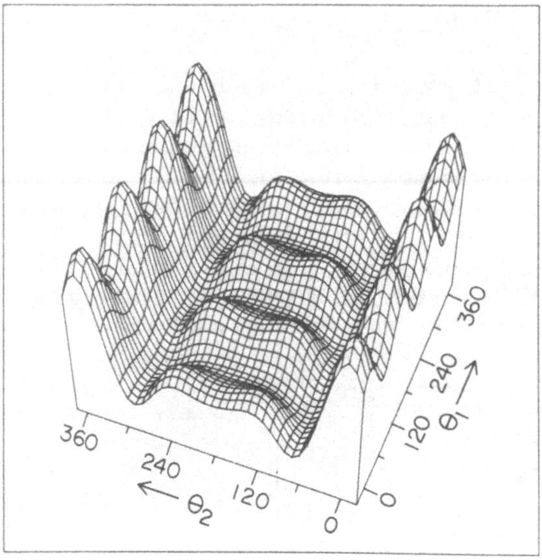


FIGURE 7. Conformational energy surface of M01. T view these plots with the perspective in which they were drawn, the eye should be a few centimeters above the plane of the paper and 20 cm from the plot origin (0,0). From ref. 47.

shows a one-fold periodicity. The characteristics of the critical points located on the fitted surface are summarized in Table V. There is a single unique minimum re-

TABLE V. Characteristics of the critical points of the M01 conformational surface.

| $\theta_1$ (deg) | $\theta_2$ (deg) | Order | Energy, hartree | Relative energy, kcal/mol |
|---|---|---|---|---|
| 60 | 77 | 0 | -189.48631 | 0.0 |
| 0 | 69 | 1 | -189.48618 | 0.1 |
| 0 | 180 | 1 | -189.48561 | 0.4 |
| 60 | 0 | 1 | -189.48480 | 0.9 |
| 60 | 180 | 1 | -189.48590 | 0.3 |
| 0 | 0 | 2 | -189.48369 | 1.6 |
| 0 | 128 | 2 | -189.48542 | 0.6 |

peated six times on the full surface. This minimum is interconnected with three distinctly different saddle points. A fourth saddle point ($\Theta_1$=0°,$\Theta_2$=180°) lies between two equivalent adjacent saddle points located at $\Theta_1$=±60° and $\Theta_2$=180°. Although the latter appear in Figure 7 as minima, they are in fact rather flat saddle points. The multitude of saddle points occur because of the existence of two nonequivalent maxima. The overall energy difference between the highest and lowest point of the surface is relatively small, less than 2 kcal/mol.

The minimum-energy conformation corresponds to a gauche arrangement of the heavy atom skeleton. A methyl hydrogen eclipses the C-H bond of the $^{\cdot}CH_2$ radical site, but the calculated barrier to methyl group rotation is only 0.1 kcal/mol. The barrier to rotation about the other C-C bond ($\Theta_2$) was computed to be 0.9 kcal/mol via the syn conformation ($\Theta_2$=0°) and 0.3 kcal/mol via the anti conformer ($\Theta_2$=180°).

MO2($T_1$). The parameters of eq. XVI, fitted to the 24 SCF data points of MO2, are summarized in Table VI. The equation representing the surface is plotted in Figure 8. Again the variation in energy along the methyl rotation reveals the characteristic three-fold periodicity of the local $C_{3v}$ symmetry. Torsion along the other C-C bond ($\Theta_2$) shows a two-fold periodicity due to the equivalence of the C-H bonds of the $-^{\cdot}CH_2$ radical. The

TABLE VI. Parameters of the MO2 analytical surface equation(47)[a].

| Term | $f(\Theta_1)$ | $g(\Theta_2)$ | Coefficient (C) | Standard error of C |
|---|---|---|---|---|
| 1 | 1.0 | 1.0 | -189480.501 | 0.027 |
| 2 | sin $3\Theta_1$ | 1.0 | -0.038 | 0.011 |
| 3 | cos $3\Theta_1$ | 1.0 | 2.352 | 0.011 |
| 4 | 1.0 | sin $2\Theta_2$ | -0.609 | 0.011 |
| 5 | 1.0 | cos $2\Theta_2$ | 0.313 | 0.011 |
| 6 | sin $3\Theta_1$ | cos $2\Theta_2$ | 0.050 | 0.016 |
| 7 | 1.0 | cos $4\Theta_2$ | -0.069 | 0.011 |

[a] Each term is a product $C_i \cdot f_i(\Theta_1) \cdot g_i(\Theta_2)$, where the coefficients give the total molecular SCF energy in millihartrees.

difference between the highest and lowest points on the surface shown in Figure 8 is 3.8 kcal/mol. Analysis of the fitted surface revealed the existence of a single minimum only slightly lower in energy than the point obtained with the fixed $\Theta_1$ and $\Theta_2$ optimization ($E(\Theta_1,\Theta_2) = E(60°,90°) = -189.48322$ hartrees(38)). In addition two non-equivalent saddle points and a single maximum where located. The characteristics of these critical points are summarized in Table VII.

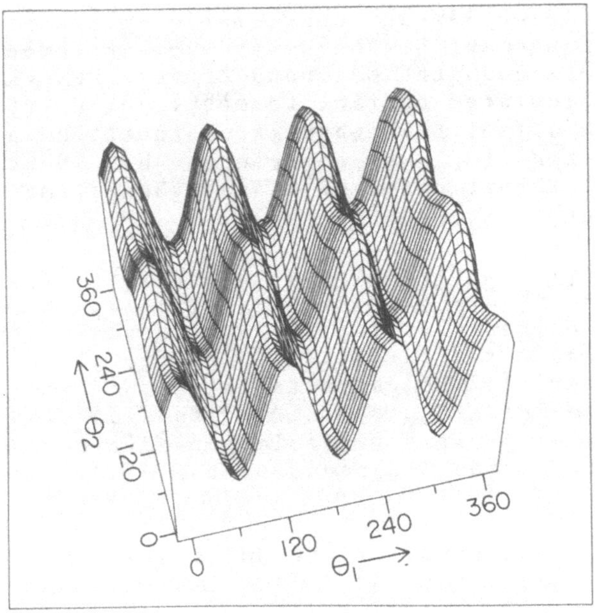


FIGURE 8. Conformational energy surface of MO2. Ref. 47.

TABLE VII. Characteristics of the critical points of the MO2 conformation energy surface.

| $\Theta_1$ (deg) | $\Theta_2$ (deg) | Order | Energy, hartree | Relative energy, kcal/mol |
|---|---|---|---|---|
| 59 | 64 | 0 | -189.48351 | 0.0 |
| 119 | 64 | 1 | -189.47880 | 3.0 |
| 60 | 145 | 1 | -189.48212 | 0.9 |
| 0 | 145 | 2 | -189.47742 | 3.8 |

The minimum-energy structure features a staggered methyl group ($\Theta_1$=60°) and the -˙$CH_2$ radical site bisects the C-O and C-C bonds ($\Theta_2$=60°). The barrier to methyl group rotation was calculated to be 0.3 kcal/mol which is similar to the ground state barriers for other saturated compounds(49,50) with the exception of methanol (1.12 kcal/mol(50)). The -˙$CH_2$ radical rotational barrier (0.9 kcal/mol) indicates essentially free rotation of that group. Note that both conformational changes follow least-motion paths; the coordinate not being rotated changes 1° or less.

The low barrier to $\Theta_2$ rotation in both MO1 and MO2 suggests that cis-trans isomerization, as might be observed in deuterium labelled methyloxirane, is energetically quite accessible in the $T_1$ state. This is consistent with the experimental finding that both cis- and trans-epoxides result from the addition of O($^3P$) to 2-butenes (28), and with the possibility of a "rapid equilibration of intermediates" during the cis-trans isomerization of irradiated diphenyloxirane(51).

MO3($T_1$). The hypersurface equation obtained by fitting the 68 MO3 SCF points to eq. XVII contains 41 terms which will not be detailed here(47). Analysis of the surface equation for critical points revealed the existence of a single unique minimum which occurs 12 times on the full surface covered by the complete rotation of all three variables. Furthermore, two unique maxima were found : an absolute maximum that occurs 24 times and a lower maximum that occurs 12 times on the full surface. Although only one unique minimum and two unique maxima were found, the surface contains six first-order and seven second-order saddle-points because the periodicity of the surface is three along $\Theta_1$, one along $\Theta_2$ with a center of inversion at (0°,180°,0°), and two along $\Theta_3$. Thus the overall surface may be decomposed into six unit cells. The topology of critical points characteristic of a one-unit cell is shown in Figure 9. Various cuts through the MO3 hypersurface are shown in Figure 10.

Starting from the minimum, the computed barrier to methyl group rotation is 0.1 kcal/mol through the saddle point (74°,141°,38°). Rotation about the internal C-O bond ($\Theta_2$) through the anti conformation (0°,180°,0°) has a computed barrier of 0.8 kcal/mol, but through the lowest syn structure (60°,0°,90°) it is 2.9 kcal/mol. Rotation of the terminal -˙$CH_2$ radical group has a calculated barrier of 1.2 or 1.8 kcal/mol via the processes already discussed. Although methyl group rotation occurs

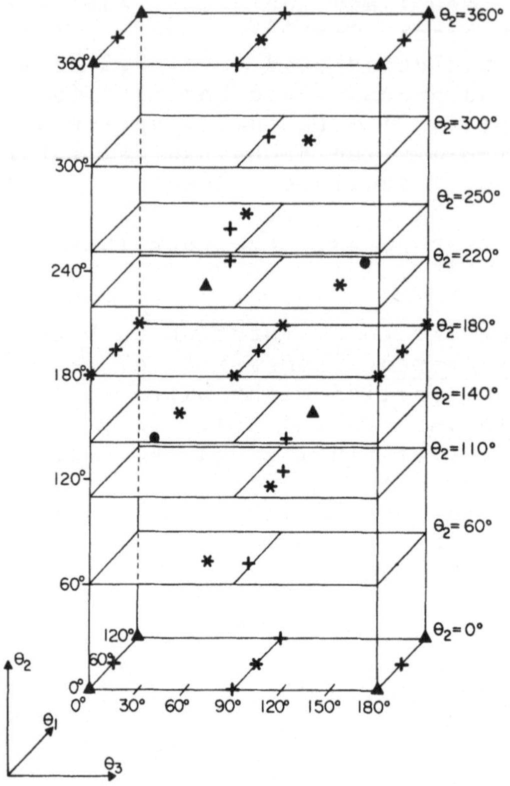


FIGURE 9. Critical point topological map for the unit cell of the MO3 hypersurface. Minima are represented by solid circles (●), first order saddle points by asterisks (✱), second order saddle points by plus signs (+) and maxima by solid triangles (▲). From ref. 47.

along least-motion paths, the two C-O bond torsions are highly coupled processes on all regions of the hypersurface.

All critical points, except those on the $\Theta_2 = 0°$ plane, lie within 2.3 kcal/mol of one another, an indication of the considerable conformation flexibility of the MO3 triplet state. The energy calculated for the (0°,0°,0°) conformation is probably too high because the

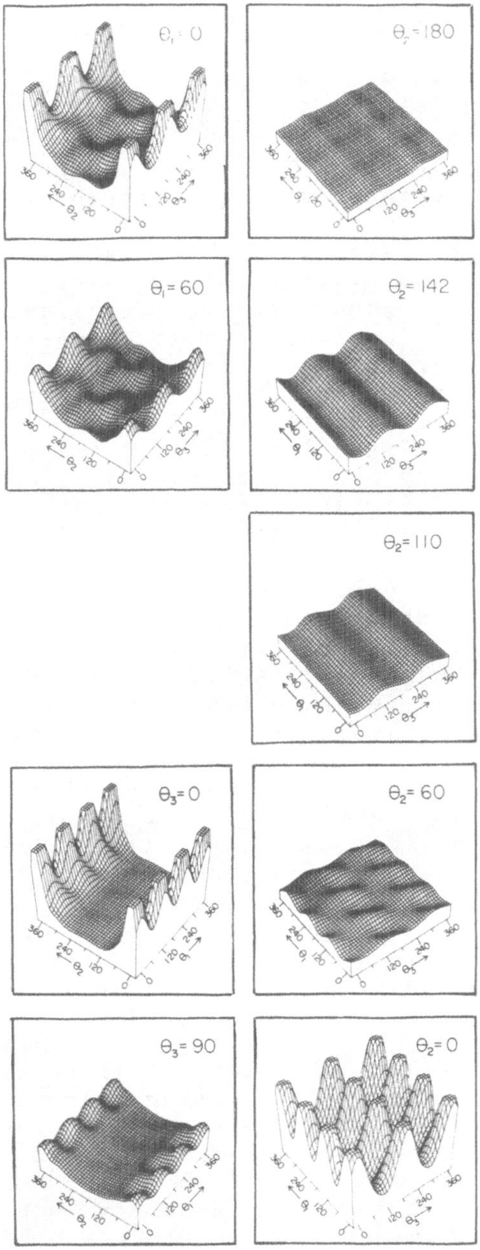


FIGURE 10. Various cross-sections of the MO3 hypersurface, all plotted with the same scale factor and all parallel to one face of the unit cell (Figure 9.). From ref. 47.

geometry was not reoptimized to reduce the non-bonded interaction of the methyl and the $-^{\bullet}CH_2$ groups.

The first-order saddle point (55°,52°,60°) and second-order saddle point (50°,66°,87°) are quite close in coordinate space, yet one would expect the $\Theta_3$ coordinate to differ by approximately 90°. One of these may be an artifact introduced by the fitting process.

$AC(T_1)$. The parameters of eq. XVII, fitted to the 47 AC data points must take into account the symmetry of the AC hypersurface : the conformations ($\Theta_1,\Theta_2,\Theta_3$) and ($\Theta_2,\Theta_1,\Theta_3$) are equivalent by superposition and must have the same energy(48). For $C_{2v}$ AC conformations, the symmetry of the open shell molecular orbitals is $5b_23b_1$, the $^3A_2$ state corresponding to the $^3n\rightarrow\pi^*$ excitation from the ground state oxygen lone pair in the molecular plane. For all $C_s$ symmetry conformations, the open shell configuration is $6a''11a'$ ($^3A''$ state), and the remaining structures belong to the $C_1$ point group, leading to a $^3A$ state.

Nine critical points were located in the $\Theta_3 \geq 0°$ portion of the conformational unit cell. The only minimum is pyramidal ($C_s$ symmetry), with both methyl groups approximately staggered relative to the C-O bond. The barrier for methyl rotation ranges from 1.1 to 2.7 kcal/mol, depending on the saddle point crossed. Note that the planar triplet acetone methyl rotation barrier is computed to be only 0.1 kcal/mol, and that the most stable conformation is predicted to be double eclipsed. For both the pyramidal and planar forms, the synchronous double methyl rotation simply doubles the barrier height. The most stable planar form is the double eclipsed (0°, 0°,0°) conformer. The barrier to rotation of $\Theta_2$ is substantially higher in the pyramidal structures than in the planar ones (about a factor of ten from the critical point determinations).

To our knowledge, there have been no other determinations of the AC triplet inversion or rotation barriers, either experimentally or theoretically. The rotation barrier in the ground state has however been computed as 1.14(52), 0.99(52) and 0.75(50) kcal/mol depending on the basis set and the assumed geometry. The corresponding experimental value(53), ranges from 0.78 to 0.83 kcal/mol. The present conformational relaxation results lower the AC energy by only 0.2 kcal/mol relative to the geometry optimized previously(38), which corresponds to the point (90°,30°,17.6°) on the hypersurface.

PA($T_1$). The parameters of eq. XVII were fitted to the 124 symmetrically unique SCF points for PA. In this case, all the linear coefficients are unique, as the hypersurface unit cell possesses no internal symmetry. Thus most of the points on the surface belong to the $C_1$ point group ($^3A$ state), while the data points with $C_s$ symmetry, the highest attainable for the PA system, correspond to $^3A''$ states, the open-shell orbitals being 6a"11a', as for AC. A total of eighteen unique critical points - three minima, seven first-order saddle points, six second-order saddle points and two maxima were located in the conformational unit cell. The three minima, all similar in energy, correspond to the two gauche plus one anti conformations for the ethyl rotation ($\Theta_2$); the methyl group is staggered in each case and the carbonyl center is pyramidal ($\Theta_3$=138°). The six barriers to ethyl rotation range from 0.8 to 1.8 kcal/mol; the three methyl rotation barriers range from 3.2 to 3.3 kcal/mol for pyramidal PA, 3.5 to 4.1 kcal/mol for planar PA. As expected, these latter values are similar to those for saturated ground state alkanes, since the triplet character is concentrated in the carbonyl portion of the molecule as shown by the Mulliken electron population analysis.

Previous calculations(54) on ground state propionaldehyde also resulted in energy differences between stable ethyl rotations of 0.4 kcal/mol, and barriers to rotation of 0.3 to 1.7 kcal/mol. Here the conformational preferences are reversed : eclipsed conformers ($\Theta_2$=0°, 120°) are favoured in the $S_o$ state while staggered conformers ($\Theta_2$=±60°,180°) are favoured in the lowest triplet state. These conformational relaxation results lower the PA energy by only 0.2 kcal/mol relative to the geometry partially optimized previously(38) which corresponds to the point (60°,180°,138.3°) on the hypersurface. Thus triplet AC is still predicted to be 2.2 kcal/mol more stable than triplet PA(38).

3.3.3. Complete geometry optimization and critical points from direct optimization for comparison with surface fitting. A very important question which arises is whether the fitted (hyper)surface equations (XVI and XVII) and the analysis for their critical points are quite reliable in revealing the true critical points of the (hyper)surfaces. For MO2, the critical points from the fitted surface (Table VII) were recalculated by direct optimization within the same geometry restrictions. The results are compared in Table VIII. It can be seen that there is no change in the overall conclusions since the values of $\Theta_1$ and $\Theta_2$ are the same for both methods and the relative energies of the critical points remain in the same order.

We have also mentioned the possibility that the imposition of $C_{2v}$ and $C_{3v}$ symmetries on portions of the triplet species may slightly affect the relative energetics. For comparison the geometry restrictions were lifted on MO2 at the single unique minimum (ie. the $CH_2$ and $CH_3$ groups were allowed to take non-$C_{2v}$ and non-$C_{3v}$ symmetries, resp.) The total energy was lowered from -189.48351 to -189.48559 hartrees, representing a stabilisation of 1.3 kcal/mol for the species without symmetry restrictions. This stabilisation is less than the energy difference of 1.8 kcal/mol between the minima in MO1 and MO2 and their relative stabilities thus remain unchanged.

These results lead us to conclude that the methods used are quite reliable

TABLE VIII. MO2 critical points from the fitted surface and by direct optimization.

| Type[a] | $\Theta_1$ (deg) | $\Theta_2$ (deg) | Order | ΔE(kcal/mol)[b] |
|---|---|---|---|---|
| Fit | 59 | 64 | 0 | 0.0 |
| Opt | 59 | 64 | 0 | 0.0 |
| Fit | 119 | 64 | 1 | 3.0 |
| Opt | 119 | 64 | 1 | 2.9 |
| Fit | 60 | 145 | 1 | 0.9 |
| Opt | 60 | 145 | 1 | 0.6 |
| Fit | 0 | 145 | 2 | 3.8 |
| Opt | 0 | 145 | 2 | 3.6 |

[a] Fit refers to results determined from the fitted equation, Opt to direct optimization.

[b] Calculated relative to the energy of the absolute minimum as determined by each method.

3.4. Summary

The main results of this quantum mechanical investigation of triplet state isomeric methyloxiranes may be summarized as follows :

1. The computed thermodynamic stabilities reveal that the $T_1$ states of only four of the $C_3H_6O$ species considered are accessible to the $O(^3P)$ + propylene adducts whereas all eight are available in the $^3Hg$ + methyloxirane reaction.

2. The Markovnikov adduct, $^{\cdot}CH(CH_3)-CH_2O^{\cdot}$, is more stable than the anti-Markovnikov addition product, $^{\cdot}OCH(CH_3)-^{\cdot}CH_2$, by 1.8 kcal/mol(47) giving a satisfactory explanation for the formation of more propionaldehyde than acetone in the addition of $O(^3P)$ to propylene.

3. The conformational energy surfaces of the ring-opened biradical species, MO1, MO2, and MO3, are generally fairly flat. This leads to low rotational barriers which indicate considerable conformational flexibility and the possibility of facile cis-trans isomerization in deuterium labeled alkene oxides.

4. The barriers to methyl rotation in triplet acetone depend strongly on the starting conformation : in planar triplet acetone the barrier is computed to be only 0.1 kcal/mol while in pyramidal acetone the barriers range from 1.1 to 2.7 kcal/mol depending on the saddle point crossed. In triplet propionaldehyde, the computed methyl rotation barriers (3.2 to 4.1 kcal/mol) are similar to those in saturated ground state alkanes.

## 4. ACKNOWLEDGEMENTS

The author thanks the Scientific Affairs Division of NATO for Research Grant 1431. He also wishes to express his deep gratitude to Professor O.P. Strausz (U. of Alberta), Professor I.G. Csizmadia(U. of Toronto) and Dr. R.K. Gosavi(U. of Alberta) for their help and collaboration in many different ways in the oxirane research projects. Very special thanks go to Mr. M.R. Peterson(U. of Toronto) for setting up MONSTERGAUSS(43) on the CDC computers at the Université Libre de Bruxelles and for having carried out an important part of the project in Brussels and Toronto. Dr. J. Olbregts (U.L.B.) is thanked for many valuable discussions and suggestions.

REFERENCES

( 1). Graedel, T.E.: 1978, "CHEMICAL COMPOUNDS IN THE ATMOSPHERE," pp. 267, 272.

( 2). Bootman, J., Lodge, D.C., and Whalley, H.E.: 1979, Muta. Res. 67, pp. 101-112.

( 3). (a) Cvetanovic, R.J.: 1955, Canad. J. Chem. 33, pp. 1684-1695; (b) Cvetanovic, R.J.: 1963, Advan. Photochem. 1, pp. 115-149; (c) Furuyama, S., Atkinson, R., Colussi, A.J., and Cvetanovic, R.J.: 1974, Intern. J. Chem. Kinetics 6, pp. 741-751.

( 4). (a) Mezey, P., Kari, R.E., Denes, A.S., Csizmadia, I.G., Gosavi, R.K., and Strausz, O.P.: 1975, Theoret. Chim. Acta 36, pp. 329-338; (b) Strausz, O.P., Gosavi, R.K., Robb, M.A., Eade, R., and Csizmadia, I.G.: 1977, Prog. Theoret. Org. Chem. 2, pp. 248-260.

( 5). (a) Scheer, M.D., and Klein, R.: 1969, J. Phys. Chem. 73, pp. 597-601; (b) Klein, R., and Scheer, M.D.: 1969, J. Phys. Chem. 73, pp. 1598-1599; (c) Scheer, M.D., and Klein, R.: 1970, J. Phys. Chem. 74, pp. 2732-2733.

( 6). (a) Special report : 1979, C & EN'S Top fifty Chemical products, Chem. Eng. News, May 7, pp. 22-27. (b) Key Chemicals, Chem. Eng. News, 1980, June 16, pp. 11, 13.

( 7). Benson, S.W.: 1964, J. Chem. Phys. 40, pp. 105-111.

( 8). Blades, A.T.: 1968, Canad. J. Chem. 46, pp. 3283-3284.

( 9). Yamaguchi, K., Yabushita, S., Fueno, T., Kato, S., and Morokuma, K.: 1980, Chem. Phys. Letters 70, pp. 27-30.

(10). Phibbs, M.K., Darwent, B.deB., and Steacie, E.W.R.: 1948, J. Chem. Phys. 16, pp. 39-44.

(11). Cvetanovic, R.J.: 1955, J. Chem. Phys. 23, pp. 1375-1380.

(12). Gomer, R., and Noyes Jr., W.A.: 1950, J. Amer. Chem. Soc. 72, pp. 101-108.

(13). Roquitte, B.C.: 1966, J. Chem. Phys. 70, pp. 2699-2702.

(14). Kawasaki, M., Ibuki, T., Iwasaki, M., and Takazaki, Y.: 1973, J. Chem. Phys. 59, pp. 2076-2082.

(15). (a) Sidhu, K.S.: 1965, Ph.D. Thesis, U. of Alberta, Edmonton, Canada; (b) Fowles, P., de Sorgo, M., Yarwood, A.J., Strausz, O.P., and Gunning, H.E.: 1967, J. Amer. Chem. Soc. 89, pp. 1352-1362.

(16). Gesser, H.: cited in reference 3(a) and private communication.

(17). Phibbs, M.K., and Darwent, B.deB.: 1950, Canad. J. Res. B28, pp. 395-402.

(18). Jakubowski, E., Ahmed, M.G., Lown, E.M., Sandhu, H.S., Gosavi, R.K., and Strausz, O.P.: 1974, J. Amer. Chem. Soc. 94, pp. 4094-4101.

(19). Lown, E.M., Sandhu, H.S., Gunning, H.E., and Strausz, O.P.: 1969, J. Amer. Chem. Soc. 90, pp. 7164-7165.

(20). Lossing, F.P.: 1957, Canad. J. Chem. 35, pp. 305-314.

(21). Calvert, J.G., and Pitts, J.N.: 1966, "PHOTOCHEMISTRY," Wiley, p. 371.

(22). Neeley, C.M.: 1969, "KINETICS OF THE MERCURY-SENSITIZED DECOMPOSITION OF ETHYLENE AND PROPYLENE OXIDES," Ph.D. Thesis, U. of Arkansas, U.S.A.

(23). (a) De Maré, G.R., and Strausz, O.P.: 1979, "THE MERCURY PHOTOSENSITIZATION OF OXIRANE IN THE GAS PHASE," in "PROBLEMS OF CHEMICAL KINETICS," collected papers from the Symposium in honour of Academician N.N. Semenov on the occasion of his 80th birthday, Moscow, April 12-14, 1976, Hayka, Moscow, pp. 38-46; (b) De Maré, G.R.: 1977, J. Photochem. 7, pp. 101-106.

(24). Kanofsky, J.R., and Gutman, D.: 1972, Chem. Phys. Letters 15, pp. 236-239.

(25). Cvetanovic, R.J.: 1955, J. Chem. Phys. 23, pp. 1208-1214, gives $\sigma^2$(oxirane) = 2.7 Å$^2$ compared to 22 Å$^2$ for $\sigma^2(C_2H_4)$ and 3.6 Å$^2$ for n-butane.

(26). De Maré, G.R., Walker, L.G., Strausz, O.P., and Gunning, H.E.: 1966, Canad. J. Chem. 44, pp. 457-460.

(27). Cvetanovic, R.J.: 1964, Prog. Reac. Kinetics 2, pp. 39-130.

(28). (a) Cvetanovic, R.J.: 1958, Canad. J. Chem. 36, pp. 623-634; (b) Hirokami, S., and Cvetanovic, R.J.: 1974, J. Amer. Chem. Soc. 96, pp. 3738-3746, and references cited therein.

(29). Orlov, V.N., and Ponomarev, A.N.: 1966, Kinet. Catal. 7, pp. 372-375.

(30). (a) Hughes, A.N., Scheer, M.D., and Klein, R.: 1966, J. Phys. Chem. 70, pp. 798-805; (b) Klein, R., and Scheer, M.D.: 1968, J. Phys. Chem. 72, pp. 616-622.

(31). Cvetanovic, R.J., and Doyle, L.C.: 1957, Canad. J. Chem. 35, pp. 605-612.

(32). Strausz, O.P., Gosavi, R.K., De Maré, G.R., and Csizmadia, I.G.: 1979, Chem. Phys. Letters 62, pp.339-340.

(33). Yamaguchi, K., Nishio, A., Yabushita, S., and Fueno, T.: 1977, Chem. Letters p. 1479; cited in reference 9.

(34). Gray, P., and Williams, A.: 1959, Trans. Far. Soc. 55, pp. 760-777.

(35). (a) Bigot, B., Sevin, A., and Devaquet, A.: 1979, J. Amer. Chem. Soc. 101, pp. 1095-1100; (b) Bigot, B., Sevin, A., and Devaquet, A.: 1979, J. Amer. Chem. Soc. 101, pp. 1101-1106.

(36). Hehre, W.J., Stewart, R.F., and Pople, J.A.: 1969, J. Chem. Phys. 51, pp. 2657-2664.

(37). Hehre, W.J., Lathan, W.A., Ditchfield, R., Newton, M.D., and Pople, J.A.: 1973, QCPE No. 236, U. of Indiana, Bloomington, Indiana, U.S.A.

(38). Strausz, O.P., Gosavi, R.K., De Maré, G.R., Peterson, M.R., and Csizmadia, I.G.: 1980, Chem. Phys. Letters 70, pp. 31-35.

(39). Barr, A.J., Goodnight, J.H., Sall, J.P., and Helwig, J.T.: "SAS76", SAS Institute Inc., P.O. Box 10066, Raleigh, NC27605, U.S.A.

(40). Powell, M.J.D.: Program VA05AD, Harwell Subroutine Library, Atomic Energy Establishment, Harwell, G.B.

(41). Davidon, W.C.: 1975, Mathematical Programming, 25, p.1. The routine is described in Technical Memos 303 and 306, Davidon, W.C., and Nazareth, L., Argonne National Laboratories, Argonne, Il., U.S.A.

(42). Schlegel, H.B.: 1975, Program FORCE, Ph.D. Thesis, Queen's University, Kingston, Ontario, Canada.

(43). Peterson, M.R., and Poirier, R.A.: 1980, Program MONSTERGAUSS, U. of Toronto, Toronto, Canada. The program incorporates the GAUSSIAN 70 (ref. 37) integral and SCF routines, analytic energy gradients (ref. 42) and automatic geometry optimization, with or without constraints, by the OC (ref. 41) or VA05AD (ref. 40) techniques.

(44). Radom, L., Lathan, W.A., Hehre, W.J., and Pople, J.A.: 1971, J. Amer. Chem. Soc. 93, pp. 5339-5342.

(45). The triplet energy of acetone is 78 kcal/mol (Schmidt, M.W., and Lee, E.K.C.: 1970, J. Amer. Chem. Soc. 92, 3579-3586). The heat of formation of acetone is -51.7 kcal/mol and that of methyl-oxirane was calculated to be -22 ± 2 kcal/mol from group additivity rules (Benson, S.W.: 1968, "THERMOCHEMICAL KINETICS," John Wiley and Sons; Janz, G.J.: 1967, "THERMODYNAMIC PROPERTIES OF ORGANIC COMPOUNDS," Academic Press).

(46). Bouma, W.J., Vincent, M.A., and Radom, L.: 1978, Intern. J. Quantum Chem. 14, pp. 767-777.

(47). De Maré, G.R., Peterson, M.R., Csizmadia, I.G., and Strausz, O.P.: 1980, J. Comp. Chem. 1, pp. 141-148.

(48). Peterson, M.R., De Maré, G.R., Csizmadia, I.G., and Strausz, O.P.; to be published.

(49). Peterson, M.R., and Csizmadia, I.G.: 1978, J. Amer. Chem. Soc. 100, pp. 6911-6916.

(50). Radom, L., Lathan, W.A., Hehre, W.J., and Pople, J.A.: 1972, Aust. J. Chem. 25, pp. 1601-1612.

(51). Albini, A., and Arnold, D.R.: 1978, Canad. J. Chem. 56, pp. 2985-2993.

(52). Allinger, N.L., and Hickey, M.J.: 1972, Tetrahedron 28, pp. 2157-2161.

(53). (a) Fately, W.G., and Miller, F.A.: 1962, Spectrochim. Acta 18, pp. 977-993; (b) Nelson, R., and Pierce, L.: 1965, J. Molec. Spect. 18, pp. 344-352; (c) Swalen, J.D., and Costain, C.C.: 1959, J. Chem. Phys. 31, pp. 1562-1574.

(54). Allinger, N.L., and Hickey, M.J.: 1973, J. Molec. Struct. 17, pp. 233-237.

# ROTATIONAL BARRIERS IN VINYL COMPOUNDS

George R. De Maré

Laboratoire de Chimie Physique Moléculaire, Faculté des Sciences CP 160, Université Libre de Bruxelles, 50 av. F.D. Roosevelt, B-1050 Brussels, Belgium.

The potential energy curve for vinyl group rotation in vinylcyclohexane, calculated using the STO-3G basis set and standard geometries from the MONSTERGAUSS program, reveals s-trans, gauche and s-cis minima with relative energies of 0.0, 5.0 and 3.2 kcal/mol, resp. The barrier to rotation from the s-trans to the gauche conformer is 5.3 kcal/mol. Similar calculations for 1,3-butadiene reveal only s-trans and gauche minima, 3.5 kcal/mol apart, with a barrier of 6.9 kcal/mol between them. Complete geometry optimization in 1,3-butadiene leads to only s-trans and s-cis minima. They are 1.83 kcal/mol apart and the barrier between them is 5.6 kcal/mol. These values are both lower than those, 2.05 and 6.73 kcal/mol, resp., reported by Radom and Pople.

+++++

Numerous experimental investigations, including Raman, IR, microwave, electron-diffraction and N.M.R. studies, have been directed at determining the number and structure of stable conformer(s) of organic compounds. While the energy differences between rotamers can often be obtained experimentally, it is more difficult to ascertain the height and position of the barriers to rotation. The latter are however accessible through *ab initio* or semi-empirical calculations (although it is well known that INDO and CNDO/2 calculations can predict the magnitude (and position) of potential barriers incorrectly(1).



*I. G. Csizmadia and R. Daudel (eds.), Computational Theoretical Organic Chemistry, 371–377.*

We have been especially interested in rotation of the vinyl group in vinylcycloalkanes(2-4) where hyperconjugation into the ring has often been invoked to explain reactivities and properties of vinylcyclopropane and vinylcyclobutane. To our knowledge, the only *ab initio* computations of the potential energy curves for rotation of the vinyl group in vinylcycloalkanes are those on vinylcyclopropanes and vinylcyclobutanes reported by Hehre(5) in 1972. He found that the unsubstituted compounds were most stable in their s-*trans* conformations and that they both possess flat secondary s-*cis* and *gauche* minima. The barrier between the s-*trans* and *gauche* minima was about 3 kcal/mol above the s-*trans*.

Since we have just completed a detailed N.M.R. investigation on vinylcyclohexane(4), we decided to carry out *ab initio* computations to determine the potential energy curve for vinyl group rotation in the axial conformer. The calculations were performed with the program MONSTERGAUSS(6), using standard geometries and the STO-3G minimal basis set(7). The variation of the relative energies with $\phi$, the $H^1$-$C^2$-$C^3$-$H^4$ dihedral angle, shown in Figure 1, is similar to that observed by Hehre

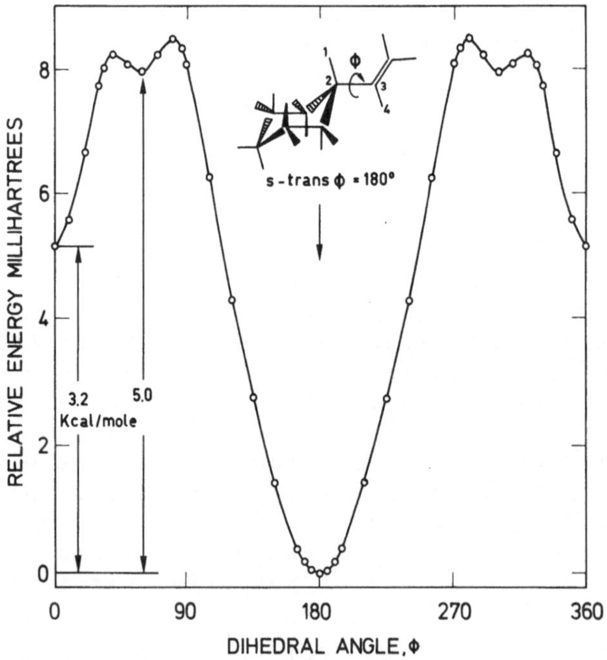


FIGURE 1. Potential for vinyl group rotation in vinylcyclohexane. E(s-*trans*) = -307.41027 hartrees.

for vinylcyclopropane(5) with the same basis set. The computations reveal the existence of minima at $\phi$ = 180, 60 and 0° or at s-<u>trans</u>, <u>gauche</u> and s-<u>cis</u> positions. Their relative energies are 0.0, 5.0 and 3.2 kcal/mol. No experimental data are available on the relative energies of vinylcyclohexane rotamers in the gas phase for direct comparison with the *ab initio* results. While the solvent may play an important role, it should be noted that the interpretation of N.M.R. spectra of vinylcyclohexane in both $C_2Cl_4$ and $CD_3COCD_3$ solutions lead to a much smaller energy difference, 0.2 kcal/mol(4), between the s-*trans* and *gauche* rotamers. It is also probable that optimization of the molecular geometry would change the relative energies somewhat. However, on a molecule such as vinylcyclohexane, with 22 atoms, this is a formidable task(computer time), even with the minimal STO-3G basis set. We therefore decided to investigate the effect of geometry optimization on rotational energy barriers by using one of Professor Csizmadia's favourite tricks - mimicking the computations on vinylcyclohexane by calculations on a smaller molecule.

First, the potential energy curve for rotation about the central C-C bond in 1,3-butadiene (dihedral angle $H^3$-$C^1$-$C^2$-$H^5$) was generated by plotting 21 STO-3G calculations in Figure 2. The plot represents rigid rotation with the standard geometries from the MONSTERGAUSS program(6). The curve reveals minima at $\phi$ = 45 and 180° (*gauche* and s-*trans*, resp.). The *gauche* minimum lies 3.5 kcal/mol above the s-*trans* one; the barrier separating the minima is 6.9 kcal/mol.

The geometry of 1,3-butadiene was then optimized at $\phi$ = 0, 90 and 180° using the analytic gradients of the SCF energy with respect to the internal geometrical parameters(6). These geometries were used to generate the potential energy curves given in Figure 3. The curves reveal the importance of geometry optimization on the results : for the potential energy scans with the s-*trans* ($\phi$ = 180°) and perpendicular optimized geometries, the s-*cis* maximum lies only slightly above the *gauche* minimum; for the scan with the s-*cis* ($\phi$ = 0°) optimized geometry, there is no s-*cis* maximum and no *gauche* minimum. The rotational barrier between the two minima is still at least 5.6 kcal/mol or about 80% of that in the standard geometry potential curve (Figure 2). The difference between the optimized minima is now only 1.83 kcal/mol, or a factor of 2 less than that obtained with the standard geometry. The above findings confirm the conclusions of Radom and Pople(8), based on optimization

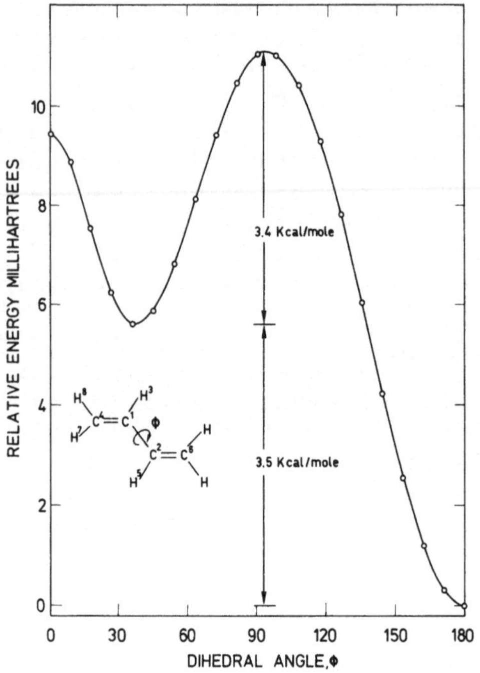


FIGURE 2. Potential for rigid rotation in 1,3-butadiene. E(s-*trans*) = -153.01412 hartrees.

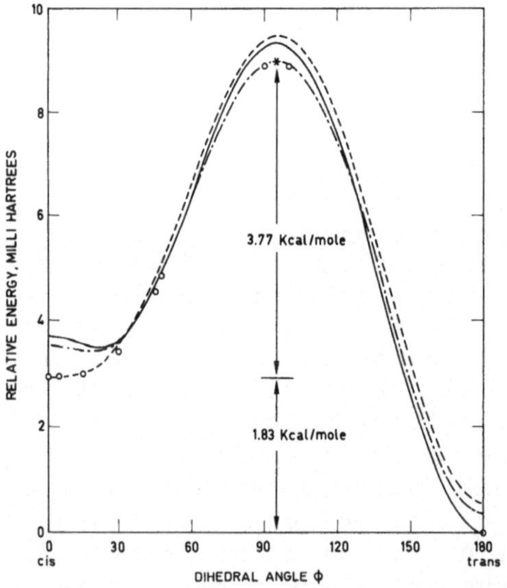


FIGURE 3. Potentials for rotation in 1,3-butadiene. Relative to E(s-*trans*) = -153.02037 hartrees.

of only the C=C-C angle during the rotational scan. They report -153.01661 hartrees for the energy of the partially optimized s-trans conformation. This is well above the -153.02037 hartrees found for the totally optimized s-trans conformation in the present study.

At fixed dihedral angles $\phi$ = 5, 15, 30, 45, 47.5, 100 and 135°, the remainder of the 1,3-butadiene geometry was optimized and the energies plotted in Figure 3. The entire geometry, including $\phi$ was optimized at the maximum and gave $\phi$ = 95.01°. The variation in the bond lengths (C=C and C-C) and the bond angles are shown in Figures 4 and 5, resp. The C-H bond lengths are almost independent of $\phi$ : as $\phi$ increases from 0 to 180°, $C^1-H^3$, $C^4-H^7$, and $C^4-H^8$ bond lengths vary between 1.0843-1.0866, 1.0809-1.0818 and 1.0812-1.0815 Å, resp. As can be seen in Figure 5, the C=C-C angles increase from 124.03 to 126.25° in going from the s-trans to the s-cis conformer. These values are slightly smaller at all $\phi$ than those found by Radom and Pople (variation from 124.2 to 126.6°(8)). It is also interesting to note that as the C=C-C angles increase, the H-C-C angles first increase slightly from 115.70° at the s-trans to 115.85° at $\phi$ = 135°, then decrease rapidly to 114.65° at the s-cis conformation.

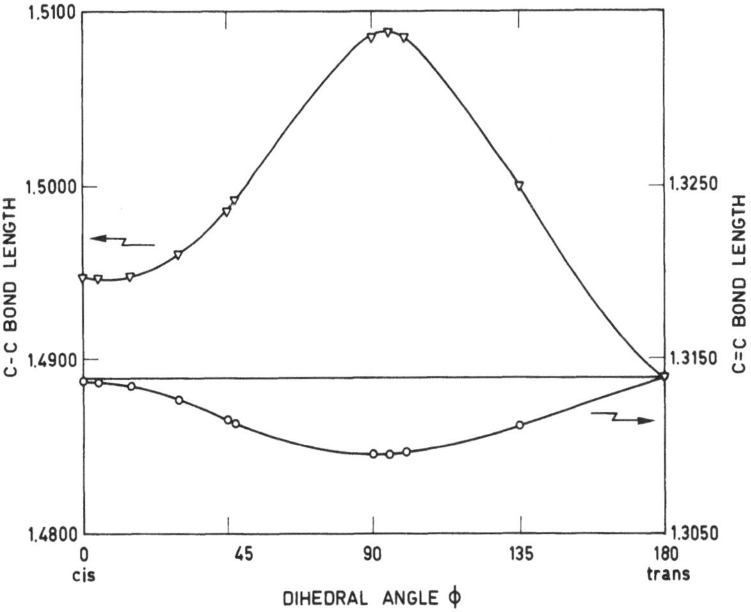


FIGURE 4. Variation of the C-C and C=C bond lengths with rotation about the central bond in 1,3-butadiene.

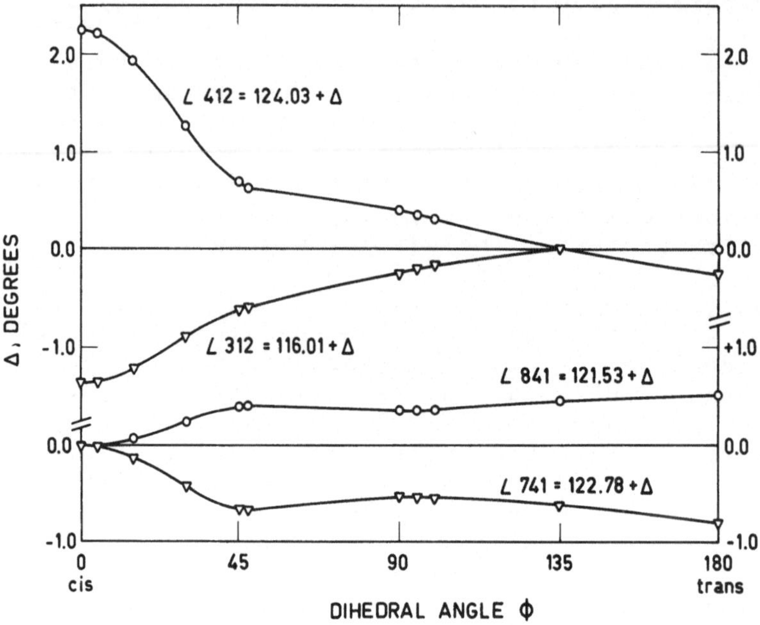


FIGURE 5. Change in the C=C-C (412), H-C-C (312) and H-C=C (741 and 841) angles with $\phi$.

The optimization of additional geometric parameters in this study as compared to that by Radom and Pople(8) has (i) reduced the energy difference between the s-trans and s-cis conformers of 1,3-butadiene from 2.05 to 1.83 kcal/mol and (ii) lowered the rotational energy barrier from 6.73 to 5.6 kcal/mol. The general conclusions which can be drawn from the above results are that geometry optimization can change rotational potential energy curves drastically and that optimization of the C=C-C bond angles appears to be the most important.

Further computations on vinyl type compounds are in progress.

ACKNOWLEDGEMENTS

The author wishes to thank Mr. D. Neisius (U.L.B) for his collaboration on this work. Helpful suggestions from, and discussions with Prof. I.G. Csizmadia (U. of Toronto), Dr. J. Olbregts (U.L.B.) and Mr. M.R. Peterson (U. of Toronto) are gratefully acknowledged.

REFERENCES

(1). Gropen, O., and Seip, H.M.: 1971, Chem. Phys. Letters 11, pp. 445-449, point out the failure of the CNDO/2 method in predicting barriers to rotation, conformations, and bond lengths in some conjugated systems.

(2). De Maré, G.R., and Martin, J.S.: 1966, J. Amer. Chem. Soc. 88, pp. 5033-5034.

(3). De Maré, G.R., Lapaille, S., Kispert, L.D., and Pittman Jr, C.U.: 1973, J. Mol. Struct. 17, pp. 417-420.

(4). De Maré, G.R., and Lapaille, S.: 1980, Org. Magn. Resonance 13, pp. 75-76.

(5). Hehre, W.J.: 1972, J. Amer. Chem. Soc. 94, pp. 6592-6597.

(6). Peterson, M.R., and Poirier, R.A.: 1980, "Program MONSTERGAUSS", U. of Toronto, Canada.

(7) Hehre, W.J., Stewart, R.F., and Pople, J.A.: 1969, J. Chem. Phys. 51, pp. 2657-2664.

(8) Radom, L., and Pople, J.A.: 1970, J. Amer. Chem. Soc. 92, pp. 4786-4795.

# THEORETICAL ASPECTS OF SMALL MOLECULE RYDBERG PHOTOCHEMISTRY

E.M. Evleth* and E. Kassab
Centre de Mécanique Ondulatoire Appliquée
C.N.R.S., 23, Rue du Maroc, 75019 Paris

Since most of the important features of the Rydberg spectroscopy of small molecules have been discussed in detail in the recent literature by Robin (1) and Sandorfy (2) we will concentrate on the theoretical aspects of their photochemical behavior. We have reviewed this subject in some detail (3) so that here we will concentrate on the construction of orbital and singlet state correlation diagrams containing Rydberg components which will permit the reader to rationalize either the photochemical behavior of some simple molecules such as water, ammonia and methane or the results of theoretical calculations on other photofragmentation reactions. The single most important feature of this article will be to show the reader how to add Rydberg orbitals and states to correlation diagrams. It will also permit predicting which reaction surfaces can be correlatively treated at the SCF level, or if configuration interaction (CI) is correlatively required, what will be the major configurational composition of the CI wave function.

## WHAT IS A RYDBERG STATE ?

Before showing how to construct these correlation diagrams we will attempt to give the reader some intuitive understanding of the theoretical nature of molecular Rydberg states as expressed in conventional orbital language. In a typical theoretical calculation the ground state wave function of a molecule is usually adequately represented by the superpositioning of a number of basis orbital functions having spacial distributions which are valence in character, i.e. most of the electronic density is within the van der Waals space of the molecule. As in the case of the excited states of atoms, certain molecular excited states will have extended electronic densities, the calculations of which would require basis orbitals having more diffuse natures. If the molecular state in question can approximately be treated as resulting from the excitation of a single electron to a highly diffuse orbital, the resulting structure can be viewed as being partitioned into n-1 electrons in valence space and one electron in Rydberg space, i.e. a cationic core, $R^+$, and a loosely bound electron in an excited atomic-like orbital. If



*I. G. Csizmadia and R. Daudel (eds.), Computational Theoretical Organic Chemistry, 379–395.*

we are dealing with a small molecule such as ammonia or ethylene (3-5) the Rydberg orbitals could have approximately 3s, 3p, 3d, 4s, etc. character. The number of Rydberg excited states would be infinite between the onset of the first transition, generally around 3 ev. (1) below the ionization potential limit, to the ionization limit. Several complications arise with this simplistic model of Rydberg states. If such states are reasonably well treated at the single configuration SCF level, the so called Rydberg orbital will have a mixed valence-diffuse composition. It is only when the diffuse composition dominates the molecular orbital in question that one is justified in calling the orbital Rydberg in character. As we will show, the orbital mix of such an MO can change along the bond rupture coordinate such that it will evolve from being nearly Rydberg to valence in character. This gives rise to a situation in which certain Rydberg states will evolve along surfaces to become the valence states of the radical pair products. Mulliken has termed the process of the radical pairs evolving from valence to Rydberg in character as Rydbergization (6). We prefer to use the term derydbergization as specifying the process in which a molecular Rydberg state involves into the valence states of the fragmentation products (3).

A reasonable view of the structure of a Rydberg state of a small molecule is obtained by removing an electron from one of the valence bonding orbitals and placing it in a large atomic-like orbital having a size which envelopes the molecule. In the case of ethylene (5) the lowest excited state is approximated as resulting from the excitation of a $\pi$-electron into a large 3s-like orbital which can be approximated by drawing a circle or ellipsoid surrounding the entire molecule resulting in a $\pi$3s state. In the cases of molecules having nonbonding electrons, such as water, ammonia, amines, alcohols, ethers, the lowest lying singlet and triplet states are conveniently given n3s electronic structures (1,3,7-9). It must be stressed, however, that these Rydberg orbitals will have irreducible representations of the molecular point group and, therefore, not fully atomic in character.

## CORRELATION DIAGRAMS, AN HISTORICAL OVERVIEW

For our development we will give an overview of the evolution of correlation diagrams. In particular, we will return to a discussion of the first generation orbital correlation diagrams incorporated in the united atom treatment (10), which has also been extended to polyatomic molecules (11). These diagrams show the origins of Rydbergization and are key elements in the understanding of this process. Likewise, the Walsh rules predate the Woodward-Hoffman rules and historically it appears that Longuet-Higgins (12) deserves the credit of stimulating these authors (13) in the their later construction and extension of orbital and state symmetry arguments (14). Finally, the development of the topicity rules by Dauben, Salem and Turro (15) directed emphasis towards certain specific photochemical problems which were not well analyzed by the Woodward-Hoffman rules.

The importance of the topicity view is that it allows an initial accounting of the number and kinds of product (or intermediate) states which will have to be eventually correlated with reactant states. Its application can be accomplished at the state level using only simple bonding diagrams. Thus if one wishes one can avoid constructing complicated orbital correlation diagrams. Likewise, the positive feature of the topicity rules is that one is forced to examine the end result first, and work backwards. This backward process may avoid leaving out an important feature in the state correlation diagram. A more sophisticated process (16) is to iron-out the problem from both directions, i.e. construct orbital correlation diagrams, followed with state correlation diagrams and double checking the results using the topicity rules. However, a final complication will arise when the diagrams are compared with actual computation. The diagrams are only convenient if the number of MO's considered is limited and simple behaving. A conventional gaussian MO level computation will have an extended basis set composition generating a large number of virtual MO's. Comparison of the two approaches will require some mental (or mathematical) truncating of the basis set composition of the virtual MO's. What we will show is how one adds Rydberg orbitals and states to the correlation diagrams and how these modify the topicity rules.

## CORRELATION DIAGRAMS FOR BOND RUPTURE REACTIONS

### 1. The Simple σ-Bond Model

The least complicated model exists in the simple σ-bond which has been constructed in MO terms from one orbital contributed each from a fragment A and B. To further simplify our argument we will deal with the case of an asymmetrically perturbed $H_2$ system in which the final orbital energies of 1s(A) and 1s(B) are different as would be the case in a heteronuclear united atom correlation diagram. This would generate the orbital correlation diagram shown in Figure 1.

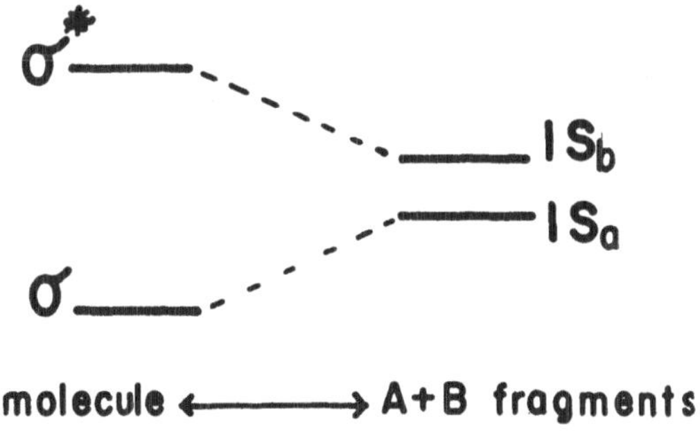


Figure 1. Simple σ-Bond Orbital Correlation, Perturbed $H_2$ Model.

Using this diagram there are three singlet configurations that can be constructed, their before-CI correlations are, $\sigma^2 \leftrightarrow (1s(A))^2$, $[Z_1(A^-, B^+)]$ ; $^1\sigma\sigma^* \leftrightarrow 1s(A)+1s(B), [^1D(A\cdot, \cdot B)]$ ; and $\sigma^{*2} \leftrightarrow (1s(B))^2$, $[Z_2(A^+, B^-)]$. The deceptive feature of these configurational correlations is that they say that the ground configuration, $\sigma^2$, will dissociate into a high energy zwitterionic state $Z_1$, while the excited state will terminate as a singlet pair of radicals ($^1D$), which is contrary to fact. This can be resolved with the construction of the state correlation diagram (Figure 2) where these configurational states are correlated by dotted lines. Crossing between the $\sigma^2$ and $^1\sigma\sigma^*$ configurations is avoided as shown by solid lines. How this crossing is computationally avoided will depend on the SCF MO basis functions employed in the CI treatment. If the functions are SCF for the $\sigma^2$ configuration the zwitterionic character of the final after-CI function for the ground state will be annihilated by equal contributions of the $Z_1$ and $Z_2$ structures. The actual case of unperturbed $H_2$, where the MO's at the dissociation limit are delocalized, is somewhat more complicated. However, in the general heteronuclear case a conceptual avoided crossing will occur (3,17). In the computational case, where the MO basis functions are not fully SCF in nature (e.g. triplet SCF-MO basis functions), the ground state after-CI function will take on mainly mixed closed shell-open shell character of the type, $\psi(S_o) = c_1(\sigma^2) + c_2\ (^1\sigma\sigma^*)$ i.e. mixed $A^-B^+$ and $^1A\cdot\cdot B$ character, which is as predicted in Figure 2 in the region of the avoided crossing. The ground state will clearly evolve to a

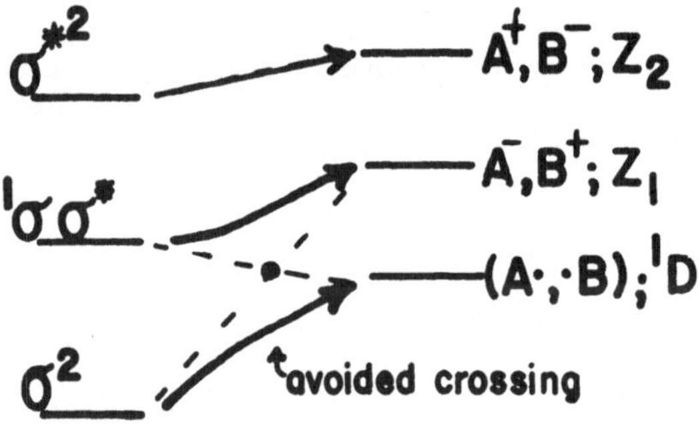


Figure 2. State Correlation Diagram for Simple σ-Bond Rupture in a AB Heteronuclear System.

diradical limit. With regard to simple σ-bond rupture, the behavior of the ground state surface could be modified to give a $Z_1$ (or $Z_2$) product if for some odd reason the zwitterion is more stable than the $^1D$ configuration. Such a modification is conceivable under strongly solvating conditions. However, a serious potential fault in this simple σ-bond rupture model is that it is basis function dependent with regard to the number and kind of excited states that can be generated. The addition of a single 2s(A) on atom A might not much alter the orbital correlation diagram but the state diagram becomes much more complicated (Figure 3).

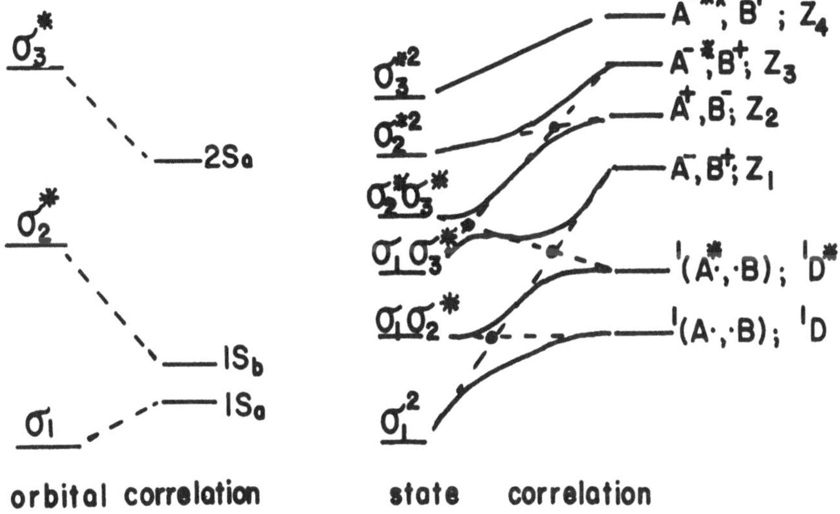


Figure 3. Orbital and State Correlation Diagram for σ-Bond Rupture Using a Simply Expanded Basis Set.

Note that the first excited state does not now correlate with the $Z_1$ state but an excited diradical state, $^1D^*$, $[A^{*\cdot}, \cdot B]$. Using the basis set employed in Figure 3 implies that one of the radical fragments is in a Rydberg (2s) state. As the basis set expands in this model 2-electron system other excited states of the reactant configuration will correlate with radical pairs in which either one or both of the radical pairs can be in various excited states (i.e. $^1D_1$, $^1D_2$, $^1D_3$, etc.), some of which will be lower in energy that the zwitterionic limit. In the two electron case these states will have Rydberg character.

2. The Four Electron $\sigma n$-Model

In larger multielectron molecules the situation becomes more complicated. One can immediately predict that various valence as well as Rydberg excited radical pairs could correlate with reactant excited states at energies less that the zwitterionic limit. Thus, one may alter the topicity rules by taking into account both excited valence and Rydberg radical pairs (i.e. $^1D^v_n$, $^1D^r_m$) (18). However, we will first demonstrate that ground-excited singlet state surface crossing is possible in certain four electron cases by examining the system :A:B in which, for the ground state, two electrons are interacting in a σ-orbital between A and B, and the other two electrons are in an nonbonding orbital n on A. The various singlet electron occupations are possible for the orbitals on A, [ψ(A), ψ'(A)], and B, [ψ(B)], where ψ'(A) represents the nonbonding orbital on A. These are, in symbol form, [:A·,·B], ($^1D_2$) ; [·A:,·B], ($^1D_1$) ; [:A$\bar{:}$,B$^+$], ($Z_1$) ; [:A$^+$,:B$^-$], ($Z_2$) and [A$\overset{+}{:}$,:B$^-$],($Z_3$). There are two state correlation diagrams possible depending on whether the energy of $D_1$ is less than or greater than $D_2$ and on whether the reaction pathway point group of the system is such that the irreducible representation of $\sigma^2n^2$ is the same or different than $^1\sigma^2n\sigma^*$. No ground-excited singlet state surface crossing will occur in the case of identical symmetries. Ammonia is an example were ground-excited state surface crossing can occur along certain pathways and not others for the NH bond rupture (19). The ammonia example is classified as a topicity 3 system (19). It is generally observed that topicity 3 or higher systems will exhibit a capacity for ground-excited state surface touchings or crossings at certain pathways (19). In most of the σn-compounds having single bonds between C,H,O and N, there is the capacity for a topicity 3 or higher reaction pathway.

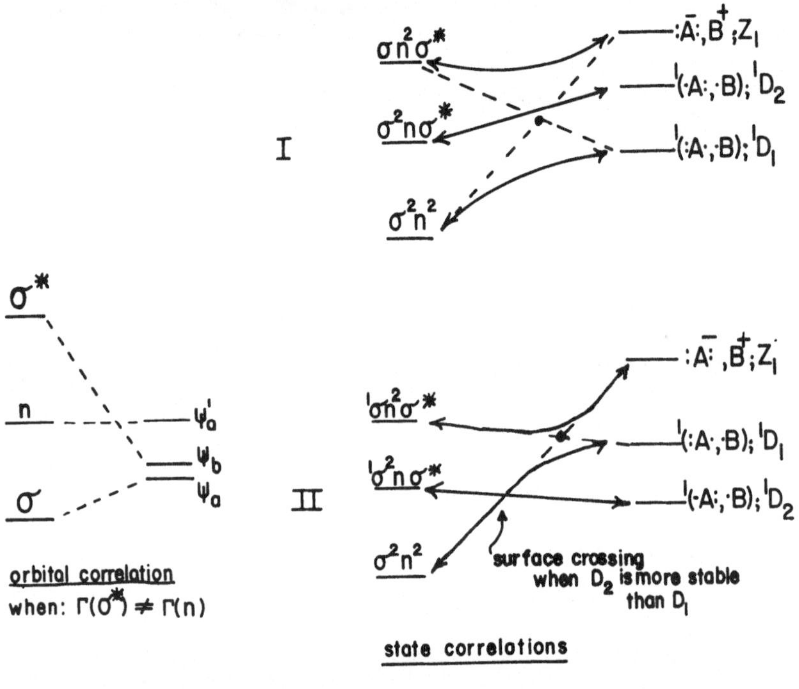


Figure 4. State and Orbital Correlation Possibilities in a σn-Four Electron Molecule.

3. Rydbergization

Examination of the homonuclear united atom diatomic correlation diagram (10) shows that the lowest $\sigma^*$ orbital, which consists of a 1s(A) - 1s(B) combination at large separations, correlates with a $2p\sigma_u$

orbital a small distances, and collapses to a 2p orbital at zero separation. While this is easily seen from the identical symmetry properties of the [1s(A) - 1s(B)] and 2p functions, this collapse can not computationally occur unless the original basis set contains either a $2p_c$ function (6) whose origin is at the center of the system, or 2s functions whose antiphasing combination [2s(A) - 2s(B)] can approximate a 2p function at A-B separations other than zero. In the former case the σ* function can be written as ; $c_1[1s(A) - 1s(B)] + c_2 2p_c$. At large distances, $c_1$ becomes large, $c_2$ small. As R(AB) → 0 a reverse situation occurs and the σ* function in $H_2$ takes on $2p_c$ character. Since the $2p_c$ function is more diffuse that the 1s functions, the σ* becomes spacially larger as the separation between the two atoms decreases. This process of the σ*-function evolving from being valence to Rydberg in character as R decreases is called Rydbergization (6). With regard to the lower energy 2s united atom function it correlates with a $2s\sigma_g$ diatomic function and eventually correlates with a [2s(A) + 2s(B)] combination of the separated atoms. This function remains Rydberg in character along the entire correlation diagram. Thus, it is seen that a particular molecular Rydberg orbital will or will not derydbergize depending how it correlates with the separated atom orbitals.

In the heteronuclear case, the gerade (g) and ungerade (u) character of the functions will disappear, a conceptual noncrossing will occur and the correlations 2pσ ↔ 2s(B) and 2sσ ↔ 1s(A) is obtained. In the hypothetical two electron three orbital example used in Figure 3 the general wavefunction, $\psi = c_1 1s(A) + c_2 1s(B) + c_3 2s(A)$ will have three solutions, $\sigma_1$, $\sigma_2^*$ and $\sigma_3^*$. At normal bonding distances, $c_1$ and $c_2$ will have comparative values in the bonding orbital, $\sigma_1$. At short distances $c_3$ will dominate in the function, $\sigma_2^*$. However, an antiphasing combination of $c_1$ and $c_2$ will dominate at larger separations and the $\sigma_2^*$ function will be valence in character. Thus, the two functions, $\sigma_2^*$, $\sigma_3^*$, will vary in valence-Rydberg character as a function of the separation of A and B. The behaviors of $\sigma_2^*$ and $\sigma_3^*$ can be conceptually viewed as occuring because of an avoided crossing (18), as will be shown by the dotted lines in the orbital correlation diagram of Figure 5.

## 4. The Four Electron nσ-Model Containing a Single Rydberg Component

In the case of a small molecule having conceptually only nonbonding and σ-bonding-antibonding orbitals one can assume that the lowest lying Rydberg orbitals lie in energy between σ-bonding and antibonding (virtual) orbitals at small internuclear distances. Therefore, Figure 4 undergoes the modification shown in Figure 5, if the irreducible representations of ψ(RM) and σ* are the same. The orbital correlations, ψ(RM) ↔ ψ(B) and σ* ↔ ψ(RA), will occur in which RM = Rydberg on the molecule, RA = Rydberg on fragment A. Note, however, if other same symmetry molecular Rydberg orbitals intervene between ψ(RM) and σ*, such as ψ(RM2), there will be a change in correlation of the σ* orbital and ψ(RM2) ↔ ψ(RA) with σ* correlating with some other atomic Rydberg orbital.

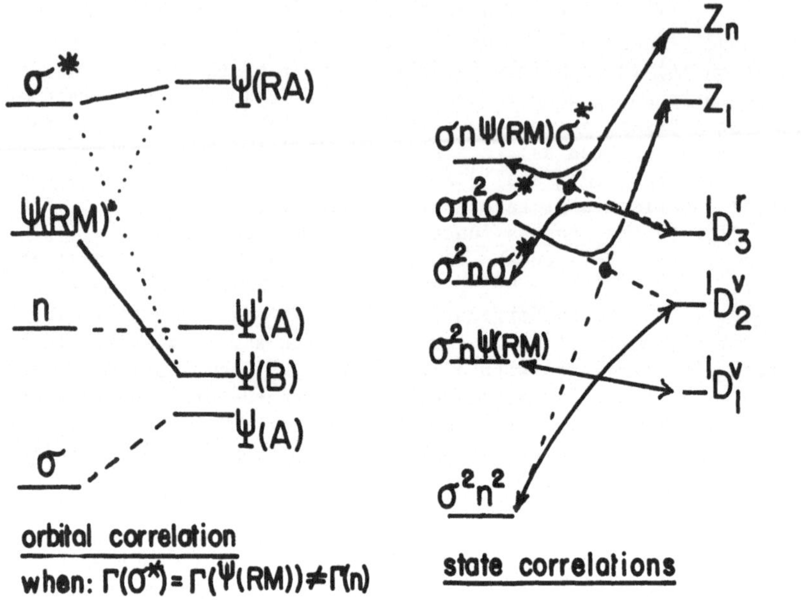


Figure 5. Orbital and State Correlation Diagrams for the σn-System Containing a Single Rydberg Orbital

Thus, while the correlation of ψ(RM1), the first molecular Rydberg orbital, will not change with the addition of other higher energy Rydberg orbitals, the behavior of the σ* orbital will be basis set dependent (18). As the Rydberg basis set increases in a small molecule calculation, a diluting effect will occur in the lowering lying valence virtual orbitals with a diminishing of their valence character. It may then be difficult to specify such an orbital as being valence in character as implied by the symbol σ*.

The state correlation diagram shown in Figure 5 shows several additional complications since the addition of a Rydberg component requires Rydberg state correlations at the reactant and product limits. If the system is symmetry restricted and the irreducible representations of the orbitals σ, σ*, and ψ(RM) are the same and different than orbital n, i.e. $\Gamma_1$ and $\Gamma_2$, respectively, the following final after CI state correlations occur : $^1\sigma^2n^2 \leftrightarrow {}^1D^v_2$, $^1\sigma^2n\psi(RM) \leftrightarrow {}^1D^v_1$, $^1\sigma^2n\sigma^* \leftrightarrow {}^1D^r_3$. This assumes that the lowest lying D state has [·A:,·B] character, as occurs in the case of the ammonia surface. Two important features are to be noted in the state correlation diagram. The $\sigma^2n\psi(RM)$ state derydbergizes without an avoided state crossing, and therefore is correlatively treatable at the single configuration SCF level. This means that CI is not correlatively necessary to treat this surface. All the other surfaces shows will require CI. In typical fashion the ground state undergoes an avoided crossing to terminate as the $D_2^v$ diradical. Whereas in Figure 4 the $\sigma^2 n\sigma^*$ state terminated as $D_2^v$ it now undergoes Rydbergization has an avoided crossing with a four open shell doubly excited state, $\sigma n\psi(RM)\sigma^*$ , and terminates as a third diradical state, $D_3^r$ which can be written as [$\cdot\dot{A}\cdot, \cdot B$] and is Rydberg in character.

As can now be seen, even in a simple four electron four orbital model complications occur. We will now concentrate on a real system, ammonia, to point out additional complications occuring with the addition of other Rydberg components. The problem of additional Rydberg components arises because once one adds a Rydberg s-orbital to a calculation, the p-components are not much higher in energy. They are of spectroscopic importance and may have photochemical importance.

## 5. Ammonia

The threshold photochemistry of $NH_3$ occurs from the lowest lying excited singlet Rydberg state (n3s). The molecule is planar in the excited state (20), exhibits virtually no fluorescence, its absorption spectrum indicating predissociation. The lowest excited state experimentally yields ground state $NH_2(^2B_1)$ and $H(^2S)$ as photoproducts, although, recently there has been reported some small amount of excited $NH_2(^2A_1)$ generation (21). The state correlation diagram first generated by Douglas (22) and cited in recent experimental work (21) argues that the derydbergization of the n3s singlet state results from an avoided state crossing between this state an a $n\sigma^*$ state, the latter correlating with $NH_2(^2B_1) + H(^2S)$. In agreement with our above analysis and Figure 5, recent theoretical work indicates that there is no avoided state crossing (8,18). In addition, the hypothesized $n\sigma^*$ state has no experimental or theoretical existence. The following state and correlation diagram will show why.

In Figure 6 is shown the expanded orbital symmetry diagram correlating the MO's of ammonia with those of $NH_2$ and H. The assumed basis set has Rydberg orbitals only on nitrogen. The key to the construction of the diagram is to place the 3s, 3p Rydberg components between the bonding and antibonding valence orbitals. Only one antibonding valence orbital is shown, the $\sigma_3^*(6a_1)$. Shown as a series of dotted lines are the correlations which would occur if the Rydberg components were not incorporated in the calculation. Note, that if no Rydberg components were included, $\sigma_3^* \leftrightarrow$ 1s(H). If only a 3s component is incorporated, $\sigma_3^* \leftrightarrow$ 3s(N). If both 3s, 3p components are included, $\sigma_3^* \leftrightarrow 3p_x$(N).

Since the $NH_3$ reaction pathway point group is $C_{2v}$, states having $A_1$, $A_2$, $B_1$ and $B_2$ symmetries are possible. Figure 7 shows the complexity of this diagram as limited to the $A_1$, $B_2$ and $B_1$ symmetries. A number of avoided crossings occur. The main feature of the diagram is that the lowest energy $1^1B_1$ state, having n3s (or $2p_z3s$) character, smoothly correlates with $^2B_1$ $NH_2$ + H (1s) without an avoided state crossing.

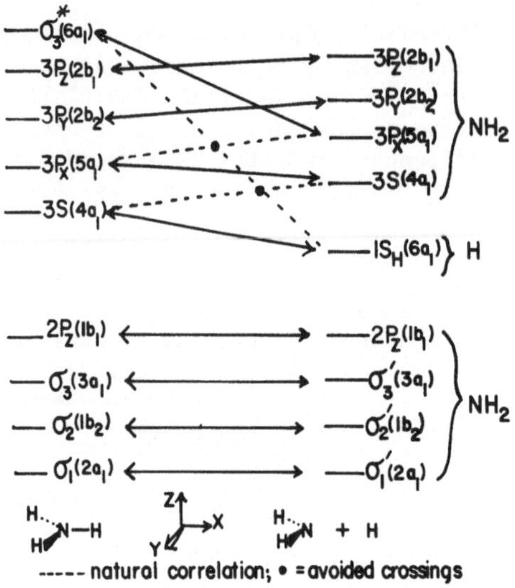


Figure 6. Orbital Correlation Diagram in $C_{2v}$ for the $NH_3 \rightarrow NH_2 + H$ Reaction Path.

The $2^1B_1$ state of $NH_3$ is clearly Rydberg ($2p_z3p_x$) and not valence ($2p_z\sigma_3^*$) in character. This latter configuration actually plays no importance in the state correlation diagram even though it appears in the orbital correlation diagram. The $2^1B_1$ state correlates with a $NH_2$ Rydberg state ($^2B_1$) + H (1s) after an avoided crossing with a higher energy four open shell $3^1B_1$ state. Finally, the correlation of the ground $1^1A_1$ state with the $^2A_1$ state of $NH_2$ + H (1s) is as anticipated in Figure 4. What is predicted in Figure 7 is actually observed computationally (18).

It is evident that a complete state correlation diagram could not be constructed without the aid of an orbital correlation diagram. The invalid aspect of the Douglas correlation diagram arises from the insufficient Rydberg basis set together with a lack of familarity with the process of Rydbergization. In addition, if the original n3s surface was computed using open shell SCF orbitals according to the generalized Brillouin theorem (23), there would be no first order CI matrix elements between the $1^1B_1$ and $2^1B_1$ configurations and no direct CI mixing in the proposed avoided crossing region (18). The Douglas diagram, however, is an important contribution in the development of state correlation diagrams because it represents an early attempt in this area. In addition, it correctly predicted the derydbergization process, excited-ground state surface crossing in the NH bond rupture reaction of planar $NH_3$ and acted as a guide to experimental work.

Figure 6 also shows that the $^1B_2$ state could arise from a

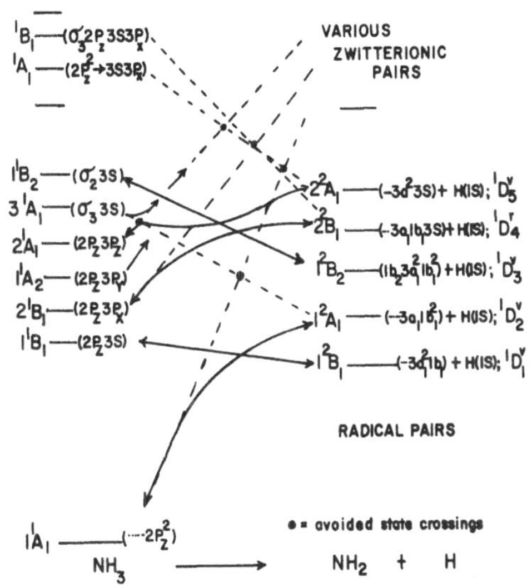


Figure 7. State Correlation Diagram for the $A_1$, $B_2$ and $B_1$ Singlet States in the $NH_3 \rightarrow NH_2 + H$, $C_{2v}$, Reaction Pathway.

simple $\sigma_2 \rightarrow 3s$ excitation. Examination of Figure 6 shows that this excitation will correlate directly at the SCF level with a theoretically predicted valence $^2B_2$ valence state of $NH_2$ + H(1s) (24).

6. Amines

The photochemistry of the amines show mainly NH bond rupture, with trimethylamine being photostable to the extent of exhibiting a large fluorescence quantum yield (25). The barrier in the n3s singlet state for NH bond rupture in ammonia and methylamine is computed to be around 10 kcal/mole while CN rupture has a value about twice this (18,26). The difference appears to arise because in the n3s state the CN bond is stabilized by some hyperconjugative interaction between the CH bonds and the half-filled $2p_z$ orbital on N (3). Since the valence space of a Rydberg state is cationic in nature, one can easily write conventional resonance structures of the type, $H_3{\equiv}C{-}NH_2(+) \leftrightarrow H_3^+{=}C{=}NH_2$. Both NH and CN bond rupture reactions in the n3s singlet state show the same rapid derydbergization as found in the $NH_3$ surface. Both photoreactions computationally yield the ground states of the radical pairs $CH_3NH$ + H and $CH_3$ + $NH_2$, respectively (26).

7. Water and Methanol

The n3s singlet states of both materials correlate directly with OH ($^2\Pi$) + H($^2S$) or $CH_3O$ ($^2E$) + H($^2S$). Likewise, for methanol, there is a direct correlation with the $CH_3$ ($^2A_1$) + OH($^2\Pi$) fragments (19). Thus, simple one electron molecule $[A\cdot B]^+$, $\psi = c_1\ [A\cdot,B+] + c_2\ [A+,\cdot B]$, there are only two solutions, one highly bonding, one highly antibonding. An expanded function, $\psi = c_1\ [A\cdot,B+] + c_2\ [A+,\cdot B] + c_3\ [A\dot{*},B+] + c_4\ [A+,\cdot B^*]$, where $A\dot{*}$ and $B\dot{*}$ refer to excited (Rydberg) components, is more general, and will have four solutions. In this latter case the second solution will not be so repulsive and will take on more and more Rydberg character as the AB distance decreases. A sort of VB avoided crossing will occur, as shown in Figure 8. In fact, the VB and MO approaches are identical in this one electron case but the system does not have to be treated in purely MO correlation terms. The essential point is that as the energy of a particularly repulsive molecular configuration rises the more likely higher energy basis orbitals will come into play.

In the cases of ammonia, water and methane, all these unfavorable configurations involve conceptual three electron configurations. For the planar approach of the $^2A_1$ $NH_2$ to an H atom, the arrangement is a bonding combination of the :A· + ·B system. The $^2B_1$ $NH_2$ + H approach is equivalent to placing ·A: + ·H, which creates, at close distances, a three electron configuration ·A:·H. In MO terms this is close to a $n\sigma^2\sigma^*$ configuration, which would also have an automatically built in a ionic VB component, $\cdot\overset{+}{A}\cdot:H^-$. In VB terms, in order to avoid this unfavorable situation we could add other same symmetry spacially expanded functions around either or both the A or B fragments. This would "allow"

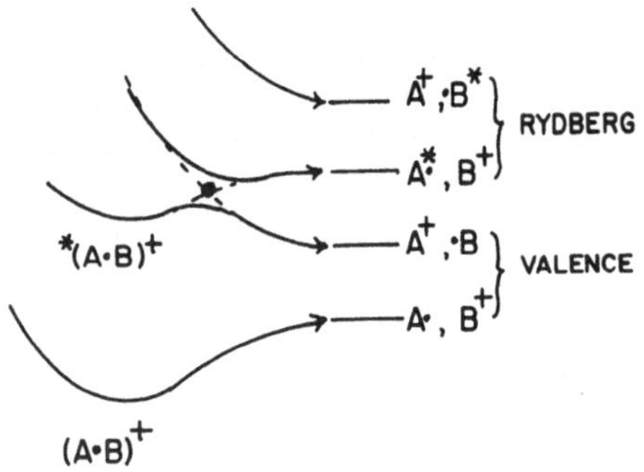


Figure 8. Behavior of a Simple one Electron σ-System on the Addition of Rydberg Functions.

one of the three electrons to escape. The resulting VB Rydberg structure can be represented by the formula, $[\cdot A\cdot\cdot B]^+(\cdot)^-$. Note that if we were dealing with an electron deficient molecule, the electron could escape into other same symmetry unoccupied or half-occupied valence AO's. But in the special cases of the approach of an H atom to either $^2B_1$ $NH_2$ or one of the degenerate pairs of OH, the odd electron will select Rydberg components to escape into. This is represented by the curves shown in Figures 9 and 10, which result from actual calculations. The process of Rydbergization is not important until the system is at close proximity. Between 1.3-1.6 times the normal NH or OH bond distance there is an abrupt change in the Rydberg-valence composition of the Rydberg both materials undergo derydbergization in the n3s state and correlate with the ground states of the radical products. Note, that the degenerate characters of both the $CH_3O$ and OH radicals permit another set of correlations linking the radical products to the ground states of water and methanol, respectively. The correlation problem is identical to that encountered in the ammonia surface if one assumes that the $^2B_1$ and $^2A_1$ states of $NH_2$ are degenerate, which they are in the linear form of the molecule (i.e. $1^2B_1$, $1^2A_1 \leftrightarrow {}^2\Pi$) (24). In contrast to ammonia and the amines, it is found that the n3s states of water and methanol show no barriers to either OH or CO bond rupture (26,27). That mainly OH rupture is observed in methanol is rationalized on the basis of the lighter mass of the H atom compared to the methyl group (3). The lack of a barrier in these surfaces could result from the higher exothermicity of the excited state reaction in the case of the n3s states of water and methanol (2-3 ev) in comparison which ammonia and the amines (1-2 ev) (3). Application of the Hammond postulate would predict a lower barrier for the more exothermic reactions (3). The quantum chemical arguments are more speculative and would require proof from a rather sophisticated partitioning of energy terms.

8. Methane

This surface has been dealt with in detail by Gordon and coworkers (28). The threshold photoreaction yields concerted departure of $H_2$ plus $CH_2$. This reaction appears deceptively different from the single bond rupture in the other materials discussed above but is correlatively similar. The lowest lying $^1\sigma 3s$ state of methane correlates directly with $CH_2(^1B_1)$ plus ground state $H_2$ for the $C_{2v}$ reaction pathway (3,28). In addition, the state undergoes simple derydbergization at the SCF level without avoided state crossings (3).

## A NON CORRELATIVE STRUCTURAL VIEW OF RYDBERGIZATION

The greatest potential fault in the above diagramatic view of Rydbergization and its consequences in photochemistry is that it does not provide much structural chemical insight. The nontheoretical chemist who normally thinks in terms of VB structures does not have to be convinced that the ground state of an A:B type molecule will separate in a fashion in which each fragment will retain an electron. This is a result which requires CI if an MO approach is used and speaks to the conceptual inferiority of the MO method at this level of thinking. In fact both methods may at times require the use of an expanded linear combination of functions in order to obtain correlatively correct result Rydbergization can often be considered to be a one electron phenomenon. In MO terms, it occurs because of the antiphasing of the valence orbitals and their tendency to spacially annihilate one another (29). The system must chose spacially larger orbitals to avoid this catastrophe and lower the energy. Somewhat equivalently, the first order valence bond wave function is insufficient. For example, in the orbital (26). With NH or OH departure the Rydberg function essentially collapses around the departing H atom. In $NH_3$, if Rydberg components are only included around the N atom, there is seen as an increase in the 1s component on the hydrogen atom and a decrease of the diffuse elements on the N atom as the H departs. If diffuse components are also used on the H atom, one sees the domination of the 3s orbital by a combination these functions which imitate a 3s(N) - like function at short distances. In this case, the departure of the H atom in $NH_3$ generates the collapse of the diffuse hydrogen functions with an increase in the importance of the 1s-like functions on the departing H atom. The situation becomes more complicated if a $3p_x$-like function is also included in the computation. But regardless of the Rydberg composition, the results are qualitatively the same, a 3s → 1s(H) derydbergization.

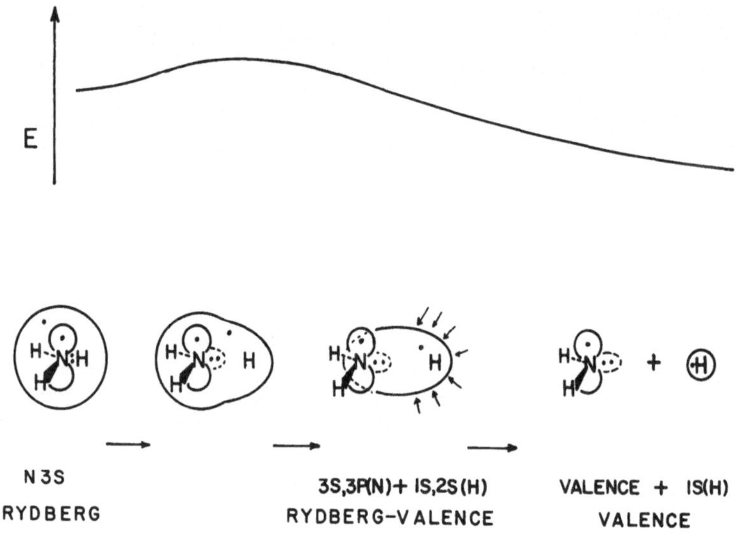


Figure 9. Conceptual View of Derydbergization in the n3s State of $NH_3$ on NH Rupture.

The case of the interaction of $^1B_1$ $CH_2$ with $H_2$ is shown in Figure 11. Here there is a placing of the $H_2$ with its two electrons into the one electron space of the hybridized orbital of $CH_2$. Because of the $\sigma(a_1)$ character of this orbital and the incoming $\sigma^2(a_1 \cdot a_1)$ orbital on the $H_2$ group, the system must chose $3s$-$3p_x$ Rydberg space to expand into. The total symmetry of the $C_{2v}$ pathway is $^1B_1$ and the odd σ-electron can not chose the half-empty $2p_x(b_1)$ orbital as the escape orbital. This would place the system in a total $^1A_1$ symmetry. The final electronic state of $CH_4$ is structurally difficult to imagine because of the degenerate point group, other than it is σ3s in character. A partial simplification is to imagine the system as being $R_2CH_2$ in $C_{2v}$. The structure can be written as $R_2C{=}H_2$, in pseudo-π terms, where in the double bond, $\Gamma(\sigma) = a_1$, $\Gamma(\pi) \doteq b_1$. Now the excited state can be viewed

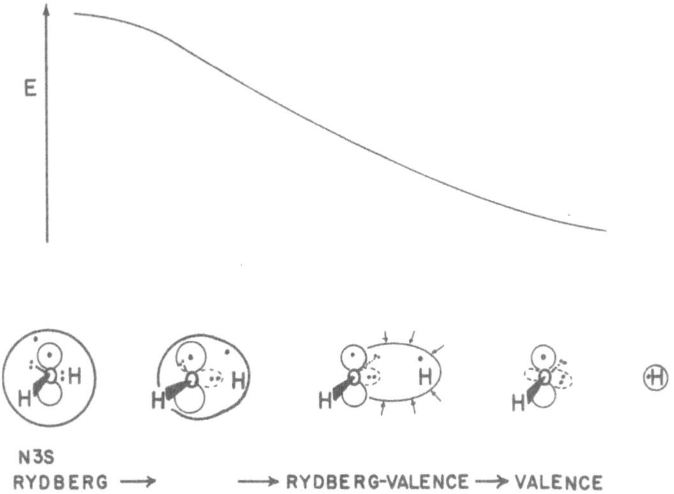


Figure 10. Conceptual View of Derydbergization in the n3s State of $H_2O$ on OH Bond Rupture.

as resulting from a $\pi \rightarrow$ 3s transition yielding an Rydberg excited state conceptually related to ethylene. The final symmetry of this excited state, $^1B_1$, will correlate with $CR_2(^1B_1) + H_2$. In this way one may see that, generally, the concerted departure of $H_2$ from either $CH_4$ or other alkanes ($R_2CH_2$) is an allowed process from an appropriate σ3s excited state.

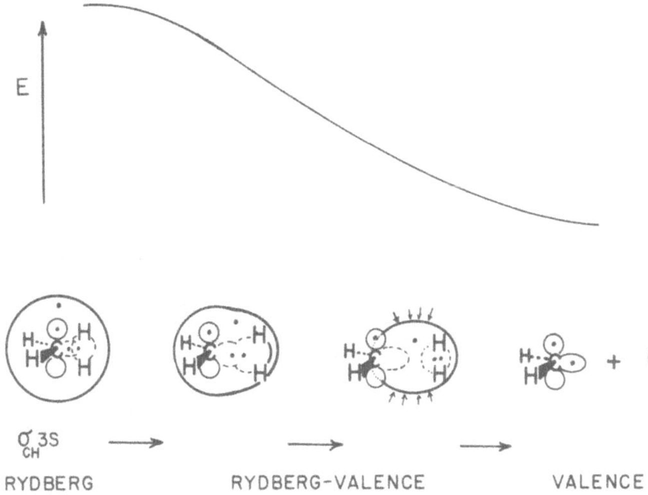


Figure 11. Conceptual view of Derydbergization on $H_2$ Departure From Excited $CH_4$.

## CONCLUSIONS

Either relatively simple or sophisticated orbital and state symmetry diagrams can be expanded to include Rydberg components. These diagrams can be used, in the case of simple molecules, to rationalize their photochemical behavior. Likewise, it is essential to use the diagrams for comparison purposes with actual theoretical calculations in order to qualitatively understand their results.

## REFERENCES


1. Robin, M. B.: 1974, 1975, "Higher Excited States in Polyatomic Molecules, Academic Press, N. Y." Volumes 1 and 2.
2. Sandorfy, C.: 1979, "Topics in Current Chemistry, Springer-Verlag, Berlin", 86, pp. 91-138.
3. Evleth, E.M., and Kassab, E.: 1980, "Quantum Theory of Chemical Activity, Daudel, R., Pullman, A., Salem, L., and Veillard, A., Ed., Reidel, Dordrecht, Holland", Volume 2.
4. McMurchie, L.E., and Davidson, E.R.: 1977, J. Chem. Phys. 67, pp. 5613.
5. Merer, A.J., and Mulliken, R.S.: 1969, Chem. Rev. 69, pp. 639.
6. a) Mulliken, R.S.: 1976, Accts. Chem. Res. 9, pp. 7.
   b) Mulliken, R.S.: 1977, Chem. Phys. Lett. 46, pp. 198.
7. Wadt, W.A., and Goddard, W.A., III: 1976, Chem. Phys. 18, pp. 1.
8. a) Horsley, J.A., and Flouquet, F.: 1970, Chem. Phys. Lett. 5, pp. 165.
   b) Runau, R., Peyerimhoff, S.D., and Buenker, R.J.: 1977, J. Mol. Spectrosc. 68, pp. 253.
   c) Müller, J., and Canuto, S.: 1980, Chem. Phys. Lett. 70, pp. 236.
9. Kassab, E., and Evleth, E.M.: to be published.
10. Herzberg, G.: 1950, "Molecular Spectra and Molecular Structure, Van Nostrand, N.Y." 1, pp. 328-329.
11. Herzberg, G.: 1966, "Molecular Spectra and Molecular Structure, Van Nostrand, N.Y." 3, Chapt. 3.
12. a) Longuet-Higgins, H.C., and Abrahamson, E.W.: 1965, J. Am. Chem. Soc. 87, pp. 2045.
    b) Hoffmann, R., and Woodward, R.B.: 1965, J. Am. Chem. Soc. 87, pp. 2046, see acknowledgement in ref. 2.
13. Woodward, R.B., and Hoffmann, R.: 1965, J. Am. Chem. Soc. 87, pp. 395.
14. Woodward, R.B., and Hoffmann, R.: 1970, "The Conservation of Orbital Symmetry, Verlag Chemie, Weinheim" see pp. 176-177.
15. Dauben, W.G., Salem, L., and Turro, N.J.: 1975, Accts. Chem. Res. 8, pp. 41.
16. Bigot, B., Roux, D., Sevin, A., and Devaquet, A.: 1979, J. Am. Chem. Soc. 101, pp. 2560.
17. Salem, L. Leforestier, C., Segal, G., and Whetmore, R.: 1975, J. Am. Chem. Soc. 97, pp. 479.
18. Evleth, E.M., Gleghorn, J.F., and Kassab, E.: submitted to Chem. Phys. Lett.

19. Evleth, E.M., and Kassab, E.: 1978, J. Am. Chem. Soc. 100, pp. 7859.
20. ref. 11, pp. 515-516.
21. a) Koda, S., and Back, R.A.: 1977, Can. J. Chem. 55, pp. 1384.
b) Donnelly, V.M., Baronavski, A.P., and McDonald, J.R.: 1979, Chem. Phys. 43, pp. 271.
22. Douglas, A.E.: 1963, Disc. Faraday Soc. 35, pp. 158.
23. Levy, B., and Berthier, G.: 1968, Int. J. Quant. Chem. 2, pp. 307.
24. Peyerimhoff, S.D., and Buenker, R.J.: 1979, Can. J. Chem. 57, pp. 3182.
25. Halpern, A.M., and Gartman, Th.: 1974, J. Am. Chem. Soc. 96, pp. 1393.
26. Kassab, E., Evleth, E.M., and Gleghorn, J.T.: in preparation.
27. Miller, K.J., Mielczarek, R.S., and Krauss, M.: 1969, J. Chem. Phys. 51, pp. 26.
28. Gordon, M.S.: 1977, Chem. Phys. Lett. 52, pp. 161.
29. Sinanoglu, O.: 1974, "Chemical Spectroscopy and Photochemistry in the Vaccum Ultraviolet, Sandorfy, C., Ausloos, P.J., and Robin, M.B., Ed. Reidel, Dordrecht" pp. 376-378.

# OUT-OF-PLANE BENDING COORDINATES FOR TETRAATOMIC MOLECULES

R.P. Steer, P.G. Mezey and A. Kapur

Department of Chemistry and Chemical Engineering
University of Saskatchewan, Saskatoon, Saskatchewan,
CANADA, S7N 0W0

## 1. INTRODUCTION

Tetraatomic molecules that are planar in their ground states often have pyramidal equilibrium conformations in some of their excited states. For tetraatomic carbonyls and thiocarbonyls (1) some of the excited state out-of-plane angles $\theta$ range from near $0^o$ to more than $40^o$ (2-5), where $\theta$ is defined in Figure 1. Out-of-plane bending coordinates have often been described by expressions in which only small deformations are assumed (6) and no specific expressions for molecules having $C_s$ ground state symmetry (with mode 6 as the out-of-plane bending) have been reported.

To gain insight into the nature of the excited state, two different but complementary approaches may be used. In the experimental approach, the time consuming step lies in the analysis of an absorption spectrum. Once the absorption spectrum has been successfully analyzed (in the present context once the out-of-plane bending mode frequencies have been successfully assigned) a potential function which is consistent with the observed frequencies is obtained. A potential function that has been used with considerable success (6,7) to describe double minimum potentials is of the form

$$V(Q) = \frac{1}{2} \lambda Q^2 + A \exp(-a^2Q^2) \qquad (1.1)$$

where $\lambda$, A and a are parameters whose values are chosen to give the best agreement with observed out-of-plane vibrational energies and Q is a mass weighted coordinate which is a function of the out-of-plane angle. Differentiation of (1.1) with respect to Q



I. G. Csizmadia and R. Daudel (eds.), Computational Theoretical Organic Chemistry, 397–402.

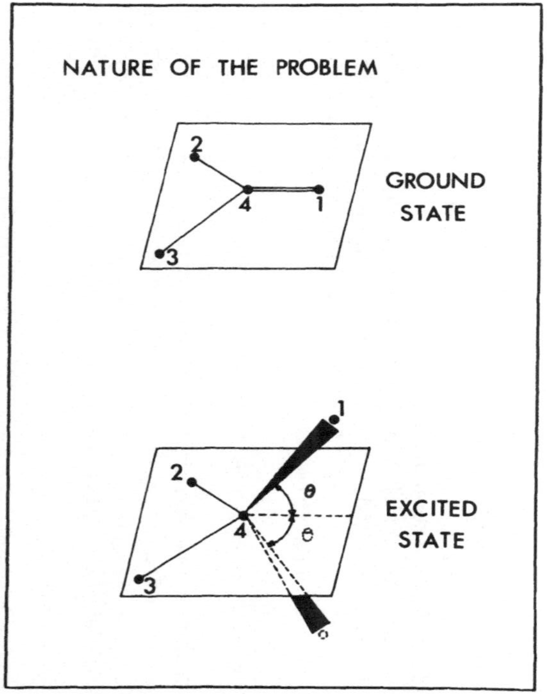


Figure 1. Geometry of model molecules.

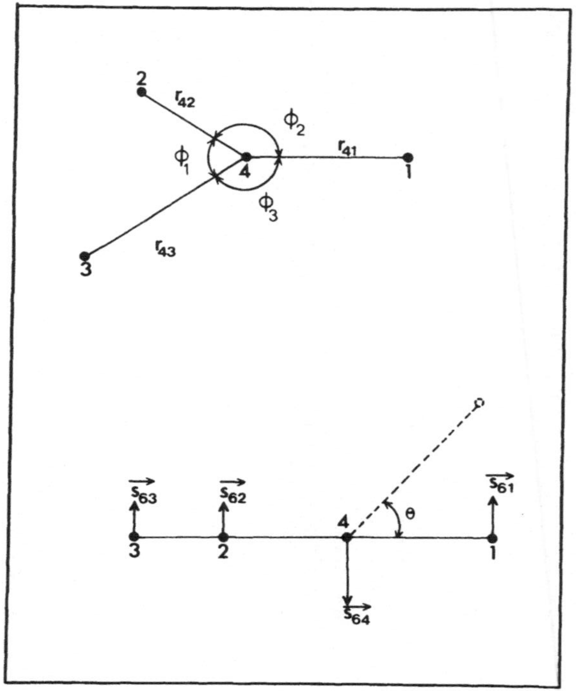


Figure 2. Definitions of structural parameters

leads to the value of $Q_m$, the mass weighted coordinate corresponding to the equilibrium conformation of the excited state. Values of the out-of-plane angle may be obtained through the use of the approximate relationship

$$Q = \mu^{\frac{1}{2}} r\theta \tag{1.2}$$

and other more exact equations. The barrier height, b, is given as the difference $V(0) - V(Q_m)$.

The experimental elucidation of the structure of excited states of tetraatomic thiocarbonyls and carbonyls thus hinges largely on a successful assignment of the out-of-plane bending mode frequencies. Since this is by no means a trivial task, it is in this area where the second, theoretical approach can be of considerable aid (8-10). Several points can be generated, using _ab initio_ calculations, along the double minimum path and then fitted to (1.1). The vibrational eigenvalue-eigenvector problem can then be solved to yield the vibrational frequencies, at least as first estimates to the experimental frequencies (11).

A crucial step in both the experimental and theoretical approaches is the transformation of the out-of-plane angle, θ, to a mass weighted coordinate. Obviously an incorrect transformation can lead to erroneous results. It is the purpose of this paper to discuss some of the problems associated with such coordinate transformations. A general formula for $C_s$ as well as $C_{2v}$ tetraatomic molecules is given.

## 2. FORMULAE FOR MASS WEIGHTED COORDINATES

One of the first formulae to be used for the transformation of the out-of-plane angle to a mass weighted coordinate is given by (6)

$$Q = \mu^{\frac{1}{2}} r\theta \tag{1.3}$$

with

$$\frac{1}{\mu} = \left(\frac{1}{M_1} + \frac{1}{M_4}\right) + \frac{2r_{41}}{M_4 z} + \left(\frac{1}{M_4} + \frac{1}{2M_2}\right)\frac{r^2_{41}}{z^2}$$

where $r_{41}$ is the distance between atoms 1 and 4, z is the perpendicular distance from atom 4 to the line joining atoms 2 and 3 and $\{M_i\}$, i = 1,4 are the masses of the respective atoms, as shown in Figure 1. Although this formula is easy to use it has two severe limitations:

(a) It is applicable for only $C_{2v}$ type of molecules.
(b) The "reduced" or effective mass does not depend on the out-of-plane angle, θ, explicitly. Thus the kinetic energy of the system is also independent of θ and this is an obvious shortcoming of the above formula.

Limitation (a) in the first formula can be removed easily by considering a transformation of the form

$$Q = \frac{\theta}{G_{66}^{\frac{1}{2}}} \tag{1.4}$$

where $G_{66}$ is the Wilson G matrix element (12) corresponding to the out-of-plane bending mode of a tetraatomic carbonyl or thiocarbonyl and has the form

$$G_{66} = \frac{1}{M_1 r_{41}^2} + \frac{1}{M_2 r_{42}^2}\left\{\frac{\sin^2\phi_2}{\cos^2\theta\sin^2\phi_1} - \frac{\tan^2\theta}{\sin^2\phi_1}\right\} + \frac{1}{M_3 r_{43}^2}\left\{\frac{\sin^2\phi_3}{\cos^2\theta\sin^2\phi_1}\right.$$
$$\left. - \frac{\tan^2\phi}{\sin^2\phi_1}\right\} + \frac{1}{M_4}\left[\frac{1}{r_{41}^2} + \frac{1}{r_{42}^2}\left\{\frac{\sin^2\phi_2}{\cos^2\theta\sin^2\phi_1} - \frac{\tan^2\theta}{\sin^2\phi_1}\right\} + \frac{1}{r_{43}^2}\right.$$
$$\left\{\frac{\sin^2\phi_3}{\cos^2\theta\sin^2\phi_1} - \frac{\tan^2\theta}{\sin^2\phi_1}\right\} + \frac{2}{r_{41}r_{42}}\left\{\frac{\cos\phi_1\cos\phi_2 - \cos\phi_3}{\cos^2\theta\sin^2\phi_1} + \frac{\tan^2\theta}{\sin^2\phi_1}\right.$$
$$\left.(\cos\phi_3 - \cos\phi_1\cos\phi_2)\right\} + \frac{2}{r_{41}r_{43}}\left\{\frac{\cos\phi_1\cos\phi_3 - \cos\phi_2}{\cos^2\theta\sin^2\phi_1}\right.$$
$$\left. + \frac{\tan^2\theta}{\sin^2\phi_1}(\cos\phi_2 - \cos\phi_1\cos\phi_3)\right\}$$
$$+ \frac{2}{r_{42}r_{43}}\left\{\frac{\cos\phi_2\cos\phi_3 - \cos\phi_1}{\cos^2\theta\sin^2\phi_1} + \frac{\tan^2\theta\cos\phi_1}{\sin^2\phi_1}\left(2 - \frac{1}{\sin^2\phi_1}\right.\right.$$
$$\left.\left.\left. - \frac{\cos^2\phi_1}{\sin^2\phi_1}\right)\right\}\right] \tag{1.5}$$

(For a definition of symbols, see Figure 2).
If $\theta = 0^o$ and the molecule under consideration belongs to the $C_{2v}$ point group, it can be shown that (1.4) and (1.3) are equivalent.

Although (1.4) does remove the first limitation of (1.3) and can be applied to $C_{2v}$ as well as $C_s$ type molecules, it still does

TABLE 1. Out-of-plane Bending Mode Frequencies for the First Triplet State of Thiocarbonyl Difluoride (in $cm^{-1}$)

| | Formula (1.3) | Formula (1.4) | Formula (1.7) | Expt. |
|---|---|---|---|---|
| $0^+-1^+$ | 565.9 | 481.3 | 543.2 | 565.9 |
| $0^--1^-$ | 565.9 | 481.3 | 543.2 | ----- |
| $0^+-2^+$ | 536.1 | 456.9 | 513.5 | 560.5 |
| $1^--2^-$ | 536.1 | 456.9 | 513.5 | 561.5 |
| $2^+-3^+$ | 497.9 | 426.5 | 474.9 | 557.2 |
| $2^--3^-$ | 498.9 | 427.0 | 476.3 | 554.7 |
| $3^+-4^+$ | 433.7 | 379.1 | 406.4 | ----- |
| $3^--4^-$ | 454.0 | 389.7 | 432.8 | ----- |

not take into account the variability of the G matrix element with the out-of-plane angle. To remove this second limitation of (1.3) the mass weighted coordinate, Q, may be written in the form

$$Q = \int_0^\theta G_{66}^{-\frac{1}{2}} \, d\theta \tag{1.6}$$

The integral shown in (1.6), when integrated by parts, leads to an infinite series and if second and higher order derivatives of $G_{66}$ with respect to $\theta$ are ignored, Q may be written as

$$Q = \sum_{i=1}^{\infty} \frac{1}{i!} \frac{1\cdot3\cdot5\cdots(2i-3)}{2^{i-1}} \theta^i G_{66}^{-(2i-1)/2} (G'_{66})^{i-1} \tag{1.7}$$

where $G'_{66} = dG_{66}/d\theta$. The first few terms in the series are

$$Q = \theta G_{66}^{-\frac{1}{2}} + \frac{1}{4}\theta^2 G_{66}^{-3/2} G'_{66} + \frac{1}{8}\theta^3 G_{66}^{-5/2}(G'_{66})^2$$
$$+ \frac{5}{64}\theta^4 G_{66}^{-7/2}(G'_{66})^3 + \ldots\ldots. \tag{1.8}$$

For $F_2CS$ it has been found that after the first four terms, higher terms do not make a significant contribution over the range $0^o \leq \theta \leq 50^o$. Table I shows the out-of-plane bending mode frequencies for the first triplet state of $F_2CS$, calculated using all three formulae.

## 3. CONCLUSIONS

Table 1 shows that good agreement can be obtained between calculated and experimental out-of-plane bending mode frequencies by using either formula (1.3) or formula (1.7). Formula (1.7) is, however, somewhat more accurate and also more general being applicable for both $C_s$ as well as $C_{2v}$ molecules. On the other hand formula (1.3) is simple to use and the numerical results are also satisfactory at small values of $\theta$. Formula (1.4) yields the least accurate results, nevertheless, even these "least accurate" calculated frequencies may aid the analysis of the experimental absorption spectrum.

## REFERENCES

1. D.R. Johnson, F.X. Powell, and W.H. Kirchoff, J. Mol. Spectrosc., 39, 136 (1971); A.J. Careless, H.W. Kroto, and B.M. Landsberg, Chem. Phys., 1, 371 (1973); K. Takagi and T. Oka, J. Phys. Soc. Japan, 18, 1174 (1963); V.W. Laurie, D.T. Pierce, and R.H. Jackson, J. Chem. Phys., 37, 2995 (1962).
2. R.H. Judge and G.W. King, Can. J. Phys., 53, 1927 (1975).
3. D.C. Moule and A.K. Mehra, J. Mol. Spectrosc., 35, 137 (1970).
4. D.A. Condirston, B.Sc. Thesis 1972, Brock University; G.L. Warkman and A.B.F. Duncan, J. Chem. Phys., 52 (1970).
5. C.R. Subramanian, Ph.D. Thesis, McMaster University.
6. J.B. Coon, N.D. Naugle, and R.D. McKenzie, J. Mol. Spectrosc., 20, 107 (1966).
7. D.C. Moule and Ch.V.S. Rao, J. Mol. Spectrosc., 45, 120 (1973).
8. A. Kapur, R.P. Steer, and P.G. Mezey, J. Chem. Phs., 69, 963 (1978).
9. A. Kapur, R.P. Steer, and P.G. Mezey, J. Chem. Phys., 70, 745 (1979).
10. A. Kapur, R.P. Steer, and P.G. Mezey, J. Chem. Phys., 71, 558 (1979).
11. For a complete discussion see A. Kapur, Ph.D. thesis, University of Saskatchewan, 1979.
12. E.B. Wilson, Jr., J.C. Decius, and P.C. Gross, "Molecular Vibrations", McGraw-Hill, New York, 1955.
13. A. Kapur, R.P. Steer, and P.G. Mezey, submitted to Chem. Phys. Lett. The derivation of (5) is quite lengthy and cumbersome, and may be obtained from the authors on request.

# THEORETICAL STUDIES ON ZEOLITE COMPOSITION AND LOEWENSTEIN'S RULE

Ernst C. Hass, Peter J. Plath[1] and Paul G. Mezey

Department of Chemistry and Chemical Engineering
University of Saskatchewan, Saskatoon, Saskatchewan, Canada S7N 0W0,
[1]Forschungsgruppe 'Angewandte Katalyse', Universität Bremen, NW2, Postfach 330440, 2800 Bremen 33, Federal Republic of Germany

## 1. INTRODUCTION

Synthetic zeolites play an important role in catalysis and, in particular, they are widely used in the petroleum industry. Zeolites are alumosilicate frameworks with a well defined geometrical structure consisting of systems of channels and cavities up to 10Å in diameter. Figure 1 shows the structural arrangement of faujasite type zeolites, the synthetic forms X and Y of which are among the most frequently used zeolite catalysts (1). Each vertex in Figure 1 represents a tetrahedrally coordinated centre occupied by a silicon or aluminum atom, while each edge represents an oxygen bridge between two tetrahedral centres. The Roman numbers indicate the possible positions of charge compensating cations.

One important property with respect to zeolite catalysis is the distribution of silicon and aluminum atoms in zeolite frameworks. To describe regularities observed in alumosilicate frameworks, such as feldspars, feldspatoids and zeolites, Loewenstein has proposed a simple rule, the so-called "Aluminum avoidance Rule": "In alumosilicate tetrahedral frameworks no oxygen should be bound to two aluminum atoms when silicon is available"(2). Thus, in zeolites containing no rings with an odd number of tetrahedral atoms, the minimum Si/Al ratio should be 1/1, and a completely ordered structure should be assumed with alternating Si and Al atoms (1).

Although the predictions of the "aluminum avoidance rule" are usually in agreement with the experimental results, several exceptions of this rule are known:



*I. G. Csizmadia and R. Daudel (eds.), Computational Theoretical Organic Chemistry, 403–408.*

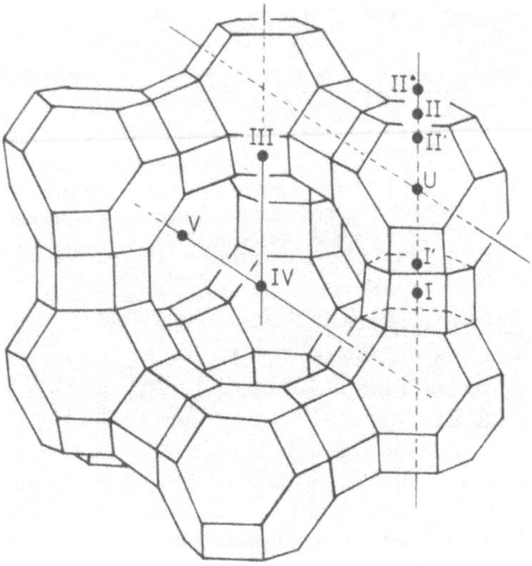


Figure 1. Structural arrangement of faujasite type zeolites.

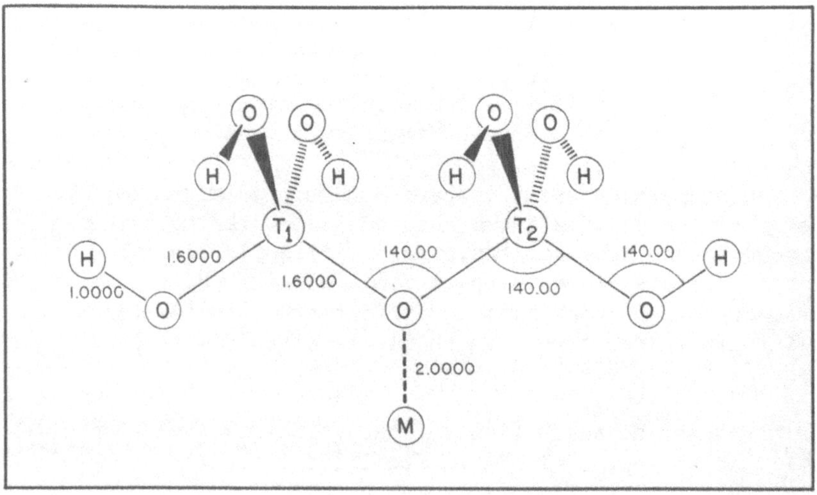


Figure 2. Geometrical parameters assumed for the starting cluster models $T_1$, $T_2$ = Si or Al, M = $Li^+$.

a) The existence of isolated Al-O-Al bridges has been established for crystalline potassium aluminate with the formula $[(OH)_3Al\text{-}O\text{-}Al(OH)_3]^{2-}\ 2K^+$ by X-ray spectroscopy (3) as well as by IR-spectroscopy (4).
b) There is evidence that tetrahedral Al atoms do share the same oxygen atoms in some feldspar minerals (5).
c) Even for zeolite lattices, particularly for crystals grown rapidly at high supersaturation, experimental data do not rule out the possibility of Al-O-Al linkages (6).

In order to get more detailed information about the relative stability of different arrangements of Si and Al tetrahedra and of the distorting effect of a Si → Al replacement we have carried out ab initio SCF MO calculations on three small model clusters $(OH)_3T_1\text{-}O\text{-}T_2(OH)_3$:

(A), $T_1 = T_2 = Si$
$(B)^-$, $T_1 = Si$ and $T_2 = Al$, and,
$(C)^{2-}$ $T_2 = T_2 = Al$

Furthermore, clusters, where an additional $Li^+$ ion was present, were also investigated.

## 2. MODEL AND METHOD

The initial geometry of the model clusters used as input for a partial geometry optimization, is shown in Figure 2. Since only the fundamental characteristics and relative stabilities have been considered throughout this study, the Si-O and Al-O bond lengths have been taken equal for the starting geometry and only roughly averaged bond lengths and bond angles (obtained from the most common zeolite geometries) have been used.

For the calculations a version of the Gaussian 70 program (7) has been employed which was modified for the 36 bit computerword DEC 2050 computer of the University of Saskatchewan, Saskatoon. Due to the large size of the clusters STO-3G basis sets were employed throughout this study (8). To obtain the equilibrium $T_1\text{-}O\text{-}T_2$ angles, partial geometry optimization has been carried out, considering only a crossection of the full energy hypersurface (7,9).

## 3. RESULTS AND DISCUSSION

Table 1 summarizes the equilibrium $T_1\text{-}O\text{-}T_2$ angles and the total energies of the three cluster models $(A)_{opt}$, $(B)^-_{opt}$ and $(C)^{2-}_{opt}$ (the index "opt" refers to the partial geometry optimization). In addition, the energy differences $\Delta E$ of the first and second replacements of one Si atom by an Al atom are listed in Table 1.

TABLE 1. Optimised $T_1$-O-$T_2$ angles, calculated total energies and energy differences of Si → Al replacements for various cluster models.

| model | $T_1$ | $T_2$ | $T_1$-O-$T_2$ angle [°] | $E_{total}$ [a.u.] | ΔE [a.u.] |
|---|---|---|---|---|---|
| $(A)_{opt}$ | Si | Si | 147.11 | -1091.67389 | |
| | | | | | 46.65467 |
| $(B)^{-}_{opt}$ | Si | Al | 147.71 | -1045.01922 | |
| | | | | | 46.83863 |
| $(C)^{2-}_{opt}$ | Al | Al | 180.00 | - 998.18059 | |

Similar calculations on the isoelectronic structures $SiH_4$ and $[AlH_4]^-$ gave total energies -287.91728 a.u. and -241.16732 a.u., for the optimum T-H bond lengths of 1.4216 Å and 1.5121 Å, respectively; the energy difference associated with the Si → Al substitution in these hydrides is 46.74996 a.u.

Comparing this ΔE value with the ΔE-values for the clusters (A), $(B)^-$ and $(C)^{2-}$ one may consider two hypothetical Si → Al exchange processes:

(i) $(HO)_3$ Si-O-Al $(OH)_3$ + $[AlH_4]^-$ → $[(HO)_3$ Si-O-Al$(OH)_3]^-$ + $SiH_4$,

$$\Delta E^{(i)} \simeq -0.095 \text{ a.u.} = -60 \text{ kcal/mole}$$

(ii) $[(HO)_3$Si-O-Al$(OH)_3]^-$ + $[AlH_4]^-$ → $[(HO)_3$Al-O-Al$(OH)_3]^{2-}$ + $SiH_4$,

$$\Delta E^{(ii)} \simeq +0.089 \text{ a.u.} = +56 \text{ kcal/mole}$$

The second process is much less favored than the first, which agrees well with the "aluminum avoidance rule". However, the calculated total energy of the doubly negative cluster models $(C)^{2-}$ and all energy differences involving this species are less reliable than the other energy values, since for some of the highest occupied MO's the orbital energies are positive, indicating an unstable HF solution. (10)

In order to obtain more physically meaningful wavefunctions for the $(C)^{2-}$ type clusters, we have carried out similar calculations for cluster models with charge compensating $Li^+$ cations, located at 2.00Å from the bridge oxygen atom (see Figure 2). In all these calculations all MO's have negative orbital energies and therefore the calculated total energies and their differences (summarized in Table 2) are comparable.

TABLE 2. Calculated total energies and energy differences of Si → Al replacements for cluster models with charge compensating $Li^+$ cations.

| Model* | $T_1$ | $T_2$ | $T_1$-O-$T_2$ angle[°] | $E_{total}$ [a.u.] | ΔE [a.u.] |
|---|---|---|---|---|---|
| $(A)_{opt} + Li^+$ | Si | Si | 147.11 | -1099.00223 | |
| | | | | | 46.45741 |
| $(B)^-_{opt} + Li^+$ | Si | Al | 147.71 | -1052.54482 | |
| | | | | | 46.62522 |
| $(C)^{2-}_{optB} + Li^+$ | Al | Al | 147.71 | -1005.91960 | |

* In the $(C)^{2-}_{opt\ B}$ cluster the Al-O-Al angle has been set equal to the optimized Si-O-Al angle, obtained for cluster model $(B)^-_{opt}$.

Based on these results, one may formulate the following hypothetical processes involving the cluster + $Li^+$ pairs and hydrides $SiH_4$ and $[AlH_4]^-$:

(iii) $(HO)_3$ Si-O-Si$(OH)_3$ + $Li^+$ +$[AlH_4]^-$ →

$$\rightarrow [(HO)_3\ Si\text{-}O\text{-}Si(OH)_3]^- + Li^+ + SiH_4,$$

$$\Delta E^{(iii)} \simeq 0.293\ \text{a.u.} = -183\ \text{kcal/mole}$$

(iv) $[(HO)_3$ Si-O-Al$(OH)_3]^-$+ $Li^+$ + $[AlH_4]^-$ →

$$\rightarrow [(HO)_3Al\text{-}O\text{-}Al(OH)_3]^{2-} + Li^+ + SiH_4,$$

$$\Delta E^{(iv)} \simeq -0.125\ \text{a.u.} = -78\ \text{kcal/mole}.$$

The two calculated "reaction heats" $\Delta E^{(iii)}$ and $\Delta E^{(iv)}$ show that both processes are exothermic, and the presence of charge compensating $Li^+$ cations has a stabilization effect, which can be estimated from (i) and (iii) as $\Delta E^{(i)} - \Delta E^{(iii)}$ = 123 kcal/mole. Nevertheless, the first Si → Al replacement in the $(HO)_3$ $T_1$-O-$T_2(OH)_3$ cluster is by 105 kcal/mole more favored than the second Si → Al substitution, supporting the "aluminum avoidance rule".

The calculated $T_1$-O-$T_2$ bending potentials (11) of cluster models (A) and $(B)^-$ show almost identical equilibrium angles of 147.11 and 147.71, respectively (see Table 1), which are also in excellent agreement with the average experimental result of 147°, obtained for silica polymorphs (12). The potential curves are very shallow in a wide interval, indicating that the first Si → Al replacement may easily be accommodated in the zeolite framework. On the other hand the potential curve obtained for the Al-O-Al in $(C)^{2-}$ type clusters is considerably steeper in the range of 140°- 150°, and indicates a minimum at 180°, e.g. a linear Al-O-Al structure is favoured. Thus, an Al-O-Al moiety should introduce a significant strain into a regular tetrahedral framework of alumosilicates.

## CONCLUSIONS

*Ab initio* SCF MO calculations on various cluster models show that Al-O-Al linkages in alumosilicate frameworks are unlikely to occur, in agreement with Loewenstein's "aluminum avoidance rule" (2). The presence of charge compensating cations, e.g. $Li^+$, may stabilize Al-O-Al arrangements, although such a stabilization is greater for moieties with Si-O-Al or Si-O-Si type bridges.

## REFERENCES

1. See e.g. D.W. Breck: "Zeolite Molecular Sieves"; Wiley (1974).
2. W. Loewenstein: Amer. Mineralogist 39, 92 (1954). see also: J.R. Goldsmith, F. Laves: Z. Kristallographie 106, 213 (1955).
3. G. Johansson: Acta Chem. Scand. 20, 505 (1966).
4. J. Haladjian, J. Roziere: J. Inorg. Nucl. Chem. 35, 3821 (1973).
5. J.V. Smith: "Feldspar Minerals", Springer, New York (1974).
6. J.V. Smith: Adv. Chem. Ser. 101, 171 (1971).
7. W.J. Hehre, W.A. Lathan, R. Ditchfield, M.D. Newton, J.A. Pople: Q.C.P.E. program 236, Indiana University, Bloomington, Indiana.
8. W.J. Hehre, R.F. Stewart, J.A. Pople: J. Chem. Phys. 51, 2657 (1969).
9. P.G. Mezey: Analysis of Conformational energy hypersurfaces; in: Progress Theoret. Org. Chem. 2, 127 (1977).
10. R. Ahlrichs: Chem. Phys. Letters 34, 570 (1975).
11. E.C. Hass, P.G. Mezey, P.J. Plath: Theochem, in press.
12. J.A. Tossel, G.V. Gibbs: Acta Cryst. A34, 463 (1978).

# SOLVENT EFFECTS - EXCITED STATE DIPOLE MOMENTS[1].

Ira Mark Brinn
Instituto de Quimica, Universidade Federal do
Rio de Janeiro, Rio de Janeiro, R.J. - 21.910, Brasil

The importance of treating solvent effects becomes apparent when one reflects that many reactions are highly sensitive[2] to the solvent used, and that the relative stability of excited states can be inverted[3] upon a change of solvent.

We propose the following simple physical model to take these effects into account ; a single solute molecule consisting of "tight" atoms within a cavity of uniform dielectric constant. (See Figure 1). Because of the tightness of the atoms, no effect of the field is noticed within each atom, however there is an effect between the atoms within a molecule. In addition, the value of the dielectric constant ($\epsilon'$) within the cavity is not equal to its value ($\epsilon$) in the bulk solvent. If the energy of a coulombic interaction is given by $E = q_1 q_2/\epsilon r$, the effect of this model on the Hamiltonian is to change the value of all diatomic integrals by $\gamma_{AB} + \gamma^0_{AB}/\epsilon'$, whereas all one center integrals remain unchanged.

CNDO/2-CI calculations[4] were carried out for formaldehyde, formamide and phenol, as a function of $\epsilon'$, which is treated as an arbitrary parameter. Figure 2 shows the calculated dipole moments of the ground state ($\mu_S 0$) and first excited singlet state ($\mu_S 1$), as a function of $\epsilon'$. In all cases $\mu_S 0$ varies by no more than 5% over the range shown, whereas $\mu_S 1$ varies by about 25%. If the results of this model are qualitatively correct, and the scale of $\epsilon'$ values chosen corresponds to a physically realistic variation of solvents, one concludes that the determination[5] of $\mu_S 1$ using the spectra of the compound of interest in solvents of different dielectric constant is subject to much greater error than previously suspected.

1. Submitted to Theor. Chim. Acta.
2. For example, the Grignard reaction.
3. Mathews, T.G., and Lytle, F.E.: 1979, J. Lumin. 21, pp. 93.



*I. G. Csizmadia and R. Daudel (eds.), Computational Theoretical Organic Chemistry, 409–411.*

4. Giessner-Prettre, T.G., and Pullman, A.: 1969, Theor. Chim. Acta 13 pp. 265.
5. a) Bayliss, N.S., and McRae, E.G.: 1954, J. Phys. Chem. 58, pp. 1002.
   b) Abe, A.: 1965, Bull. Chem. Soc., Japan, 38, pp. 1314.

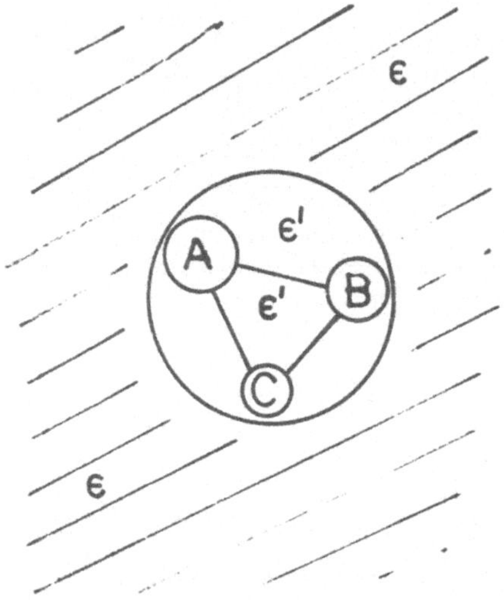


Figure 1. The Model : "tight" atoms A, B, and C within the molecule ABC, situated in a cavity of uniform dielectric constant ($\epsilon'$), whose value is different from that of the bulk solvent.

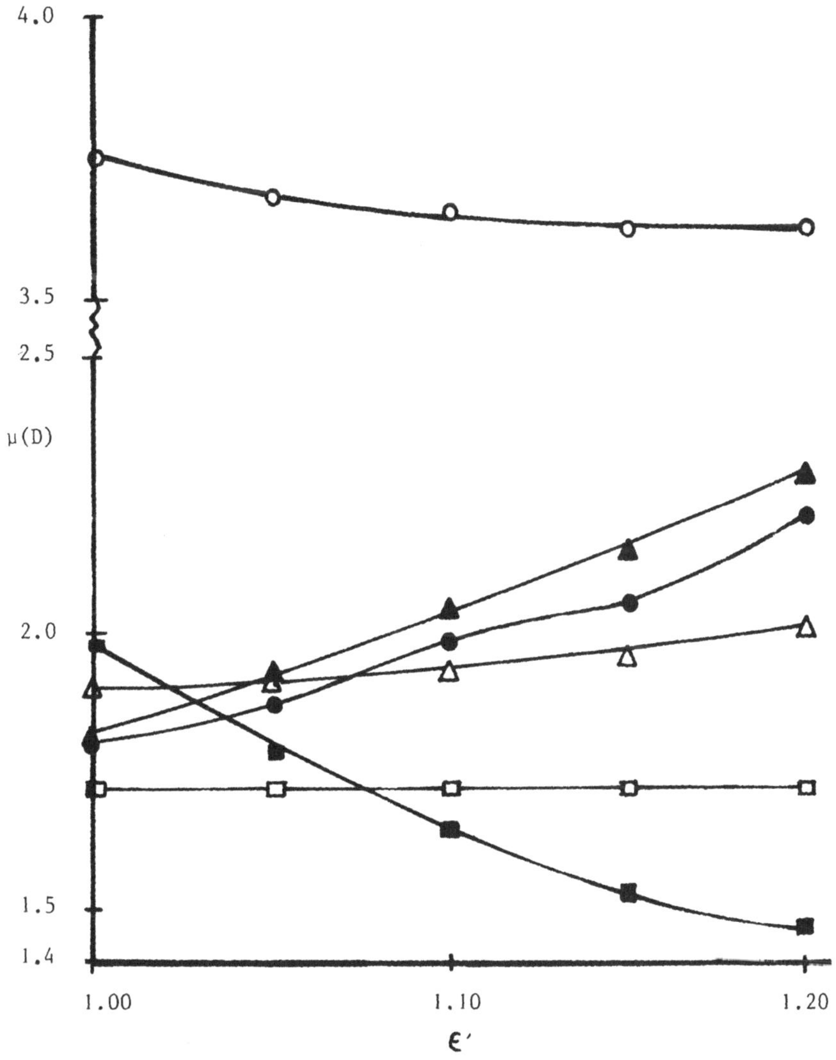


Figure 2. Calculated dipole moments as a function of ε' for formaldehyde (triangles), formamide (circles) and phenol (squares). $S_0$ - open figures, $S_1$ - shaded figures.

# COMPUTATIONAL LABORATORY PROJECTS

Michael R. Peterson

Department of Chemistry, University of Toronto,
Toronto, Ontario M5S 1A1, Canada.

A booklet containing 25 computational project suggestions was prepared in advance by Dr. John Goddard, Michael Peterson and Dr. Raymond Poirier. Most of these problems fell into the domain of theoretical organic chemistry, but some inorganic and spectroscopic chemistry applications were also included.

Each project consisted of an outline of the problem, a brief set of leading references, and a step-by-step suggested plan of attack. This latter plan was not mandatory; in fact, the participants were encouraged to try every available technique or option. In addition, the computer cards for the first suggested calculation had all been prepared and checked in advance. Some had even been run completely to ensure that every participant in the computational laboratory would have at least one successful calculation on their project.

The problems were designed to illustrate many of the new techniques available in modern theoretical organic chemistry, most of which are described in detail in other sections of this volume. These include automatic simultaneous optimization of molecular geometries using analytic gradient (force) methods, transition state determination, direct configuration interaction (CI) calculations, quantitative perturbational molecular orbital (PMO) analysis, one-electron properties (e.g., multipole moments), a variety of basis sets, and gradient optimization of molecular basis sets.

Several projects were concerned with one of the traditional areas of application of MO theory, conformational analysis: rotational barriers in $CH_3$-$CH_3$, $CH_3$-$NH_2$, $CH_3$-OH and HO-OH, and the inversion



*I. G. Csizmadia and R. Daudel (eds.), Computational Theoretical Organic Chemistry, 413–416.*

barriers in substituted amines R-$NH_2$. Quantitative PMO analysis was included in these projects to assist in the rationalization of the results.

Several other projects considered intramolecular rearrangements, and the effect of different basis sets: HCN $\rightleftharpoons$ HNC, HCP $\rightleftharpoons$ HPC, HC(=O)F $\rightleftharpoons$ H-C-O-F, and H-N=C=X $\rightleftharpoons$ N≡C-X-H for X = O and S. The 1,3-dipolar addition reaction of a model nitrone ($CH_2=NH^+-O^-$) to acetylene constituted another project.

Three molecular complexes were included: $C_2H_7^+(CH_4+CH_3^+)$, $N_2H_7^+(NH_3+NH_4^+)$ and $HN_4F(NH_3+HF)$. Another project concerned the carbonium ions $C_2H_5^+$ and $C_2H_7^+$, including the effect of CI on the relative stabilities.

Two other proposed projects involved the determination of the sites of protonation of enamine ($CH_2=CH-NH_2$) and formamide ($HC(=O)NH_2$). The proton affinities of $H_2O$ and $H_2S$ were the subject of another project.

Rounding out the list of projects concerned primarily with gradient optimization of geometries were the inorganic complexes $Al+C_2H_2$ and $Al+C_2H_4$, and the model Grignard reagents $CH_3BeF$ and $CH_3MgF$.

A series of problems were designed to illustrate CI calculations, as applied to spectroscopy. The first area was singlet-triplet energy gaps, for the systems $XH_2$ (X = C, $B^-$, $N^+$, $P^+$), formaldehyde ($H_2C=O$) and formimine ($H_2C=NH$). The other major area was ionization potentials, illustrated by water and methane.

A final pair of projects were to study the effect of basis set optimization in the molecule on various one-electron properties for the first-row hydrides.

All the calculations were performed with the MONSTERGAUSS ab initio program [1], which incorporates the integral and SCF packages of GAUSSIAN 76 [2], together with many other features. These include automatic geometry optimization (in internal coordinates) using the analytic gradient [3] of the SCF energy, employing powerful variable metric simultaneous optimization techniques [4,5]. Transition state structures may also be determined using the VA05AD optimization routine [6], which computes the Hessian matrix (second derivatives or force constant matrix) by gradient differences to confirm the order of the critical point located.

The direct CI calculations [7] are possible only for closed shell singlet states at present. Other MONSTERGAUSS features include a standard geometry program [8], a quantitative PMO analysis routine [9], and a Boy's localization routine [10]. The one-electron properties program includes the multipole moments (dipole, quadrupole and octupole), the potential (1/r), the electric field ($1/r^2$), the electric field gradient ($1/r^3$), the charge density and the optical rotatory strength. The analytic SCF gradient with respect to the gaussian orbital exponents and contraction coefficients may also be computed [3]. It may be combined with the geometry gradient for simultaneous molecular geometry and basis set optimizations.

Most of these features were illustrated by a geometry optimization of water, with the calculation of the localized orbitals, many one-electron properties and the all singles and doubles CI energy at the final point. All computational laboratory participants also received a copy of this job.

We thank the Centre National de Recherche Scientifique (CNRS) of France for making this laboratory possible with a generous grant of computer funds. All calculations were performed on the IBM 370/168 and 3032 computers at the Centre Inter Régional de Calcul Electronique (CIRCE) at Orsay, France, near Paris. The jobs were actually submitted via a terminal in Nice - we thank Dr. Earl Evleth and Mlle. Peggy Evleth for providing a daily courier service between the Nice terminal and the Menton meeting site.

References

1. Program MONSTERGAUSS, a program to perform ab initio molecular orbital calculations, M.R. Peterson and R.A. Poirier, Dept. of Chemistry, University of Toronto, Toronto, Ontario M5S 1A1, Canada.
2. GAUSSIAN 76: J.S. Binkley, R.A. Whitehead, P.C. Hariharan, R. Seeger, J.A. Pople, W.J. Hehre and M.D. Newton, Quantum Chemistry Program Exchange (QCPE), Program No. 368, Dept. of Chemistry, Indiana University, Bloomington, Indiana 47405, U.S.A.
3. H.B. Schlegel, Ph.D. Thesis, Queen's University, 1975. See also Dr. Schlegel's lecture notes elsewhere in this volume.
4. Optimally Conditioned (OC) Optimization Method: W.C. Davidon and L. Nazareth, Technical Memos 303 and 306, 1977, Applied Mathematics Division, Argonne National Laboratories, Argonne, Illinois 60439, U.S.A. The algorithm was described in W.C. Davidon, Mathematical Programming, 9, 1 (1975).
5. Broyden-Fletcher-Goldfarb-Shanno (BFGS) Optimization Method: M.J.D. Powell, Subroutine VA13AD, Atomic Energy Research Establishment (AERE) Subroutine Library, Harwell, Didcot,

Berkshire, U.K. The algorithm is described in R. Fletcher, Comp. J., 13, 317 (1970).
6. M.J.D. Powell, Subroutine VA05AD, AERE Subroutine Library, Harwell, Didcot, Berkshire, U.K. The algorithm is very similar to one described by M.J.D. Powell in "Numerical Methods for Nonlinear Algebraic Equations", P. Rabinowitz (Ed.), Gordon and Breach, London, 1970, Ch. 6 and 7.
7. N.C. Handy, J.D. Goddard and H.F. Schaefer III, J. Chem. Phys., 71, 426 (1979). See also Dr. Goddard's lecture notes elsewhere in this volume.
8. M.R. Peterson and I.G. Csizmadia, unpublished work.
9. M-H. Whangbo, H.B. Schlegel and S. Wolfe, J. Am. Chem. Soc., 99, 1296 (1975). D. Kost, H.B. Schlegel, M-H. Whangbo, D.J. Mitchell and S. Wolfe, Can. J. Chem., 57, 729 (1979). See also the lecture notes, elsewhere in this volume, of Dr. Whangbo and Dr. Bernardi.
10. The localization is performed by an extensively modified version of Program BOYLOC: D. Peeters, Q.C.P.E., Program No. 330, Dept. of Chemistry, Indiana University, Bloomington, Indiana 47405, U.S.A. The algorithm is described in R.C. Haddon and G.R.J. Williams, Chem. Phys. Letters, 42, 453 (1976).

INDEX OF SUBJECT